Hou · Guo Homogeneous Denumerable Markov Processes

Hou Zhenting Guo Qingfeng

Homogeneous Denumerable Markov Processes

Springer-Verlag
Berlin Heidelberg New York
London Paris Tokyo

Science Press
Beijing

Hou Zhenting
Chang Sha Railway Institute
Chang Sha
The People's Republic of China

Guo Qingfeng
Department of Mathematics
Xiang Tan University
Xiang Tan
The People's Republic of China

Revised edition of the original Chinese edition published by Science Press Beijing 1978 as the second volume in the Series in Pure and Applied Mathematics.

Distribution rights throughout the world, excluding The People's Republic of China, granted to Springer-Verlag Berlin Heidelberg New York London Paris Tokyo

Mathematics Subject Classification (1980): 60-XX, 82-XX, 90-XX, 93-XX

ISBN-13: 978-3-642-68129-5 e-ISBN-13: 978-3-642-68127-1
DOI: 10.1007/978-3-642-68127-1

Softcover reprint of the hardcover 1st edition 1988

Typesetting: Science Press, Beijing, The People's Republic of China

2141/3140-543210 Printed on acid-free paper

Preface

Markov processes occupy an important position in the study of probability theory. The homogeneous denumerable Markov process is one of the main branches of Markov processes and has a wide range of applications in various fields of science and technology (for example, in physics, cybernetics, queueing theory, dynamical programming theory, etc.).

The theoretical problems in homogeneous denumerable Markov processes are generally of two types: first, the study of various properties of any given process; second, the qualitative study of Q-processes and the construction of all Q-processes for any given Q-matrix. This book is a summary of the research results in the above two types of problems obtained by the authors in recent years. Most of the results are published here for the first time.

This book consists of five parts. Parts I and II deal with theoretical foundations. Parts III and IV are devoted to the study of the first type of problems and Part V to that of the second type of problems.

Two mathematical methods are used throughout this book. The first is called "the method of minimal nonnegative solution", and the second "the limit transition method" (i.e., to approximate arbitrary processes by processes with simpler sample functions). Generally speaking, these two methods are coordinated in this way: The method of minimal nonnegative solution is used to investigate the processes with simpler sample functions (Markov chain, the minimal Q-process and the Q-process of order one); on the basis of this, the limit transition method is used to investigate arbitrary processes by means of the "construction theorem" shown in Part I. The former is due to the authors; the latter was established by Professor Wang Zikun in his paper "Structure of Birth and Death Processes" and is extended by the authors.

Chapters I and II, being somewhat harder, form the framework of the whole book. Their contents will not be used until Chapter XI. Readers may first quickly glance over them to get an impression of the main features, and then consult them in detail whenever needed.

For their encouragement and help, comments and advice, we are indebted to many comrades, particularly to Professors Miao Bangjun, Yang Xiangqun, Wang Zikun, Yue Minyi, Han Jiye, Messrs. Mo Wenchuan, Chen Mufa and Professor Li Junxian. Thanks are also due to Professor K. L. Chung.

It is difficult to avoid making mistakes and shortcomings on account of our modest ability and limited knowledge. Criticism and suggestions are cordially requested.

February 1988

Hou Zhenting
Guo Qingfeng

Contents

PART I

CONSTRUCTION THEORY OF SAMPLE FUNCTIONS OF HOMOGENEOUS DENUMERABLE MARKOV PROCESSES

CHAPTER I

The First Construction Theorem

§ 1.1 Introduction

Let $(\Omega, \mathscr{F}, P)$ be a complete probability space, and $E=(1, 2, \cdots)$. $\sigma(\omega)$ is a random variable taking values in $[0, \infty)$. $x(t, \omega)$ is a function taking values in E and defined for all $\omega\in\Omega$, $t\in[0, \sigma(\omega))$. For every $t\geqslant 0$, let

$$\tilde{x}(t, \omega)=\begin{cases} x(t, \omega), & t<\sigma(\omega); \\ \Delta, & t\geqslant\sigma(\omega). \end{cases}$$

We call $X(\omega)=\{x(t, \omega), t<\sigma(\omega)\}$ a homogeneous denumerable Markov process defined on the complete probability space $(\Omega, \mathscr{F}, P)$ with the minimal state space E if $\hat{X}(\omega)=\{\hat{x}(t, \omega), t\geqslant 0\}$ is a continuous parameter Markov chain (abbreviated to c.p.M.c.) defined on the complete probability space $(\Omega, \mathscr{F}, P)$ with the minimal state space $E\cup\{\Delta\}$ and stationary transition probabilities defined as in [1, II, §4], and $\sigma(\omega)$ is called a living time of $x(\cdot, \omega)$. We shall take the corresponding concepts and notation about c.p.M.c. from [1, II] for the homogeneous denumerable Markov chain $X(\omega)$ defined above. For example, $p_{ij}(t)=p(x(t+s)=j|x(s)=i)$ is called a transition probability matrix of $X(\omega)$; $p_{ij}(t)$ is called standard if $\lim_{t\to 0}p_{ij}(t)=\delta_{ij}$; etc. Futhermore, by the relation between $X(\omega)$ and $\hat{X}(\omega)$, it is easy to establish that most of the results of [1, II] are still true for $X(\omega)$. So we shall frequently use the concepts, notations and some suitable results from [1, II].

Let $X(\omega)=\{x(t, \omega), t<\sigma(\omega)\}$ be a homogeneous denumerable Markov process defined on a complete probability space $(\Omega, \mathscr{F}, P)$, with a minimal state space $E=(1, 2, \cdots)$, and a standard transition probability matrix $(p_{ij}(t), i, j\in E)_{t\geqslant 0}$, its Q matrix satisfies the relation

$$q_{ij}\geqslant 0 \ (i\neq j), \qquad \sum_{j\neq i} q_{ij}=-q_{ii}\equiv q_i<+\infty. \tag{1.1.1}$$

For simplicity, we always suppose, by [1, II, Theorem 5.1, Theorem 5.7 and §7], without the affecting transition probability matrix, that $X(\omega)$ has the following properties (D):

(D_1) It is right continuous, therefore well-separable, Borel measurable and possesses the strong Markov property.

(D_2) $\{\omega: \mu(t: x(t, \omega) = +\infty) = 0\} = \Omega$, where μ stands for Lebesgue measure.

(D_3) $\{\omega$: for any $i \in E$ and any $t \in [0, \sigma(\omega))$, there only exists a finite number of i-intervals of $x(\cdot, \omega)$ in $[0, t)\} = \Omega$.

It is well known that the processes with $Q = (q_{ij})$ as their density matrix (all processes with the same transition probability matrix are identified as one and the same process) are not necessarily unique and they are generally referred to as Q-processes.

How difficult it is to study the properties of Q-processes mostly depends on whether the structures of their sample functions are complicated or simple. In what follows we will first distinguish "the minimal Q-processes", with the simplest structure of sample functions, and the so-called "Q-processes of order one" with comparatively simple structure of sample functions. They lay the foundation for studying arbitrary Q-processes.

Definition 1.1.1 *τ is called a flying point of $x(\cdot, \omega)$ for $\tau \leqslant \sigma(\omega)$ if $\tau = \sigma(\omega)$ or $\tau < \sigma(\omega)$, and $x(\cdot, \omega)$ has infinitely many jump points in $(\tau - \varepsilon, \tau + \varepsilon)$ for any $\varepsilon > 0$.*

Evidently, a limiting point of the flying points is also a flying point, hence the set of all the flying points of $x(\cdot, \omega)$ is a closed set. So there exists the least flying point of $x(\cdot, \omega)$, denoted by τ_1. Sometimes, τ_1 is called the first flying point of $x(\cdot, \omega)$.

Definition 1.1.2 *If Q-process $X(\omega) = \{x(t, \omega), t < \sigma(\omega)\}$ satisfies the condition*

$$\tau_1(\omega) = \sigma(\omega) \qquad (\omega \in \Omega), \tag{1.1.2}$$

then we say that $X(\omega)$ is a minimal Q-process or a Q-process of order zero.

Definition 1.1.3 *A Q-process $X(\omega) = \{x(t, \omega), t < \sigma(\omega)\}$ is said to be a Q-process of order one if it satisfies the following two conditions:*

$$\text{(1)} \qquad \{\omega: x(\tau_1(\omega)) = \infty\} = \varnothing \tag{1.1.3}$$

(2) *$x(\cdot, \omega)$ has at most a finite number of flying points in $[0, t)$ for any $\omega \in \Omega$ and any $t \in [0, \sigma(\omega))$.*

By Definitions 1.1.2 and 1.1.3, we know that a Q-process of order zero is the degeneration of a Q-process of order one.

In [2], Wang Zikun strictly proved the construction theorem for the sample functions of birth and death processes, which was successfully used in a series of studies of birth and death processes in [2] and [3] by Wang Zikun, as well as in [4] and [5] by Yang Chaoqun. The aim of this chapter is to establish the construction theorem 1.5.1 for sample functions of Q-processes so that we may go on to study the properties of the processes according to the following procedure: first, to investigate minimal Q-processes with the simplest structure of sample functions; then to consider Q-processes of order one with comparatively simple structure of sample functions; and finally, to deal with any Q-process by using the "limit transition method" provided by the construction theorem.

§ 1.2 Definition of transformation g_n

Let $X(\omega)=\{x(t,\omega),\ t<\sigma(\omega)\}$ be an arbitrary Q-process, and $D_n=(1,\cdots,n)$. Let

$$\beta_0^{(n)}(\omega)\equiv 0, \tag{1.2.1}$$

let

$$\sigma_1^{(n)}(\omega)\ \text{be the first flying point of}\ x(\cdot,\omega), \tag{1.2.2}$$

and let

$$\beta_1^{(n)}(\omega)=\begin{cases}\inf\{t:\ \sigma_1^{(n)}(\omega)\leqslant t<\sigma(\omega),\ x(t,\omega)\in D_n\},\\ \sigma(\omega),\quad \text{if the above set is empty.}\end{cases} \tag{1.2.3}$$

Suppose that $\sigma_{m-1}^{(n)}(\omega)$ and $\beta_{m-1}^{(n)}(\omega)$ have been defined. Then we stipulate that $\sigma_m^{(n)}(\omega)=\beta_m^{(n)}(\omega)=\sigma(\omega)$, if $\beta_{m-1}^{(n)}(\omega)=\sigma(\omega)$, otherwise let

$$\sigma_m^{(n)}(\omega)\ \text{be the first flying point of}\ x(\cdot,\omega)\ \text{after}\ \beta_{m-1}^{(n)}(\omega), \tag{1.2.4}$$

and let

$$\beta_m^{(n)}(\omega)=\begin{cases}\inf\{t:\ \sigma_m^{(n)}(\omega)\leqslant t<\sigma(\omega),\ x(t,\omega)\in D_n\}\\ \sigma(\omega),\quad \text{if the above set is empty.}\end{cases} \tag{1.2.5}$$

Thus we have

$$\begin{aligned}0\equiv\beta_0^{(n)}(\omega)&\leqslant\sigma_1^{(n)}(\omega)\leqslant\beta_1^{(n)}(\omega)\leqslant\cdots\\ &\leqslant\sigma_m^{(n)}(\omega)\leqslant\beta_m^{(n)}(\omega)\leqslant\cdots\leqslant\sigma(\omega).\end{aligned} \tag{1.2.6}$$

Let

$$\tau_k^{(n)}(\omega)=\begin{cases}0, & k=0,\\ \sum_{s=1}^{k}\left(\sigma_s^{(n)}(\omega)-\beta_{s-1}^{(n)}(\omega)\right), & k>0,\end{cases} \tag{1.2.7}$$

$$\sigma^{(n)}(\omega)=\sum_{s=1}^{\infty}\left(\sigma_s^{(n)}(\omega)-\beta_{s-1}^{(n)}(\omega)\right)=\lim_{k\to\infty}\tau_k^{(n)}(\omega), \tag{1.2.8}$$

$$\alpha_t^{(n)}(\omega)=\begin{cases}\beta_{k-1}^{(n)}(\omega)+\left(t-\tau_{k-1}^{(n)}(\omega)\right), & \text{if}\ \tau_{k-1}^{(n)}(\omega)\leqslant t<\tau_k^{(n)}(\omega),\\ \sigma(\omega); & \text{if}\ t\geqslant\sigma^{(n)}(\omega).\end{cases} \tag{1.2.9}$$

For any $\omega\in\Omega$, $t<\sigma^{(n)}(\omega)$, let

$$x^{(n)}_{(t,\,\omega)} = x(\alpha_t^{(n)}(\omega),\,\omega). \tag{1.2.10}$$

We denote by g_n the transformation from $X(\omega) = \{x(t,\,\omega),\ t < \sigma(\omega)\}$ to $X^{(n)}(\omega) = \{x^{(n)}(t,\,\omega),\ t < \sigma^{(n)}(\omega)\}$, i.e.,

$$g_n(X(\omega)) = X^{(n)}(\omega). \tag{1.2.11}$$

§ 1.3 Convergence of the sequence $X^{(n)}(\omega)$ $(n \geqslant 1)$

Lemma 1.3.1

$$\lim_{k\to\infty} \beta_k^{(n)}(\omega) = \beta^{(n)}(\omega) = \sigma(\omega) \qquad (\omega \in \Omega). \tag{1.3.1}$$

Proof. By (1.2.6), the limit $\lim_{k\to\infty} \beta_k^{(n)}(\omega) = \beta^{(n)}(\omega)$ $(\omega \in \Omega)$ exists. Let

$$\Delta = \{\omega\colon \beta^{(n)}(\omega) < \sigma(\omega)\}. \tag{1.3.2}$$

Hence

$$\beta^{(n)}(\omega) < +\infty \qquad (\omega \in \Delta). \tag{1.3.3}$$

For every $\omega \in \Delta$, from Property (D_1) and $x(\beta_{k-1}^{(n)}(\omega)) \in D_n$ it follows that

$$\beta_{k-1}^{(n)}(\omega) < \sigma_k^{(n)}(\omega) \leqslant \sigma(\omega),$$

if $\beta_{k-1}^{(n)}(\omega) < \sigma(\omega)$. Thus we have

$$\begin{aligned} 0 \equiv \beta_0^{(n)}(\omega) < \sigma_1^{(n)}(\omega) \leqslant \beta_1^{(n)}(\omega) < \sigma_2^{(n)}(\omega) \leqslant \cdots \leqslant \beta_{k-1}^{(n)}(\omega) \\ < \sigma_k^{(n)}(\omega) \leqslant \beta_k^{(n)}(\omega) < \cdots < \sigma(\omega) \qquad (\omega \in \Delta). \end{aligned} \tag{1.3.4}$$

Hence

$$0 \equiv \beta_0^{(n)}(\omega) < \beta_1^{(n)}(\omega) < \cdots < \beta_k^{(n)}(\omega) < \cdots < \sigma(\omega) \qquad (\omega \in \Delta). \tag{1.3.5}$$

Since $D_n = (1,\,2,\,\cdots,\,n)$ contains a finite number of elements and $x(\beta_k^{(n)}(\omega)) \in D_n$, there must be an $i \in D_n$ such that for any $\omega \in \Delta$,

$$x(\beta_k^{(n)}(\omega)) = i \text{ holds for an infinite number of } k. \tag{1.3.6}$$

So

$$x(\cdot, \omega) \text{ has an infinite number of } i\text{-intervals in } [0, \beta^{(n)}(\omega)). \tag{1.3.7}$$

Therefore we see $\Delta = \varnothing$ by (1.3.3) and Property (D_3). □ [1]

Lemma 1.3.2 *For any $\omega \in \Omega$ and $t < \sigma(\omega)$, we have*

$$\lim_{n \to \infty} \mu(B_t^{(n)}(\omega)) = 0, \tag{1.3.8}$$

where

$$B_t^{(n)}(\omega) = \bigcup_{k=1}^{\infty} [\sigma_k^{(n)}(\omega), \beta_k^{(n)}(\omega)) \cap [0, t). \tag{1.3.9}$$

Proof. Let

$$A(\omega) = \{t: x(t, \omega) \neq +\infty, t < \sigma(\omega)\}, \tag{1.3.10}$$

$$B(\omega) = \{t: x(t, \omega) = +\infty, t < \sigma(\omega)\}, \tag{1.3.11}$$

$$A^{(n)}(\omega) = \bigcup_{k=1}^{\infty} [\beta_{k-1}^{(n)}(\omega), \quad \sigma_k^{(n)}(\omega)), \tag{1.3.12}$$

$$B^{(n)}(\omega) = \bigcup_{k=1}^{\infty} [\sigma_k^{(n)}(\omega), \quad \beta_k^{(n)}(\omega)). \tag{1.3.13}$$

It is easy to show by Lemma 1.3.1 that

$$A(\omega) \cup B(\omega) = A^{(n)}(\omega) \cup B^{(n)}(\omega) = [0, \sigma(\omega)), \tag{1.3.14}$$

$$A^{(n)}(\omega) \subseteq A^{(n+1)}(\omega), \tag{1.3.15}$$

$$B^{(n)}(\omega) \supseteq B^{(n+1)}(\omega), \tag{1.3.16}$$

$$A(\omega) = \bigcup_{n=1}^{\infty} A^{(n)}(\omega), \tag{1.3.17}$$

$$B(\omega) = \bigcap_{n=1}^{\infty} B^{(n)}(\omega). \tag{1.3.18}$$

Hence if we let

---

1 The symbol □ indicates the end of the proof or the fact that the proof is omitted.

$$A_t(\omega) = A(\omega) \bigcap [0, t), \tag{1.3.19}$$

$$B_t(\omega) = B(\omega) \bigcap [0, t), \tag{1.3.20}$$

$$A_t^{(n)}(\omega) = A^{(n)}(\omega) \bigcap [0, t), \tag{1.3.21}$$

then it follows from (1.3.9) and (1.3.13)—(1.3.18) that

$$B_t^{(n)}(\omega) = B^{(n)}(\omega) \bigcap [0, t), \tag{1.3.22}$$

$$A_t(\omega) \bigcup B_t(\omega) = A_t^{(n)}(\omega) \bigcup B_t^{(n)}(\omega) = [0, t), \tag{1.3.23}$$

$$A_t^{(n)}(\omega) \subseteq A_t^{(n+1)}(\omega), \tag{1.3.24}$$

$$B_t^{(n)}(\omega) \supseteq B_t^{(n+1)}(\omega), \tag{1.3.25}$$

$$A_t(\omega) = \bigcup_{n=1}^{\infty} A_t^{(n)}(\omega), \tag{1.3.26}$$

$$B_t(\omega) = \bigcap_{n=1}^{\infty} B_t^{(n)}(\omega). \tag{1.3.27}$$

From

$$\mu(B_t^{(n)}(\omega)) \leqslant t < +\infty, \tag{1.3.28}$$

$$\mu(B_t(\omega)) \leqslant \mu(B(\omega)) = 0, \tag{1.3.29}$$

and (1.3.27), we get

$$\lim_{n \to \infty} \mu(B_t^{(n)}(\omega)) = \mu(B_t(\omega)) = 0. \quad \square \tag{1.3.30}$$

Lemma 1.3.3 *For any $\omega \in \Omega$ and $t < \sigma(\omega)$, we have*

$$\alpha_t^{(n)}(\omega) \downarrow t \qquad (n \uparrow +\infty). \tag{1.3.31}$$

Proof. We know from Lemma 1.3.2 that

$$\lim_{n \to \infty} \sigma^{(n)}(\omega) = \sigma(\omega). \tag{1.3.32}$$

Hence, since $t < \sigma(\omega)$ there exists $N > 0$ such that

$$t < \sigma^{(n)}(\omega) \qquad (n \geqslant N). \tag{1.3.33}$$

It follows from the definition of $\alpha_t^{(n)}(\omega)$ that

$$\alpha_t^{(n)}(\omega) = t + \mu(B^{(n)}_{\alpha_t^{(n)}(\omega)}(\omega)) \qquad (n \geqslant N), \tag{1.3.34}$$

$$\alpha_t^{(n)}(\omega) \geqslant \alpha_t^{(n+1)}(\omega) \qquad (n = 1, 2, \cdots). \tag{1.3.35}$$

From the definition of $B_t^{(n)}(\omega)$ and (1.3.35), we have

$$0 \leqslant \mu(B^{(n)}_{\alpha_t^{(n)}(\omega)}(\omega)) \leqslant \mu(B^{(n)}_{\alpha_t^{(N)}(\omega)}(\omega)) \qquad (n \geqslant N). \tag{1.3.36}$$

From the definition of $\alpha_t^{(n)}(\omega)$ and (1.3.33), we see that

$$\alpha_t^{(N)}(\omega) < \sigma(\omega) \leqslant +\infty. \tag{1.3.37}$$

Therefore we have

$$\lim_{n\to\infty} \mu(B^{(n)}_{\alpha_t^{(N)}(\omega)}(\omega)) = 0 \tag{1.3.38}$$

by Lemma 1.3.2,(1.3.31) follows immediately from (1.3.34), (1.3.35), (1.3.36) and (1.3.38). □

Theorem 1.3.1

$$\lim_{n\to\infty} x^{(n)}(t, \omega) = x(t, \omega), \tag{1.3.39}$$

for every point t in $[0, \sigma(\omega))$ $(\omega\in\Omega)$.

Proof. Suppose $t < \sigma(\omega)$. Then there exists $N>0$ such that

$$x^{(n)}(t, \omega) = x(\alpha_t^{(n)}(\omega), \omega) \qquad (n \geqslant N). \tag{1.3.40}$$

The theorem now follows immediately from Property (D_1) and Lemma 1.3.3. □

§ 1.4. Further properties of $X^{(n)}(\omega)$ $(n \geqslant 1)$

Let $X(\omega)=\{x(t, \omega), t<\sigma(\omega)\}$ be any Q process. N_t stands for a σ-algebra in the space $\Omega_t=(\sigma>t)$ generated by the sets of the form $(x_u=i, \sigma>t)$ $(u\leqslant t, i\in E\bigcup\{+\infty\})$. Then we have

$$u \leqslant t, \qquad A\in N_t \Longrightarrow A\bigcap\Omega_t\in N_t. \tag{1.4.1}$$

We define an optional random variable according to [6] as follows:

Definition 1.4.1 *The function* $\delta(\omega)$ $(\omega\in\Omega)$ *is called an optional random* *variable with respect to the process*

$$X(\omega)=\{x(t,\omega),\ t<\sigma(\omega)\},$$

if

(1) $$0 \leqslant \delta(\omega) \leqslant \sigma(\omega) \qquad (\omega \in \Omega) \tag{1.4.2}$$

and

(2) $$(\delta \leqslant t < \sigma) \in N_t \tag{1.4.3}$$

for all $t \geqslant 0$. *Obviously, Condition* (2) *can be replaced by the following condition:*

(2)′ $$(\delta > t) \in N_t \tag{1.4.4}$$

for all $t \geqslant 0$.

Lemma 1.4.1 *If* $\delta(\omega)$ *is an optional random variable with respect to the process*

$$X(\omega)=\{x(t,\omega),\ t<\sigma(\omega)\},$$

then

$$(\delta > u,\ \sigma > t) \in N_t \qquad (u \leqslant t), \tag{1.4.5}$$

$$(\delta \geqslant u,\ \sigma > t) \in N_t \qquad (u \leqslant t), \tag{1.4.6}$$

$$(\delta = u,\ \sigma > t) \in N_t \qquad (u \leqslant t), \tag{1.4.7}$$

$$(\delta \leqslant u,\ \sigma > t) \in N_t \qquad (u \leqslant t), \tag{1.4.8}$$

and

$$(\delta < u,\ \sigma > t) \in N_t \qquad (u \leqslant t). \tag{1.4.9}$$

Proof. Equations (1.4.1) and (1.4.4) yield (1.4.5) and (1.4.8). We deduce from (1.4.1) and (1.4.5) that

$$(\delta \geqslant u,\ \sigma > t) = \bigcap_{n=1}^{\infty}\left(\delta > u - \frac{1}{n},\ \sigma > t\right) \in N_t. \tag{1.4.10}$$

So we obtain (1.4.6), then (1.4.9). Furthermore (1.4.7) follows from (1.4.5) and (1.4.6). □

Lemma 1.4.2 *Let δ_1, δ_2 be the optional random variables with respect to the process*

$$X(\omega) = \{x(t,\ \omega),\ t < \sigma(\omega)\}$$

and $\delta_1 \leqslant \delta_2$. Then

$$(\delta_2 - \delta_1 < r,\ \delta_2 \leqslant t,\ \sigma > t) \subset N_t \tag{1.4.11}$$

and

$$(\delta_2 - \delta_1 > r,\ \delta_2 \leqslant t,\ \sigma > t) \in N_t. \tag{1.4.12}$$

Proof.
(i) Proof of (1.4.11).
If $r \leqslant 0$, then

$$(\delta_2 - \delta_1 < r,\ \delta_2 \leqslant t,\ \sigma > t) = \varnothing \in N_t. \tag{1.4.13}$$

If $r > 0$, and R denotes the set of rational number, then by Lemma 1.4.1,

$$\begin{aligned}
&(\delta_2 - \delta_1 < r,\ \delta_2 \leqslant t,\ \sigma > t)\\
&= (\delta_2 < \delta_1 + r,\ \delta_2 < t,\ \sigma > t) \bigcup (\delta_2 < \delta_1 + r,\ \delta_2 = t,\ \sigma > t)\\
&= \left\{ \bigcup_{\substack{r_1, r_2 \in R \\ r_1, r_2 < t \\ r_2 < r_1 + r}} [(\delta_2 < r_2,\ \sigma > t) \bigcap (\delta_1 > r_1,\ \sigma > t)] \right\}\\
&\quad \bigcup [(\delta_1 > t - r,\ \sigma > t) \bigcap (\delta_2 = t,\ \sigma > t)] \in N_t.
\end{aligned} \tag{1.4.14}$$

Equation (1.4.11) follows from (1.4.13) and (1.4.14).

(ii) Proof of (1.4.12).
If $r < 0$, then

$$(\delta_2 - \delta_1 > r,\ \delta_2 \leqslant t,\ \sigma > t) = (\delta_2 \leqslant t,\ \sigma > t) \in N_t. \tag{1.4.15}$$

If $r \geqslant 0$, then by Lemma 1.4.1,

$$(\delta_2-\delta_1>r,\ \delta_2\leqslant t,\ \sigma>t)$$

$$=\left\{\bigcup_{\substack{r_1,r_2\in R\\ r_1,r_2<t\\ r_2>r_1+r}}[(\delta_2>r_2,\ \sigma>t)\cap(\sigma_2<t,\ \sigma>t)\cap(\delta_1<r_1,\ \sigma>t)]\right\}$$

$$\bigcup[(\delta_1<t-r,\ \sigma>t)\cap(\delta_2=t,\ \sigma>t)]\in N_t. \tag{1.4.16}$$

Equation (1.4.12) follows from (1.4.15) and (1.4.16). □

Lemma 1.4.3 *For any* $s\in[0,+\infty)$, $\alpha_s^{(n)}(\omega)$ *is an optional random variable with respect to the Q-process* $X(\omega)=\{x(t,\omega),\ t<\sigma(\omega)\}$.

Proof. In the proof of this lemma, we shall write $\alpha_s^{(n)}(\omega)$ as $\alpha(\omega)$ for short, and note that $\beta_0^{(n)}(\omega)$, $\sigma_k^{(n)}(\omega)$, $\beta_k^{(n)}(\omega)$ $(k=1,2,\cdots)$ are optional random variables with respect to the Q-process $X(\omega)$.

We now prove it for the following two cases:

(i) $t<s$.

If $\omega\in(\sigma>t)$ and $s\geqslant\sigma^{(n)}(\omega)$, then $\alpha(\omega)=\sigma(\omega)>t$; if $\omega\in(\sigma>t)$ and $s<\sigma^{(n)}(\omega)$, then $\alpha(\omega)\geqslant s>t$. Hence $(\alpha>t)\supseteq(\sigma>t)$, On the other hand, $(\alpha>t)\subseteq(\sigma>t)$. Therefore, we have

$$(\alpha>t)=(\sigma>t) \tag{1.4.17}$$

when $t<s$.

(ii) $t\geqslant s$.

$(\alpha>t)=(\alpha>t,\ \sigma>t)$, since $(\alpha>t)\subseteq(\sigma>t)$. Hence it follows from Lemma 1.3.1 that

$$(\alpha>t)=\left[\bigcup_{k=1}^{\infty}(\beta_{k-1}^{(n)}\leqslant t<\sigma_k^{(n)},\ \alpha>t,\ \sigma>t)\right]$$

$$\bigcup\left[\bigcup_{k=1}^{\infty}(\sigma_k^{(n)}\leqslant t<\beta_k^{(n)},\ \alpha>t,\ \sigma>t)\right]. \tag{1.4.18}$$

When $\beta_0^{(n)}\leqslant t<\sigma_1^{(n)}$, we have $\alpha\leqslant s\leqslant t$. Therefore

$$(\beta_0^{(n)}\leqslant t<\sigma_1^{(n)},\ \alpha>t,\ \sigma>t)=\varnothing. \tag{1.4.19}$$

From (1.3.22) and the fact that $\alpha>t$ if and only if $\mu(B_t^{(n)})>t-s$, it follows that

$$(\beta_{k-1}^{(n)}\leqslant t<\sigma_k^{(n)},\ \alpha>t,\ \sigma>t)$$

$$= (\beta_{k-1}^{(n)} \leqslant t < \sigma_k^{(n)},\ \mu(B_t^{(n)}) > t-s,\ \sigma > t)$$

$$= \left[\left(\sum_{j=1}^{k-1} (\beta_j^{(n)} - \sigma_j^{(n)})\right) > t-s,\ \beta_{k-1}^{(n)} \leqslant t,\ \sigma > t\right] \bigcap (\sigma_k^{(n)} > t)$$

$$= \left\{ \bigcup_{\substack{r_j \in R \\ \sum_{j=1}^{k-1} r_j > t-s}} \left[\bigcap_{j=1}^{k-1} (\beta_j^{(n)} - \sigma_j^{(n)} > r_j,\ \beta_j^{(n)} \leqslant t,\ \sigma > t\right]\right\} \bigcap (\sigma_k^{(n)} > t)\ (k>1).$$

(1.4.20)

From (1.3.21) and the fact that $\alpha > t$ if and only if $\mu(A_t^{(n)}) < s$, it follows that

$$(\sigma_k^{(n)} \leqslant t < \beta_k^{(n)},\ \alpha > t,\ \sigma > t)$$

$$= (\sigma_k^{(n)} \leqslant t < \beta_k^{(n)},\ \mu(A_t^{(n)}) < s,\ \sigma > t)$$

$$= \left[\left(\sum_{j=1}^{k} \sigma_j^{(n)} - \beta_{j-1}^{(n)}\right) < s,\ \sigma_k^{(n)} \leqslant t,\ \sigma > t\right] \bigcap (\beta_k^{(n)} > t)$$

$$= \left\{ \bigcup_{\substack{r_j \in R \\ \sum_{j=1}^{k} r_j < s}} \left[\bigcap_{j=1}^{k} (\sigma_j^{(n)} - \beta_{j-1}^{(n)} < r_j,\ \sigma_j^{(n)} \leqslant t,\ \sigma > t)\right]\right\} \bigcap (\beta_k^{(n)} > t). \quad (1.4.21)$$

From Definition 1.4.1, Lemma 1.4.2 and (1.4.17)—(1.4.21), we see that $(\alpha > t) \in N_t$. Thus this lemma follows from $0 \leqslant \alpha(\omega) \leqslant \sigma(\omega)$ $(\omega \in \Omega)$ and Definition 1.4.1. □

Theorem 1.4.1 $X^{(n)}(\omega) = \{x^{(n)}(t, \omega),\ t < \sigma^{(n)}(\omega)\}$ *is a Q-process of order one.*

Proof. Obviously, we only need to prove that $X^{(n)}(\omega)$ is a homogeneous Markov process. Suppose

$$0 \leqslant t_1 \leqslant t_2 \leqslant \cdots \leqslant t_{k-1} < t_k, \qquad i_1, i_2, \cdots, i_k \in E. \qquad (1.4.22)$$

Let

$$\Delta_t^{(n)} = \{\omega\colon \alpha_t^{(n)}(\omega) < \sigma(\omega)\}. \qquad (1.4.23)$$

Then, clearly

$$\Delta_t^{(n)} = \{\omega\colon t < \sigma^{(n)}(\omega)\}, \qquad (1.4.24)$$

and

$$\Delta_{t_1}^{(n)} \supseteq \Delta_{t_2}^{(n)} \supseteq \cdots \supseteq \Delta_{t_k}^{(n)}. \qquad (1.4.25)$$

From Lemma 1.4.3, $\alpha^{(n)}_{t_j}$ $(j=1, 2, \cdots, k)$ is an optional random variable with respect to the process $X(\omega)$. Let

$$\tilde{x}(t, \omega) = x(\alpha^{(n)}_{t_{k-1}}(\omega) + t, \omega) \qquad (\omega \in \Delta^{(n)}_{t_{k-1}}, t < \sigma(\omega) - \alpha^{(n)}_{t_{k-1}}(\omega)). \quad (1.4.26)$$

From the strong Markov property, the fact that $\alpha^{(n)}_{t_{k-1}}(\omega)$ is an optional random variable with respect to the process $X(\omega)$, and $P(x(\alpha^{(n)}_{t_{k-1}}(\omega)) = +\infty) = 0$, we see that

$$\tilde{X}(\omega) = \{\tilde{x}(t, \omega), t < \sigma(\omega) - \alpha^{(n)}_{t_{k-1}}(\omega)\} \qquad (\omega \in \Delta^{(n)}_{t_{k-1}})$$

is a Q-process with the same transition probabilities as $X(\omega)$. Let $\tilde{\alpha}^{(n)}_t(\omega)$ $(\omega \in \Delta^{(n)}_{i_{k-1}})$ be a random variable of the Q-process $\tilde{X}(\omega)$ defined in the same way as $\alpha^{(n)}_t(\omega)$ in (1.2.9). It is apparent that

$$\alpha^{(n)}_{t_k}(\omega) - \alpha^{(n)}_{t_{k-1}}(\omega) = \tilde{\alpha}^{(n)}_{t_k - t_{k-1}}(\omega) \qquad (\omega \in \Delta^{(n)}_{t_k}). \quad (1.4.27)$$

Hence

$$\begin{aligned}
&P(x^{(n)}(t_k) = i_k | x^{(n)}(t_1) = i_1, x^{(n)}(t_2) = i_1, \cdots, x^{(n)}(t_{k-1}) = i_{k-1}) \\
&= P(x(\alpha^{(n)}_{t_k}) = i_k | x(\alpha^{(n)}_{t_1}) = i_1, x(\alpha^{(n)}_{t_2}) = i_2, \cdots, x^{(n)}(\alpha_{t_{k-1}}) = i_{k-1}) \\
&= P(x(\alpha^{(n)}_{t_k}) = i_k | x(\alpha^{(n)}_{t_{k-1}}) = i_{k-1}) \\
&= P(\tilde{x}(\alpha^{(n)}_{t_k} - \alpha^{(n)}_{t_{k-1}}) = i_k | \tilde{x}(0) = i_{k-1}) \\
&= P(\tilde{x}(\tilde{\alpha}^{(n)}_{t_{k-1}t_{k-1}}) = i_k | \tilde{x}(0) = i_{k-1}) \\
&= P(x(\alpha^{(n)}_{t_{k-1}t_{k-1}}) = i_k | x(0) = i_{k-1}) \\
&= P(x^{(n)}(t_k - t_{k-1}) = i_k | x^{(n)}(0) = i_{k-1}).
\end{aligned} \quad (1.4.28)$$

Therefore $X^{(n)}(\omega)$ is a homogeneous Markov process. □

Remark 1.4.1 The proof of Theorem 1.4.1 can be further simplified if we use the notation and results of [7]. In fact, it is easily seen that $\alpha^{(n)}_t(\omega)$ satisfies Conditions (3.i)–(3.v) of [7]. (In [7] the original process is assumed to be honest. As a matter of fact, this assumption is not essential.) Thus we know immediately from Theorem 2.1 of [7] that $X^{(n)}(\omega)$ is a homogeneous Markov process, so Theorem 1.4.1 holds.

§ 1.5 The first construction theorem

Summing up the above results, we obtain

Theorem 1.5.1 (The First Construction Theorem). *Suppose* $X(\omega) = \{x(t, \omega), t < \sigma(\omega)\}$ *is any Q-process, Let*

$$X^{(n)}(\omega) = g_n(X(\omega)). \tag{1.5.1}$$

Then

(i) $X^{(n)}(\omega) = \{x^{(n)}(t, \omega), t < \sigma^{(n)}(\omega)\}$ *is a Q-process of order one;*

(ii)
$$\lim_{n \to \infty} x^{(n)}(t, \omega) = x(t, \omega) \tag{1.5.2}$$

for every t in $[0, \sigma(\omega))$ $(\omega \in \Omega)$. □

CHAPTER II
The Second Construction Theorem

§ 2.1 Introduction

Suppose $(\Omega, \mathcal{F}, P)$ is a complete probability space,

$$X^{(n)}(\omega) = \{x^{(n)}(t, \omega), t < \sigma^{(n)}(\omega)\}$$

is a sequence of Q-processes of order one on it, and

$$g_n(X^{(n+1)}(\omega)) = X^{(n)}(\omega). \tag{2.1.1}$$

Then we have

$$g_n(X^{(m)}(\omega)) = X^{(n)}(\omega) \qquad (m \geqslant n). \tag{2.1.2}$$

For definitions of the Q-process, the Q-process of order one, and the transformation g_n, see Chapter I. It is noted that all the Q-processes of this chapter satisfy (1.1.1) and Property (D) in Chapter I.

No misunderstanding should arise when ω is omitted from the related notation in §§2.2—2.4 since the problem of convergence of $x^{(n)}(t, \omega)$ $(n \geqslant 1)$ as $n \to +\infty$ is concerned with only a fixed $\omega \in \Omega$.

In this chapter we will demonstrate the construction theorem 2.5.1 which will be used mainly to construct all Q-processes for any given Q-matrix.

§ 2.2 The mapping T_{mn}

For each fixed natural number n, we define $\tau_k^{(n)}$ $(k = 0, 1, \cdots)$ in the following way:

$$\tau_0^{(n)} = 0, \tag{2.2.1}$$

$$\tau_1^{(n)} \text{ is the first flying point of } x^{(n)}(t). \tag{2.2.2}$$

Suppose $\tau_{k-1}^{(n)}$ is defined. Then let $\tau_k^{(n)} = \sigma^{(n)}$ if $\tau_{k-1}^{(n)}(\omega) = \sigma^{(n)}$. Otherwise, let

$$\tau_k^{(n)} \text{ be the first flying point of } x^{(n)}(t) \text{ after } \tau_{k-1}^{(n)}. \tag{2.2.3}$$

Since $x^{(n)}(t)$ is a Q-process of order one, it can be seen that the above $\tau_k^{(n)}(k=0, 1, \cdots)$ are determined uniquely, and

$$\lim_{k\to\infty} \tau_k^{(n)} = \sigma^{(n)}. \tag{2.2.4}$$

For $m \geqslant n$, let

$$\beta_0^{(m,n)} = 0, \tag{2.2.5}$$

let

$$\sigma_1^{(m,n)} \text{ be the first flying point of } x^{(m)}(t), \tag{2.2.6}$$

and let

$$\beta_1^{(m,n)} = \begin{cases} \inf\{t: \sigma_1^{(m,n)} \leqslant t < \sigma^{(m)},\ x^{(m)}(t) \in D_n\}, \\ \sigma^{(m)}, \quad \text{if the above set is empty,} \end{cases} \tag{2.2.7}$$

where $D_n = (1, 2, \cdots, n)$. Suppose $\sigma_{k-1}^{(m,n)}$, $\beta_{k-1}^{(m,n)}$ are defined. Then let $\sigma_k^{(m,n)} = \beta_k^{(m,n)} = \sigma^{(m)}$, if $\beta_{k-1}^{(m,n)} = \sigma^{(m)}$; otherwise, let

$$\sigma_k^{(m,n)} \text{ be the first flying point of } x^{(m)}(t) \text{ after } \beta_{k-1}^{(m,n)} \tag{2.2.8}$$

and let

$$\beta_k^{(m,n)} = \begin{cases} \inf(t: \sigma_k^{(m,n)} \leqslant t < \sigma^{(m)},\ x^{(m)}(t) \in D_n), \\ \sigma^{(m)}, \quad \text{if the above set is empty.} \end{cases} \tag{2.2.9}$$

It is clear that

$$\sigma^{(1)} \leqslant \sigma^{(2)} \leqslant \cdots \leqslant \sigma^{(m)} \leqslant \cdots, \tag{2.2.10}$$

$$0 = \beta_0^{(m,n)} \leqslant \sigma_1^{(m,n)} \leqslant \cdots \leqslant \sigma_k^{(m,n)} \leqslant \beta_k^{(m,n)} \leqslant \cdots \leqslant \sigma^{(m)}, \tag{2.2.11}$$

$$\sigma_k^{(n,n)} \leqslant \sigma_k^{(n+1,n)} \leqslant \cdots \leqslant \sigma_k^{(n+s,n)} \leqslant \cdots \qquad (k = 1, 2, \cdots), \tag{2.2.12}$$

$$\beta_k^{(n,n)} \leqslant \beta_k^{(n+1,n)} \leqslant \cdots \leqslant \beta_k^{(n+s,n)} \leqslant \cdots \qquad (k = 0, 1, \cdots), \tag{2.2.13}$$

$$\sigma_k^{(n,n)} = \beta_k^{(n,n)} = \tau_k^{(n)} \qquad (k = 1, 2, \cdots), \tag{2.2.14}$$

$$\sigma_k^{(m,n)} - \beta_{k-1}^{(m,n)} = \sigma_k^{(n,n)} - \beta_{k-1}^{(n,n)} = \tau_k^{(n)} - \tau_{k-1}^{(n)} \qquad (m \geqslant n), \tag{2.2.15}$$

$$\sum_{k=1}^{\infty} \sigma_k^{(m,n)} - \beta_{k-1}^{(m,n)} = \sum_{k=1}^{\infty} \tau_k^{(n)} - \tau_{k-1}^{(n)} = \sigma^{(n)} \qquad (m \geqslant n). \tag{2.2.16}$$

By virtue of (2.2.10), (2.2.12) and (2.2.13), there exist the (finite or infinite) limits

$$\lim_{m\to\infty} \sigma^{(m)} = \sigma, \tag{2.2.17}$$

$$\lim_{m\to\infty} \sigma_k^{(m,n)} = \sigma_k^{(n)}, \tag{2.2.18}$$

and

$$\lim_{m\to\infty} \beta_k^{(m,n)} = \beta_k^{(n)}. \tag{2.2.19}$$

By (2.2.11), we have

$$0 = \beta_0^{(n)} \leqslant \sigma_1^{(n)} \leqslant \cdots \leqslant \sigma_k^{(n)} \leqslant \beta_k^{(n)} \leqslant \cdots \leqslant \sigma. \tag{2.2.20}$$

From (2.2.15), if $\beta_{k-1}^{(n)} < +\infty$ then it can be deduced that

$$\sigma_k^{(n)} - \beta_{k-1}^{(n)} = \sigma_k^{(m,n)} - \beta_{k-1}^{(m,n)} = \tau_k^{(n)} - \tau_{k-1}^{(n)} \qquad (m \geqslant n). \tag{2.2.21}$$

Let

$$R^{(m,n)} = \bigcup_{k=1}^{\infty} [\beta_{k-1}^{(m,n)}, \sigma_k^{(m,n)}). \tag{2.2.22}$$

We now define the mapping T_{mn} from $[0, \sigma^{(n)})$ onto $R^{(m,n)}$: let

$$T_{mn}t = \beta_{k-1}^{(m,n)} + t - \tau_{k-1}^{(n)}, \tag{2.2.23}$$

when $t \in [\tau_{k-1}^{(n)}, \tau_k^{(n)})$.

Clearly, if $t_1 < t_2$ and $t_1, t_2 \in [0, \sigma^{(n)})$, then

$$T_{mn}t_1 < T_{mn}t_2. \tag{2.2.24}$$

Therefore the inverse T_{mn}^{-1} of T_{mn} exists and is unique. From (2.1.2),

$$x^{(n)}(t) = x^{(m)}(T_{mn}t), \qquad t \in [0, \sigma^{(n)}), \tag{2.2.25}$$

$$x^{(m)}(t) = x^{(n)}(T_{mn}^{-1}t), \qquad t \in R^{(m,n)}. \tag{2.2.26}$$

Obviously, if $k \geqslant m \geqslant n$, then

$$T_{km}T_{mn}t = T_{kn}t, \qquad t \in [0, \sigma^{(n)}), \tag{2.2.27}$$

$$T_{mn}^{-1}T_{km}^{-1}t = T_{kn}^{-1}t, \qquad t \in R^{(k,n)}. \tag{2.2.28}$$

In particular,

$$T_{mn}\tau_k^{(n)} = \beta_k^{(m,n)},$$

$$x^{(n)}\left(\tau_k^{(n)}\right) = x^{(m)}\left(\beta_k^{(m,n)}\right). \tag{2.2.29}$$

§ 2.3 The mapping W_n

Suppose $t \in [0, \sigma^{(n)})$. Since

$$t = T_{nn}t \leqslant T_{n+1,n}t \leqslant \cdots \leqslant T_{n+k,n}t \leqslant \cdots, \tag{2.3.1}$$

there exists the (finite or infinite) limit:

$$\lim_{k\to\infty} T_{n+k,n}t = \dot{t}. \tag{2.3.2}$$

Let

$$\hat{\sigma}^{(n)} = \sup\left\{t\colon t \in [0, \sigma^{(n)}),\ \lim_{k\to\infty} T_{n+k,n}t < +\infty\right\}. \tag{2.3.3}$$

It is easily seen that

$$\lim_{k\to\infty} T_{n+k,n}t < +\infty \tag{2.3.4}$$

if and only if $t \in [0, \hat{\sigma}^{(n)})$, and there exists

$$\lim_{n\to\infty} \hat{\sigma}^{(n)} = \hat{\sigma}. \tag{2.3.5}$$

Let

$$A^{(n)} = \left\{\dot{t}\colon \text{there exists } t \in [0, \hat{\sigma}^{(n)}), \text{ such that } \lim_{k\to\infty} T_{n+k,n}t = \dot{t}\right\}. \tag{2.3.6}$$

Obviously, there exists $1 \leqslant s_n \leqslant +\infty$ such that

$$[0, \hat{\sigma}^{(n)}) = \bigcup_{k=1}^{s_n} [\tau_{k-1}^{(n)}, \tau_k^{(n)}) \tag{2.3.7}$$

and

$$A^{(n)} = \bigcup_{k=1}^{s_n} [\beta_{k-1}^{(n)}, \sigma_k^{(n)}). \tag{2.3.8}$$

And if $s_n < +\infty$, then

$$\beta_0^{(n)} < \beta_1^{(n)} < \cdots < \beta_{s_n-1}^{(n)} < \beta_{s_n}^{(n)} = \sigma. \tag{2.3.9}$$

Let

$$J_n = \begin{cases} \{1, 2, \cdots, s_n\}, & \text{if } s_n < +\infty, \\ \{1, 2, \cdots\}, & \text{if } s_n = +\infty. \end{cases} \tag{2.3.10}$$

We define the mapping W_n from $[0, \hat{\sigma}^{(n)})$ onto $A^{(n)}$:

$$W_n t = \lim_{k \to \infty} T_{n+k,n} t, \quad t \in [0, \hat{\sigma}^{(n)}). \tag{2.3.11}$$

It is apparent that

$$T_{n+k,n} t_2 - T_{n+k,n} t_1 \geqslant t_2 - t_1, \tag{2.3.12}$$

if $t_1 < t_2$, $t_1, t_2 \in [0, \hat{\sigma}^{(n)})$.

Then

$$W_n t_2 - W_n t_1 \geqslant t_2 - t_1. \tag{2.3.13}$$

Thus, the inverse W_n^{-1} of W_n exists and is unique. If

$$t = \tau_{k-1}^{(n)} + \alpha_1, \quad 0 \leqslant \alpha_1 < \tau_k^{(n)} - \tau_{k-1}^{(n)}, \quad k \in J_n, \tag{2.3.14}$$

then

$$W_n t = \beta_{k-1}^{(n)} + \alpha_1; \tag{2.3.15}$$

conversely, if

$$\dot{t} = \beta_{k-1}^{(n)} + \alpha_2, \quad 0 \leqslant \alpha_2 < \sigma_k^{(n)} - \beta_{k-1}^{(n)}, \quad k \in J_n, \tag{2.3.16}$$

then

$$W_n^{-1} \dot{t} = \tau_{k-1}^{(n)} + \alpha_2, \tag{2.3.17}$$

and in particular

$$W_n \tau_k^{(n)} = \beta_k^{(n)}, \quad k \in J_n. \tag{2.3.18}$$

Let

$$B^{(n)} = [0, \sigma) \setminus A^{(n)}, \tag{2.3.19}$$

$$A_t^{(n)} = A^{(n)} \cap [0, t), \tag{2.3.20}$$

$$B_t^{(n)} = B^{(n)} \cap [0, t). \tag{2.3.21}$$

Hence we have

$$A_\sigma^{(n)} = A^{(n)}, \quad B_\sigma^{(n)} = B^{(n)}, \tag{2.3.22}$$

$$t + \mu(B^{(n)}_{W_n t}) = W_n t, \qquad t \in [0, \hat{\sigma}^{(n)}), \tag{2.3.23}$$

$$t - \mu(B^{(n)}_t) = W_n^{-1} t, \qquad t \in A^{(n)}, \tag{2.3.24}$$

where μ denotes Lebesgue measure.

We can prove the following two lemmas without difficulty.

Lemma 2.3.1 *If*

$$t_1 \in [0, \hat{\sigma}^{(n)}), \qquad t_2 \in A^{(n)}, \tag{2.3.25}$$

then

$$W_n t_1 = W_m T_{mn} t_1 \qquad (m \geqslant n), \tag{2.3.26}$$

$$W_n^{-1} t = T_{mn}^{-1} W_m^{-1} t_2 \qquad (m \geqslant n). \ \square \tag{2.3.27}$$

Lemma 2.3.2 *For any $s \in [0, \sigma)$, we have*

$$A_s^{(n)} \subseteq A_s^{(n+1)} \qquad (n = 1, 2, \cdots). \tag{2.3.28}$$

Hence

$$B_s^{(n)} \supseteq B_s^{(n+1)} \qquad (n = 1, 2, \cdots). \ \square \tag{2.3.29}$$

Let

$$A_t = \bigcup_{n=1}^{\infty} A_t^{(n)}, \qquad 0 \leqslant t \leqslant \sigma, \tag{2.3.30}$$

$$B_t = \bigcap_{n=1}^{\infty} B_t^{(n)}, \qquad 0 \leqslant t \leqslant \sigma, \tag{2.3.31}$$

$$A = A_\sigma, \tag{2.3.32}$$

$$B = B_\sigma. \tag{2.3.33}$$

Then

$$A_t \cap B_t = \varnothing, \qquad A_t \cup B_t = [0, t), \tag{2.3.34}$$

$$A \cap B = \varnothing, \qquad A \cup B = [0, \sigma). \tag{2.3.35}$$

Lemma 2.3.3 *$\hat{\sigma} = \sigma$ if and only if*

$$\mu(B) = 0. \tag{2.3.36}$$

Before proving the lemma, let us dispose of some preliminaries:

Proposition 2.3.1 *If $0 \leqslant t \leqslant \sigma$, $t < +\infty$, then*

$$\lim_{m\to\infty} \mu(B_t^{(m)}) = \mu(B_t). \square \tag{2.3.37}$$

Proposition 2.3.2 *If* $0 \leqslant s \leqslant \sigma$ *and*

$$0 \leqslant s_k \uparrow s \qquad (k\uparrow +\infty), \tag{2.3.38}$$

then

$$\lim_{k\to\infty} \mu(B_{s_k}^{(m)}) = \mu(B_s^{(m)}), \tag{2.3.39}$$

and

$$\lim_{k\to\infty} \mu(B_{s_k}) = \mu(B_s). \square \tag{2.3.40}$$

Proposition 2.3.3 *If*

$$\beta_k^{(n)} < +\infty, \tag{2.3.41}$$

then

$$\mu(B_{\beta_k^{(n)}}) = 0. \tag{2.3.42}$$

Proof. From (2.3.23) and (2.3.41), we have

$$\beta_k^{(m,n)} + \mu(B^{(m)}_{W_m \beta_k^{(m,n)}}) = W_m \beta_k^{(m,n)} \qquad (m \geqslant n), \tag{2.3.43}$$

but

$$\beta_k^{(m,n)} = T_{mn} \tau_k^{(n)} \qquad (m \geqslant n). \tag{2.3.44}$$

Hence we confirm on the basis of Lemma 2.3.1 and (2.3.18) that

$$W_m \beta_k^{(m,n)} = W_n \tau_k^{(n)} = \beta_k^{(n)}. \tag{2.3.45}$$

Substituting (2.3.45) in (2.3.43), we get

$$\beta_k^{(m,n)} + \mu(B^{(m)}_{\beta_k^{(n)}}) = \beta_k^{(n)}. \tag{2.3.46}$$

From (2.2.19), (2.3.41) and Proposition (2.3.1) we get

$$\beta_k^{(n)} + \mu(B_{\beta_k^{(n)}}) = \beta_k^{(n)}; \tag{2.3.47}$$

thus we obtain immediately (2.3.42). This proposition is proved. $\square$

Lemma 2.3.4

$$W_n t \downarrow t \quad (n \to \infty) \qquad \text{for } t \in [0, \hat{\sigma}). \tag{2.3.48}$$

Proof. Let

$$W_m t - t = g(t)$$

for $m \geqslant n$. Then

$$W_m t - t = g(t) \leqslant g(T_{mn}(t)) = W_m(T_{mn}(t)) - T_m(t) = W_n t - T_{mn} t.$$

Letting $n \to \infty$, we obtain (2.3.48). □

Proof of the Lemma 2.3.3.

Suppose $\hat{\sigma} = \sigma$. We have

$$\lim_{k \to \infty} \beta_k^{(n)} = \sigma, \qquad n = 1, 2, \cdots. \tag{2.3.49}$$

In fact that for any $n > 0$,

$$\beta_k^{(n)} \geqslant \beta_k^{(m,n)} \geqslant \sigma_k^{(m,n)} \geqslant \tau_k^{(m)}. \tag{2.3.50}$$

So

$$\lim_{k \to \infty} \beta_k^{(n)} \geqslant \lim_{k \to \infty} \tau_k^{(m)} = \sigma^{(m)} \qquad (m \geqslant n). \tag{2.3.51}$$

Thus

$$\lim_{k \to \infty} \beta_k^{(m)} \geqslant \lim_{m \to \infty} \sigma^{(m)} = \sigma. \tag{2.3.52}$$

Clearly,

$$\lim_{k \to \infty} \beta_k^{(m)} \leqslant \sigma. \tag{2.3.53}$$

Then we have (2.3.49).

If $\sigma < \infty$, we have

$$\mu(B_{\beta_k^{(n)}}) = 0 \tag{2.3.54}$$

by Proposition 2.3.3. It follows that

$$\mu(B) = \lim_{k \to \infty} \mu(B_{\beta_k^{(n)}}) = 0. \tag{2.3.55}$$

If $\sigma = \infty$ for fixed t there exists n which is sufficiently large such that $t < \hat{\sigma}^{(n)}$. We have

$$W_n(t)\downarrow t, \qquad W_n^{-1}(t)\uparrow t \qquad (n\to\infty)$$

by Lemma 2.3.4, therefore

$$\lim_{n\to\infty} \mu(A_t^{(n)}) = \lim_{n\to\infty} W_n^{-1}(t) = t. \tag{2.3.56}$$

So

$$\lim_{n\to\infty} \mu(\beta_t^{(n)}) = 0. \tag{2.3.57}$$

Since t is arbitrary, we have

$$\mu(B) = 0. \tag{2.3.58}$$

We proceed to prove the other assertion. If $\sigma < \infty$, then we have $\sigma^{(n)} \to \infty$ for every n and

$$\lim_{m\to\infty} T_{mn}(t) \leqslant t + \sigma < \infty \quad \text{for } t < \sigma^{(n)}. \tag{2.3.59}$$

Thus $\hat{\sigma}^{(n)} = \sigma^{(n)}$ $(n = 1, 2, \cdots)$. It follows that

$$\hat{\sigma} = \sigma. \tag{2.3.60}$$

If $\hat{\sigma} < \sigma = \infty$, then for any $t \in (\hat{\sigma}, \infty)$ we have

$$t \notin A^{(n)}, \qquad n = 1, 2, \cdots.$$

So, $t \in B$. It follows that

$$\mu(B) = \mu([\hat{\sigma}, \infty)) > 0.$$

But this contradicts (2.3.36). Hence

$$\hat{\sigma} = \infty = \sigma. \square$$

§ 2.4 Constructing auxiliary functions

Let

$$x(t) = \begin{cases} x^{(n)}(W_n^{-1} t), & \text{if } t \in A^{(n)}, \\ +\infty, & \text{if } t \in B. \end{cases} \tag{2.4.1}$$

From (2.2.26) and Lemma 2.3.1 we get

$$x^{(n+1)}(W_{n+1}^{-1}t) = x^{(n)}(T_{n+1,n}^{-1}W_{n+1}^{-1}t) = x^{(n)}(W_n^{-1}t), \qquad t\in A^{(n)}. \tag{2.4.2}$$

Thus $x(t)$ is determined uniquely in $[0, \sigma)$.

Lemma 2.4.1 *$x(t)$ is right continuous in $[0, \hat{\sigma})$, and*

$$\mu\{t: t\in[0, \hat{\sigma}), x(t) = +\infty\} = 0. \tag{2.4.3}$$

Proof. Equation (2.4.3) follows from Lemma 2.3.3. It is easy to see from the right continuity of $x^{(n)}(t)$ that $x(t)$ is right continuous at every point of A. We shall prove that $x(t)$ is also right continuous at every point of B.

Suppose $t\in B$. Noting Lemma 2.3.3, we can readily prove that there exists a sequence of intervals $[\sigma_{k_n}^{(n)}, \beta_{k_n}^{(n)})$ $(n = 1, 2, \cdots)$, such that

$$t\in[\sigma_{k_n}^{(n)}, \beta_{k_n}^{(n)})\in B^{(n)} \qquad (n = 1, 2, \cdots), \tag{2.4.4}$$

$$[\sigma_{k_n}^{(n)}, \beta_{k_n}^{(n)}) \supset [\sigma_{k_{n+1}}^{(n+1)}, \beta_{k_{n+1}}^{(n+1)}) \qquad (n = 1, 2, \cdots), \tag{2.4.5}$$

and

$$\lim_{n\to\infty} \sigma_{k_n}^{(n)} = \lim_{n\to\infty} \beta_{k_n}^{(n)} = t. \tag{2.4.6}$$

Clearly, the values of $x(t)$ in the interval $[\sigma_{k_n}^{(n)}, \beta_{k_n}^{(n)})$ are greater than n, so

$$\lim_{s\to t} x(s) = +\infty = x(t). \tag{2.4.7}$$

Therefore $x(t)$ is right continuous at every point of B. □

Lemma 2.4.2 *For any $t\in[0, \sigma)$ and any $i\in E$, there is only a finite number of i-intervals of $x(\cdot)$ in $[0, t)$.*

Proof. Since the number of i-intervals of $x(\cdot)$ in $[0, t)$ is not more than the number of i-interval of $x^{(i)}(\cdot)$, and $x^{(i)}(\cdot)$ has only a finite number of them in $[0, t)$, we get this lemma immediately. □

Lemma 2.4.3 *If $\hat{\sigma} = \sigma$, then*

$$\lim_{n\to\infty} x^{(n)}(t) = x(t), \qquad t\in[0, \sigma). \tag{2.4.8}$$

Proof. By (2.3.5), we may take $t\in[0, \hat{\sigma}^{(n)})$ if n is sufficiently large. So we have

$$x^{(n)}(t) = x(W_n t). \tag{2.4.9}$$

Equation (2.4.8) follows from (2.4.9) and Lemmas 2.3.4 and 2.4.1. □

§ 2.5 The second construction theorem

Theorem 2.5.1 (The Second Construction Theorem).
Let

$$X^{(n)}(\omega) = \{x^{(n)}(t, \omega), t < \sigma^{(n)}(\omega)\} \qquad (n \geqslant 1) \tag{2.5.1}$$

be a sequence of Q-processes of order one. If

$$g_n(X^{(n+1)}(\omega)) = X^{(n)}(\omega) \tag{2.5.2}$$

and

$$\hat{\sigma}(\omega) = \sigma(\omega), \qquad \omega \in \Omega, \tag{2.5.3}$$

then

(i) $$\lim_{n\to\infty} x^{(n)}(t, \omega) = x(t, \omega) \qquad (\omega \in \Omega) \tag{2.5.4}$$

exists for any $t \in [0, \sigma(\omega))$, *and* $X(\omega) = \{x(t, \omega), t < \sigma(\omega)\}$ *is a Q-process;*

(ii) $$g_n(X(\omega)) = X^{(n)}(\omega). \tag{2.5.5}$$

Proof. In view of §2.4, we need only prove the following two points:
(i) $X(\omega)$ is a homogeneous Markov process.

(ii) $$P\{\omega: x(t, \omega) = +\infty\} = 0, \tag{2.5.6}$$

for any $t \in [0, +\infty)$.

This proof can be established in the same way as for Theorems 6.2 and 6.3 of [2]. □

§ 2.6 Summary

Definition 2.6.1 *The Q-process sequence* $X^{(n)}(\omega)$ $(n \geqslant 1)$ *is called a basic sequence of Q-processes if satisfying* (2.5.2) *and*

$$\sigma(\omega) = \hat{\sigma}(\omega) \qquad (\omega \in \Omega), \tag{2.6.1}$$

where $\sigma(\omega) = \lim_{n\to\infty} \sigma^{(n)}(\omega)$.

Theorem 2.6.1 *Let $X(\omega)=\{x(t,\omega),\ t<\sigma(\omega)\}$ be a Q-process. If we let*

$$g_n(X(\omega)) = X^{(n)}(\omega), \tag{2.6.2}$$

then

(i) $X^{(n)}(\omega) = \{x^{(n)}(t,\omega),\ t<\sigma^{(n)}(\omega)\}$ $(n\geqslant 1)$ *is a basic sequence of Q-processes.*

(ii)
$$\lim_{n\to\infty} x^{(n)}(t,\omega) = x(t,\omega) \qquad (\omega\in\Omega) \tag{2.6.3}$$

holds for any $t\in[0,\sigma(\omega))$. □

Theorem 2.6.2 *Let $X^{(n)}(\omega)=\{x^{(n)}(t,\omega),\ t<\sigma^{(n)}(\omega)\}$ $(n\geqslant 1)$ be a basic sequence of Q-processes. Then*

(i)
$$\lim_{n\to\infty} x^{(n)}(t,\omega) = x(t,\omega) \qquad (\omega\in\Omega) \tag{2.6.4}$$

exists for any $t\in[0,\sigma(\omega))$, and $X(\omega)=\{x(t,\omega),\ t<\sigma(\omega)\}$ is a Q-process, where

$$\sigma(\omega) = \lim_{n\to\infty} \sigma^{(n)}(\omega) \qquad (\omega\in\Omega); \tag{2.6.5}$$

(ii)
$$g_n(X(\omega)) = X^{(n)}(\omega). \ \square \tag{2.6.6}$$

§ 2.7 Two notes

Note 2.7.1 Suppose $X^{(n)}(\omega)=\{x^{(n)}(t,\omega),\ t<\sigma^{(n)}(\omega)\}$ is a sequence of Q-processes of order one and $g_n(X^{(n+1)}(\omega))=X^{(n)}(\omega)$. $X(\omega)$ is defined by (2.4.1). Let

$$\eta_i^{(n)}(\omega) = \begin{cases} \inf\{0<t<\sigma^{(n)}(\omega),\ \ x^{(n)}(t,\omega)=i\}, \\ \infty \quad \text{if the above set is empty,} \end{cases} \tag{2.7.1}$$

$$\eta_i(\omega) = \begin{cases} \inf\{0<t<\sigma(\omega),\ \ x(t,\omega)=i\}, \\ \infty, \quad \text{if the above set is empty.} \end{cases} \tag{2.7.2}$$

$$f_{ij}^{(n)} = P_i(\eta_i^{(n)}<\infty), \tag{2.7.3}$$

$$f_{ij} = P_i(\eta_i<\infty), \tag{2.7.4}$$

$$\Phi_{ij}^{(n)}(\lambda) = E_i e^{-\lambda\eta_j^{(n)}} \quad . \tag{2.7.5}$$

The calculation of $\Phi_{ij}^{(n)}(\lambda)$ will be introduced in Chapter X of this book.

For describing Theorem 2.5.1, we have the following result:

$$\hat{\sigma}=\sigma \text{ if and only if } f_{ij}^{(n)} \to f_{ij} \quad (n\to\infty) \qquad \text{for any } i, j\in E, \tag{2.7.6}$$

i.e.,

$$\lim_{n\to\infty}\lim_{\lambda\to 0}\Phi_{ij}^{(n)}(\lambda)=\lim_{\lambda\to 0}\lim_{n\to\infty}\Phi_{ij}^{(n)}(\lambda). \tag{2.7.7}$$

Note 2.7.2 The first counterexample for Theorem 2.5.1 in the Chinese edition of this book was given by Xiong Daguo. The theorem was revised by Zou Jiezhong and Tang Lingqi here.

PART II

THEORY OF MINIMAL NONNEGATIVE SOLUTIONS FOR SYSTEMS OF NONNEGATIVE LINEAR EQUATIONS

CHAPTER III

General Theory

§ 3.1 Introduction

The purpose of this chapter is to give an exposition of the general theory of minimal nonnegative solutions for systems of nonnegative linear equations.

The results or basic idea in §§2, 3 and 6 of this chapter come from [8, Chapter I, §2]. Thus, most of the proofs for the conclusions in this chapter are omitted.

§ 3.2 Definition of a system of nonnegative linear equations and definition, existence and uniqueness of its minimal nonnegative solution

Definition 3.2.1 *A system of linear equations*

$$x_i = \sum_{k \in E} c_{ik} x_k + b_i \qquad (i \in E) \tag{3.2.1}$$

is called a system of nonnegative linear equations if

$$0 \leqslant c_{ik} < +\infty \quad (i,\ k \in E), \qquad 0 \leqslant b_i \leqslant +\infty \quad (i \in E)^1,$$

where E is the finite set $\{1, 2, \cdots, n\}$ or the denumerable set $\{1, 2, \cdots\}$.

In what follows we always assume that (3.2.1) is a system of nonnegative linear equations.

Definition 3.2.2 *The nonnegative solution $0 \leqslant x_i^* \leqslant +\infty$ $(i \in E)$ of (3.2.1) is said to be its minimal nonnegative solution if for any nonnegative solution $0 \leqslant x_i \leqslant +\infty$ $(i \in E)$ of (3.2.1), we have*

$$x_i^* \leqslant x_i \qquad (i \in E). \tag{3.2.2}$$

Please note that henceforth we often write V-$\lim_{n \to \infty} a_n = a$ *for* $a_n \uparrow a \quad (n \uparrow +\infty)$.

---

1 In this part of our book we always take the set of nonnegative numbers to contain $+\infty$.

Theorem 3.2.1 *The minimal nonnegative solution of* (3.2.1) *exists and is unique. If we let*

$$\left.\begin{aligned} x_i^{(0)} &\equiv 0 \qquad (i \in E), \\ x_i^{(n+1)} &= \sum_{k \in E} c_{ik} x_k^{(n)} + b_i \qquad (n \geqslant 0,\ i \in E), \end{aligned}\right\} \tag{3.2.3}$$

then the limit

$$\text{V-}\lim_{n \to \infty} x_i^{(n)} = x_i^* \qquad (i \in E) \tag{3.2.4}$$

exists and hence x_i^* $(i \in E)$ *is the minimal nonnegative solution of* (3.2.1).□

In this part, we always denote the minimal nonnegative solution[1] of (3.2.1) by x_i^* $(i \in E)$, and (3.2.2) is referred to as the minimality of x_i^* $(i \in E)$.

Corollary 3.2.1 *If* (3.2.1) *is homogeneous, i.e.,* $b_i \equiv 0$ $(i \in E)$, *then*

$$x_i^* \equiv 0 \qquad (i \in E). \ \square \tag{3.2.5}$$

Theorem 3.2.2 *Suppose*

$$b_i^{(n)} \geqslant 0 \qquad (n \geqslant 1,\ i \in E), \tag{3.2.6}$$

$$\text{V-}\lim_{n \to \infty} b_i^{(n)} = b_i \qquad (i \in E). \tag{3.2.7}$$

If we let

$$\left.\begin{aligned} \tilde{x}_i^{(1)} &= b_i^{(1)} \qquad (i \in E), \\ \tilde{x}_i^{(n+1)} &= \sum_{k \in E} c_{ik} \tilde{x}_k^{(n)} + b_i^{(n+1)} \qquad (n \geqslant 1,\ i \in E), \end{aligned}\right\} \tag{3.2.8}$$

then

$$\text{V-}\lim_{n \to \infty} \tilde{x}_i^{(n)} = x_i^* \qquad (i \in E). \ \square \tag{3.2.9}$$

Corollary3.2.2 *Suppose*

$$a_i^{(n)} \geqslant 0 \qquad (n \geqslant 1,\ i \in E), \tag{3.2.10}$$

---

1 This method for finding the solution is henceforth called the successive approximation method.

$$\sum_{n=1}^{\infty} a_i^{(n)} = b_i \qquad (i \in E). \tag{3.2.11}$$

If we let

$$\left.\begin{aligned} \tilde{y}_i^{(1)} &= a_i^{(1)} \qquad (i \in E), \\ \tilde{y}_i^{(n+1)} &= \sum_{k \in E} c_{ik} \tilde{y}_k^{(n)} + a_i^{(n+1)} \qquad (n \geqslant 1,\ i \in E), \end{aligned}\right\} \tag{3.2.12}$$

then

$$\sum_{n=1}^{\infty} \tilde{y}_i^{(n)} = x_i^* \qquad (i \in E). \tag{3.2.13}$$

Corollary 3.2.3 *If we let*

$$\left.\begin{aligned} y_i^{(1)} &= b_i \qquad (i \in E), \\ y_i^{(n+1)} &= \sum_{k \in E} c_{ik} y_k^{(n)} \qquad (n \geqslant 1,\ i \in E), \end{aligned}\right\} \tag{3.2.14}$$

then

$$\sum_{n=1}^{\infty} y_i^{(n)} = x_i^* \qquad (i \in E). \ \square \tag{3.2.15}$$

Theorem 3.2.3 *The unique nonnegative solution of* (3.2.1),*satisfying the inequality* $0 \leqslant \bar{x}_i \leqslant p x_i^*$ $(i \in E,\ p \geqslant 1,\ p$ *a constant), is* x_i^* $(i \in E)$, *which can be obtained by starting from any initial value* $\bar{x}_i^{(0)}$ $(i \in E)$ *satisfying the condition* $0 \leqslant \bar{x}_i^{(0)} \leqslant p x_i^*$ $(i \in E)$, *using the successive approximation method.* □

Theorem 3.2.4 *If* $x_i^* < +\infty$ $(i \in E)$, *then the unique solution of* (3.2.1), *satisfying the inequality* $|\bar{x}_i| \leqslant p x_i^*$ $(i \in E, p \geqslant 1, p$ *a constant), is* x_i^* $(i \in E)$, *which can be obtained by successive approximation, starting from any initial value* $\bar{x}_i^{(0)}$ $(i \in E)$ *satisfying the condition* $|\bar{x}_i^{(0)}| \leqslant p x_i^*$ $(i \in E)$. □

Corollary 3.2.4 *If* x_i^* $(i \in E)$ *possesses the property*

$$0 < \inf_{i \in E} x_i^* \leqslant \sup_{i \in E} x_i^* < +\infty \qquad (i \in E), \tag{3.2.16}$$

then the corresponding system of homogeneous equations

$$x_i = \sum_{k \in E} c_{ik} x_k \qquad (i \in E) \tag{3.2.17}$$

of (3.2.1) *has no nonzero bounded solution.*

Proof. If the system of homogeneous equations does have a nonzero bounded solution

$$\bar{x}_i \quad (i \in E), \qquad |\bar{x}_i| < K < +\infty \quad (i \in E).$$

then, putting $\inf x_i^* = A$,

$$\frac{|\bar{x}_i|}{x_i^*} < \frac{|\bar{x}_i|}{A} \leqslant \frac{K}{A} \qquad (i \in E), \tag{3.2.18}$$

i.e.,

$$|\bar{x}_i| \leqslant \frac{K}{A} x_i^* \qquad (i \in E). \tag{3.2.19}$$

But it is easy to see that $x_i^* + \bar{x}_i$ is the solution of (3.2.1), and

$$|x_i^* + \bar{x}_i| \leqslant x_i^* + |\bar{x}_i| \leqslant \left(1 + \frac{K}{A}\right) x_i^* = p x_i^* \qquad (i \in E). \tag{3.2.20}$$

So the equality

$$x_i^* + \bar{x}_i = x_i^* \qquad (i \in E) \tag{3.2.21}$$

holds, owing to Theorem 3.2.4. Hence

$$\bar{x}_i \equiv 0 \qquad (i \in E). \tag{3.2.22}$$

We are led to a contradiction. Therefore (3.2.17) has no nonzero bounded solution. □

§ 3.3 Comparison theorem and linear combination theorem

Definition 3.3.1 *An inequality system*

$$X_i \geqslant \sum_{k \in E} C_{ik} X_k + B_i \qquad (i \in E) \tag{3.3.1}$$

is called a major system of (3.2.1) *if the following inequalities hold:*

$$c_{ik} \leqslant C_{ik} \qquad (i,\ k \in E), \tag{3.3.2}$$

$$b_i \leqslant B_i \qquad (i \in E). \tag{3.3.3}$$

Theorem 3.3.1 *Let $X_i \ (i \in E)$ be an arbitrary nonnegative solution of the major system* (3.3.1) *of* (3.2.1). *Then*

$$x_i^* \leqslant X_i \qquad (i \in E). \ \square \tag{3.3.4}$$

Corollary 3.3.1 *A necessary and sufficient condition for x_i^* $(i \in E)$ to possess the property*

$$x_i^* < s_i \leqslant +\infty \qquad (i \in G) \tag{3.3.5}$$

is, that a certain major system of (3.2.1) *has a nonnegative solution X_i $(i \in E)$ with the property*

$$X_i < s_i \qquad (i \in G), \tag{3.3.6}$$

where $G \subset E$ and s_i $(i \in G)$ are positive numbers. $\square$

Corollary 3.3.2 *Let $a_k \geqslant 0$ $(k \in \{0\} \bigcup E)$ and $s > 0$. Then the inequality*

$$a_0 + \sum_{k \in E} a_k x_k^* < s \tag{3.3.7}$$

holds iff a certain major system of (3.2.1) *has a nonnegative solution X_k $(k \in E)$ with the property*

$$a_0 + \sum_{k \in E} a_k X_k < s. \ \square \tag{3.3.8}$$

Theorem 3.3.2 *Let G be a finite or denumerable set, $s \in G$. If $x_i^{(s)*}$ $(i \in E)$ is the minimal nonnegative solution of the system of nonnegative linear equations*

$$x_i = \sum_{k \in E} c_{ik} x_k + b_i^{(s)} \qquad (i \in E), \tag{3.3.9}$$

then

$$\sum_{s \in G} a_s x_i^{(s)*} \qquad (i \in E) \tag{3.3.10}$$

is the minimal nonnegative solution of the system of nonnegative linear equations

$$x_i = \sum_{k \in E} c_{ik} x_k + \Big(\sum_{s \in G} a_s b_i^{(s)}\Big) \qquad (i \in E), \tag{3.3.11}$$

where $a_s \geqslant 0$ $(i \in E)$. $\square$

Theorem 3.3.3 *If $a_i > 0$ $(i \in G)$, then $a_i x_i^*$ $(i \in E)$ is the minimal nonnegative solution of the system of nonnegative linear equations*

$$x_i = \sum_{k \in E} c_{ik} \frac{a_i}{a_k} x_k + a_i b_i \qquad (i \in E).\square \tag{3.3.12}$$

Corollary 3.3.3 *ax_i^* $(i \in E)$ is the minimal nonnegative solution of the system of nonnegative linear equations*

$$x_i = \sum_{k \in E} c_{ik} x_k + ab_i \qquad (i \in E), \tag{3.3.13}$$

provided $a \geqslant 0$. $\square$

§ 3.4 Localization theorem

Theorem 3.4.1 *Let G be a nonempty subset of E. The minimal nonnegative solution of the system of nonnegative linear equations*

$$x_i = \sum_{k \in G} c_{ik} x_k + \Big(\sum_{k \in E \setminus G} c_{ik} x_k^* + b_i \Big) \qquad (i \in G) \tag{3.4.1}$$

is $\tilde{x}_i^* = x_i^*$ $(i \in G)$. $\square$

Corollary 3.4.1 *If*

$$c_{ik} = 0, \qquad i \in G, \qquad k \in E \setminus G, \tag{3.4.2}$$

then the minimal nonnegative solution of the system of nonnegative linear equations

$$x_i = \sum_{k \in G} c_{ik} x_k + b_i \qquad (i \in G) \tag{3.4.3}$$

is $\tilde{x}_i^* = x_i^*$ $(i \in G)$, and the minimal nonnegative solution of the system of nonnegative linear equations

$$x_i = \sum_{k \in E \setminus G} c_{ik} x_k + \Big(\sum_{k \in G} c_{ik} x_k^* + b_i \Big) \qquad (i \in E \setminus G) \tag{3.4.4}$$

is $\tilde{x}_i^* = x_i^*$ $(i \in E \setminus G)$. $\square$

§ 3.5 Connecting property of the minimal nonnegative solution

Definition 3.5.1 *Let $A = (a_{ij},\ i, j \in E)$ be a nonnegative matrix. If there exists a finite subset $\{i, j_1, j_2, \cdots, j_s, j\}$ of E such that*

$$a_{ij_1}a_{j_1j_2}\cdots a_{j_sj} > 0, \tag{3.5.1}$$

then we say that j can be reached from i in A, and write $i \underset{A}{\leadsto} j$; otherwise, we say that j cannot be reached from i and write $i \underset{A}{\not\leadsto} j$. We say that i communicates with j in A, and write $i \underset{A}{\leftrightsquigarrow} j$, if $j \underset{A}{\leadsto} i$ and $i \underset{A}{\leadsto} j$. We say that J can be reached from I in A, and write $I \underset{A}{\leadsto} J$, if I, J are subsets of E and there exist $i \in I$, $j \in J$ such that $i \underset{A}{\leadsto} j$; otherwise we say that J cannot be reached from I in A and write $I \underset{A}{\not\leadsto} J$. We say that J can be strongly reached from I in A, denoted by $I \underset{A}{\rightrightarrows} J$, if $\{i\} \underset{A}{\leadsto} J$ for all $i \in I$.

Remark 3.5.1 Obviously, if $i \underset{A}{\leadsto} j$ and $j \underset{A}{\leadsto} l$, then $i \underset{A}{\leadsto} l$

Remark 3.5.2 If $i \neq j$, then whether $i \underset{A}{\leadsto} j$ or not is independent of a_{ki} $(k \in E)$, in view of (3.5.1).

Remark 3.5.3 From now on we shall write $i \underset{A}{\leadsto} J$ for $\{i\} \underset{A}{\leadsto} J$, etc.; and if G is a subset of E, letting $A_G = (a_{ik}, i, k \in E\setminus G)$, we shall write $i \overset{G}{\underset{A}{\leadsto}} j$ for $i \underset{A_G}{\leadsto} j$, etc.

Theorem 3.5.1 *Let $C = (c_{ik}, i, k \in E)$ denote the coefficient matrix on the right side of* (3.2.1). *If $i \underset{C}{\leadsto} j$, then*
(i) $x_j^* = 0$ *if* $x_i^* = 0$;
(ii) $x_j^* < +\infty$ *if* $x_i^* < +\infty$.

Proof. Suppose

$$x_i^* = 0. \tag{3.5.2}$$

From $i \underset{C}{\leadsto} j$, we know there exists a finite subset $\{i, j_1, j_2, \cdots, j_s, j\}$ of E such that

$$c_{ij_1}c_{j_1j_2}\cdots c_{j_sj} > 0. \tag{3.5.3}$$

By means of (3.5.2), (3.5.3) and

$$x_i^* = \sum_{k \neq j_1} c_{ik}x_k^* + c_{ij_1}x_{j_1}^* + b_i, \tag{3.5.4}$$

we obtain immediately $x_{j_1}^* = 0$. Similarly, from $x_{j_1}^* = 0$, we may obtain $x_{j_2}^* = 0, \cdots$; finally, from $x_{j_s}^* = 0$ we obtain $x_j^* = 0$, thus proving (i). The validity of (ii) can be established in the same way. □

Corollary 3.5.1 *If $i \underset{C}{\leftrightsquigarrow} j$, then x_i^* and x_j^* are simultaneously equal to zero or greater than zero, finite or infinite.* □

Remark 3.5.4 Obviously, if the minimal nonnegative solution x_i^* $(i \in E)$ in Theorem 3.5.1 is changed to an arbitrary nonnegative solution, the conclusion of the theorem is still true.

Definition 3.5.2 *Any element of E is called a subscript of the matrix $A = (a_{ij}, i, j \in \mathrm{E})$ and E is called the subscript set.*

Definition 3.5.3 *$i \in E$ is called an essential subscript of the matrix $A = (a_{ij}, i,$*

$j \in E$), if for any $k \in E$, $i \underset{A}{\sim} k$ implies $k \underset{A}{\sim} i$; otherwise, i is called a nonessential subscript of A.

§ 3.6 Limit theorem

Theorem 3.6.1 *Let $E = (1, 2, \cdots)$ and*

$$c_{ik}^{(N)} \geqslant 0 \quad (i,\ k,\ N \in E), \qquad \text{V-}\lim_{N \to +\infty} c_{ik}^{(N)} = c_{ik} \quad (i,\ k \in E), \tag{3.6.1}$$

$$b_i^{(N)} \geqslant 0 \quad (i,\ N \in E), \qquad \text{V-}\lim_{N \to +\infty} b_i^{(N)} = b_i \quad (i \in E), \tag{3.6.2}$$

$$\left.\begin{aligned} y_i^{(N,1)} &= b_i^{(N)} \qquad (i,\ N \in E), \\ y_i^{N,n+1} &= \sum_{k \in E} c_{ik}^{(N)} y_k^{(N,n)} \qquad (n \geqslant 1,\ i,\ N \in E), \end{aligned}\right\} \tag{3.6.3}$$

$$\left.\begin{aligned} y_i^{(1)} &= b_i \qquad (i \in E), \\ y_i^{(n+1)} &= \sum_{k \in E} c_{ik} y_k^{(n)} \qquad (n \geqslant 1,\ i \in E), \end{aligned}\right\} \tag{3.6.4}$$

and let $\tilde{x}_i^{N*}$ $(i \in E)$ be the minimal nonnegative solution of the system of nonnegative linear equations

$$x_i = \sum_{k \in E} c_{ik}^{(N)} x_k + b_i^{(N)} \qquad (i \in E). \tag{3.6.5}$$

Then

$$\text{V-}\lim_{N \to \infty} y_i^{(N,n)} = y_i^{(n)} \qquad (n \geqslant 1,\ i \in E), \tag{3.6.6}$$

$$\text{V-}\lim_{N \to \infty} \tilde{x}_i^{(N)*} = x_i^* \qquad (i \in E). \ \square \tag{3.6.7}$$

Corollary 3.6.1 *With definition of E, $c_{ik}^{(N)}$, $b_i^{(N)}$ as in Theorem 3.6.1, if $\bar{x}^{(N)*}$ $(i = 1, 2, \cdots, N)$ denotes the minimal nonnegative solution of the system of linear equations*

$$x_i = \sum_{k=1}^{N} c_{ik}^{(N)} x_k + b_i^{(N)} \qquad (i = 1,\ 2,\ \cdots,\ N), \tag{3.6.8}$$

then

$$\text{V-}\lim_{N \to \infty} \bar{x}_i^{(N)*} = x_i^* \qquad (i \in E). \ \square \tag{3.6.9}$$

Corollary 3.6.2 *Let $E = (1, 2, \cdots)$. If $x_i^{(N)*}$ $(i = 1, 2, \cdots, N)$ is the minimal nonnegative solution of the system of nonnegative linear equations*

$$x_i = \sum_{k=1}^{N} c_{ik} x_k + b_i \qquad (i \in E), \tag{3.6.10}$$

then

$$\text{V-}\lim_{N \to \infty} x_i^{(N)*} = x_i^* \qquad (i \in E). \ \square \tag{3.6.11}$$

Remark 3.6.1 In Corollaries 3.6.1 and 3.6.2, the calculation of the minimal nonnegative solution of systems of nonnegative linear equations of denumerable dimensions is reduced to the calculation of the minimal nonnegative solution of systems of nonnegative linear equations of finite dimensions, which will form the main subject of Chapters IV and V.

§ 3.7 Matrix representation

Theorem 3.7.1 *Let $C = (c_{ij},\ i, j \in E)$, $B = (b_i,\ i \in E)$ and $X^* = (x_i^*,\ i \in E)$ be column vectors. Then*

$$X^* = \left(\sum_{n=0}^{\infty} C^n \right) B, \tag{3.7.1}$$

i.e.,

$$x_i^* = \sum_{k \in E} \left(\sum_{n=0}^{\infty} c_{ik}^{(n)} \right) b_k = b_i + \sum_{k \in E} \left(\sum_{n=1}^{\infty} c_{ik}^{(n)} \right) b_k \qquad (i \in E), \tag{3.7.2}$$

where $c_{ij}^{(n)}$ $(i, j \in E)$ is determined uniquely by $C^n = (c_{ij}^{(n)},\ i, j \in E)$.

§ 3.8 Dual theorem

Suppose C, D, B are nonnegative matrices defined on $E \times E$, and the entries of both C and D are finite. O stands for the null matrix defined on $E \times E$. Let

$$\left.\begin{aligned} X^{(0)} &= O, \\ X^{(n+1)} &= CX^{(n)} + B \qquad (n \geqslant 0), \end{aligned}\right\} \tag{3.8.1}$$

$$\left.\begin{aligned} \tilde{X}^{(0)} &= O, \\ \tilde{X}^{(n+1)} &= \tilde{X}^{(n)} D + B \qquad (n \geqslant 0). \end{aligned}\right\} \tag{3.8.2}$$

It is obvious that the limits $X^* = \text{V-}\lim_{n\to\infty} X^{(n)}$ and $\tilde{X}^* = \text{V-}\lim_{n\to\infty} \tilde{X}^{(n)}$ exist.

Remark 3.8.1 In this section B, $X^{(n)}$, $\tilde{X}^{(n)}$, X^*, $\tilde{X}^*$ all denote the matrices, this differs from the notation in §3.7.

Theorem 3.8.1 *If*

$$CB = BD, \tag{3.8.3}$$

then

$$X^* = \tilde{X}^*. \tag{3.8.4}$$

Proof. It is not hard to show that

$$X^* = \sum_{n=0}^{\infty} C^n B, \tag{3.8.5}$$

$$\tilde{X}^* = \sum_{n=0}^{\infty} BD^n. \tag{3.8.6}$$

If (3.8.3) holds, then

$$C^n B = BD^n \qquad (n = 0, 1, \cdots). \tag{3.8.7}$$

Therefore from (3.8.5), (3.8.6) and (3.8.7) we readily obtain (3.8.4). □

CHAPTER IV
Calculation

§ 4.1 Some lemmas

Lemma 4.1.1 *Let $A = (a_{ij})$ be a primitive matrix of order n, and let r be its maximal eigenvalue. If*

$$r \geqslant 1, \tag{4.1.1}$$

then

$$\sum_{k=0}^{\infty} a_{ij}^{(k)} = +\infty \qquad (i, j = 1, 2, \cdots, n), \tag{4.1.2}$$

where $a_{ij}^{(k)}$ is determined uniquely by $A^k = (a_{ij}^{(k)})$.

Proof. The equality

$$\lim_{k\to\infty} \frac{A^k}{r^k} = \left.\frac{C(\lambda)}{\psi'(\lambda)}\right|_{\lambda=r} = \frac{C(r)}{\psi'(r)} \tag{4.1.3}$$

holds, owing to [9, Chapter XIII, (84)], where $C(\lambda)$ is the derived additional matrix of the matrix A, and $\psi'(\lambda)$ is the derivative of the minimal polynomial $\psi(\lambda)$ of A. Thus we deduce

$$C(r) > 0 \tag{4.1.4}$$

from [9, Chapter XIII, (53)]. Since the coefficient of the first term of $\psi(\lambda)$ is 1, and r is a simple zero of order 1 and the maximal real zero point, we obtain readily

$$\psi'(r) > 0. \tag{4.1.5}$$

So

$$\lim_{k\to\infty} \frac{A^k}{r^k} > 0. \tag{4.1.6}$$

Equation (4.1.2) follows by (4.1.1) and (4.1.6). □

Lemma 4.1.2 *Let $A = (a_{ij})$ be an irreducible nonnegative matrix of order n and let r be its maximal eigenvalue. If*

$$r \geqslant 1, \tag{4.1.7}$$

then

$$\sum_{k=0}^{\infty} a_{ij}^{(k)} = +\infty \qquad (i, j = 1, 2, \cdots, n). \tag{4.1.8}$$

Proof. Let h be the nonprimitive index of A. Based on Corollary 2 in [9, Chapter XIII, §5], we may assume, without loss of generality, that A^h has the form

$$A^h = \begin{pmatrix} A_1 & & 0 \\ & A_2 \ddots & \\ 0 & & A_h \end{pmatrix}, \tag{4.1.9}$$

where $A_i\ (i = 1, 2, \cdots, h)$ are primitive matrices, with r as their maximal eigenvalue. It is easy to prove that (4.1.8) holds by the irreducibility of A and Lemma 4.1.1. □

Lemma 4.1.3 *Let $A = (a_{ij})$ be a nonnegative matrix of order n, and let r be its maximal eigenvalue. If*

$$r < 1, \tag{4.1.10}$$

then the series

$$\sum_{k=0}^{\infty} A^k \tag{4.1.11}$$

converges, and vice versa.

Proof. Using the fact that the modulus of r is the maximal one of all the eigenvalues of A, and [10, Theorem 3.7], we get this lemma immediately. □

Lemma 4.1.4 *Let $A = (a_{ij})$ be a nonnegative matrix of order n, and let r be its maximal eignvalue. If*

$$r < 1, \tag{4.1.12}$$

or equivalently

$$\sum_{k=0}^{\infty} A^k \tag{4.1.13}$$

converges, then $(A - I)^{-1}$ exists, and

$$(A - I)^{-1} = \sum_{k=0}^{\infty} A^k, \tag{4.1.14}$$

where I is the unit matrix of order n.

Proof. This lemma follows readily from [10, Theorem 3.7].

Lemma 4.1.5 *Let $A = (a_{ij})$ be a nonnegative matrix of order n, and let r be its maximal eigenvalue. Then*

$$r < 1 \tag{4.1.15}$$

iff

$$(-1)^k \begin{vmatrix} a_{11}-1 & a_{12} & \cdots & \cdots \\ a_{21} & a_{22}-1 & \cdots & \cdots \\ \cdots & \cdots & \cdots & \cdots \\ a_{k1} & a_{k2} & \cdots & a_{kk}-1 \end{vmatrix} > 0 \ (k = 1, 2, \cdots, n). \tag{4.1.16}$$

Proof. This lemma follows immediately from [9, Chapter XIII, Theorem 4].
□

§ 4.2 Reduction of the problems

Suppose

$$x_i = \sum_{k \in E} c_{ik} x_k + b_i \qquad (i \in E) \tag{4.2.1}$$

is a system of nonnegative linear equations. Let

$$t_{ik} = \begin{cases} c_{ik}, & i, \ k \in E, \\ b_i & i \in E, \quad k = 0, \\ 0, & i = 0, \quad k \in E, \\ 1, & i = k = 0. \end{cases} \tag{4.2.2}$$

We define a matrix

$$T = (t_{ik}, \ i, \ k \in \{0\} \cup E), \tag{4.2.3}$$

and let

$$E_1 = \{i: \ i \in E, \ i \underset{\mathrm{T}}{\not\widetilde{\to}} 0\}, \tag{4.2.4}$$

$$E_2 = E \setminus E_1 = \{i: \ i \in E, \ i \underset{\mathrm{T}}{\widetilde{\to}} 0\}. \tag{4.2.5}$$

Definition 4.2.1 *If* $E_1 = \varnothing$, *then* (4.2.1) *is called a system of strictly nonhomogeneous nonnegative linear equations or, for brevity, a system of strictly nonhomogeneous equations.*

Remark 4.2.1 If $E_2 = \varnothing$, then (4.2.1) is obviously a system of homogeneous nonnegative linear equations, referred to simply as a system of homogeneous equations.

Theorem 4.2.1 *The minimal nonnegative solution* x_i^* $(i \in E)$ *of the system of nonnegative linear equations* (4.2.1) *is determined uniquely as follows:*

(i) $\{x_i^*, \ i \in E_1\}$ *is the minimal nonnegative solution of the system of homogeneous equations*

$$x_i = \sum_{k \in E_1} c_{ik} x_k \qquad (i \in E_1). \tag{4.2.6}$$

Hence

$$x_i^* = 0 \qquad (i \in E_1). \tag{4.2.7}$$

(ii) $\{x_i^*,\ i \in E_2\}$ *is the minimal nonnegative solution of the system of strictly nonhomogeneous equations*

$$x_i = \sum_{k \in E_2} c_{ik} x_k + b_i \qquad (i \in E_2), \tag{4.2.8}$$

and

$$x_i^* > 0 \qquad (i \in E_2) \tag{4.2.9}$$

Proof. The proof is easily completed in a manner similar to the proof of Theorem 3.5.1. □

Corollary 4.2.1 *A necessary and sufficient condition for the minimal nonnegative solution*

$$x_i^* \equiv 0 \qquad (i \in E) \tag{4.2.10}$$

of the system of nonnegative linear equations (4.2.1) *is that* (4.2.1) *is homogeneous, and*

$$x_i^* > 0 \qquad (i \in E) \tag{4.2.11}$$

iff (4.2.1) *is strictly nonhomogeneous.* □

Let

$$r_{ik} = \begin{cases} c_{ik}, & i,\ k \in E, \\ 1, & b_i = +\infty, \quad i \in E \quad k = 0 \text{ or } i = k = 0, \\ 0, & b_i < +\infty, \quad i \in E, \quad k = 0 \text{ or } i = 0, \quad k \in E. \end{cases} \tag{4.2.12}$$

We define a matrix

$$R = (r_{ik},\ i,\ k \in \{0\} \cup E), \tag{4.2.13}$$

and let

$$\left.\begin{aligned} G_1 &= \{i\colon\ i \in E,\ i \underset{R}{\not\rightsquigarrow} 0\}, \\ G_2 &= E \backslash G_1 = \{i\colon\ i \in E,\ i \underset{R}{\rightsquigarrow} 0\}. \end{aligned}\right\} \tag{4.2.14}$$

Definition 4.2.2 *If*

$$G_2 = \varnothing, \tag{4.2.15}$$

then (4.2.1) *is called an ordinary system of nonnegative linear equations or simply, an ordinary system of equations. If*

$$G_1 = \varnothing, \tag{4.2.16}$$

then (4.2.1) *is called a system of nonnegative linear equationswith essentially infinite constant terms, or simply, a system of equations with essentially infinite constant terms.*

Theorem 4.2.2 *The minimal nonnegative solution* x_i^* $(i \in E)$ *of the system of nonnegative linear equations* (4.2.1) *is determined uniquely as follows:*

(i) $\{x_i^*, i \in G_1\}$ *is the minimal nonnegative solution of the ordinary system of equations*

$$x_i = \sum_{k \in G_1} c_{ik} x_k + b_i \qquad (i \in E). \tag{4.2.17}$$

(ii) $\{x_i^*, i \in G_2\}$ *is the minimal nonnegative solution of the system of equations with essentially infinite constant terms*

$$x_i = \sum_{k \in G_2} c_{ik} x_k + \left(\sum_{k \in G_1} c_{ik} x_k^* + b_i \right) \qquad (i \in G_2), \tag{4.2.18}$$

and

$$x_i^* = +\infty \qquad (i \in G_2). \tag{4.2.19}$$

Proof. It is easy to prove this theorem by referring to the proof of Theorem 3.5.1. □

Corollary 4.2.2 *If* (4.2.1) *is a system of equations with the essentially infinite constant terms, then*

$$x_i^* = +\infty \qquad (i \in E). \ \square \tag{4.2.20}$$

§ 4.3 Ordinary systems of strictly nonhomogeneous equations with dimension n

In Corollaries 3.6.1 and 3.6.2, the calculation of the minimal nonnegative solution of systems of nonnegative linear equations with denumerable dimension is reduced to the calculation of the minimal nonnegative solution of systems of nonnegative linear equations with finite dimensions. In Theorems 4.2.1 and 4.2.2, the calculation of the minimal nonnegative solution of systems of nonnegative linear equations with dimension n is reduced to the calculation of the minimal nonnegative solution of ordinary systems of strictly nonhomogeneous equations with dimension n. In what follows, we shall present a method to calculate the minimal nonnegative solutions of the latter kind of systems of equations. The proof of the following theorem is omitted since it may be easily deduced from Lemmas 4.1.1—4.1.5.

Let

$$x_i = \sum_{k=1}^{n} c_{ik} x_k + b_i \qquad (i = 1, 2, \cdots, n) \tag{4.3.1}$$

be an ordinary system of strictly nonhomogeneous equations with dimension n. According to [9, Chapter XIII, §4], $C = (c_{ik})$ may be transformed to

$$\begin{pmatrix} C_1 & 0 & \cdots & 0 & 0 & \cdots & 0 & 0 \\ 0 & C_2 & \cdots & 0 & 0 & \cdots & 0 & 0 \\ \cdots & \cdots & \cdots & \cdots & \cdots & \cdots & \cdots & \cdots \\ 0 & 0 & \cdots & C_g & 0 & \cdots & 0 & 0 \\ C_{g+1,1} & C_{g+1,2} & \cdots & C_{g+1,g} & C_{g+1} & \cdots & 0 & 0 \\ \cdots & \cdots & \cdots & \cdots & \cdots & \cdots & \cdots & \cdots \\ C_{l1} & C_{l2} & \cdots & \cdots & \cdots & \cdots & C_{l,l-1} & C_l \end{pmatrix} \tag{4.3.2}$$

after permuting the rows and columns in the same way if necessary, where C_i $(i = 1, 2, \cdots, l)$ are irreducible nonnegative matrices. For each fixed j $(j = g + 1, g + 2, \cdots, l)$, at least one of the matrices C_{jk} $(1 \leqslant k < j)$ is not zero. Let E_i, r_i $(i = 1, 2, \cdots, l)$ denote the subscript set and the maximal eigenvalue of C_i, respectively. Let

$$R_1 = \{i\colon\ i \in (1, 2, \cdots, l),\ r_i < 1\}, \tag{4.3.3}$$

$$R_2 = \{i\colon\ i \in (1, 2, \cdots, l),\ r_i \geqslant 1\}, \tag{4.3.4}$$

$$R_1^{(1)} = \{i\colon\ i \in R_1,\ E_i \underset{C}{\not\rightsquigarrow} \bigcup_{j \in R_2} E_j\}, \tag{4.3.5}$$

$$R_1^{(2)} = \{i\colon\ i \in R_1,\ E_i \underset{C}{\rightsquigarrow} \bigcup_{j \in R_2} E_j\}. \tag{4.3.6}$$

Theorem 4.3.1 *The minimal nonnegative solution x_i^* $(i = 1, 2, \cdots, n)$ of the ordinary system of strictly nonhomogeneous equation* (4.3.1) *with dimension n is determined uniquely as follows:*

(i) $\{x_i^*,\ i \in \bigcup_{j \in R_1^{(1)}} E_j\}$ *is the unique finite solution of the system of nonnegative linear equations with finite dimension*

$$x_i = \sum_{k \in \bigcup_{j \in R_1^{(1)}} E_j} c_{ik} x_k + b_i \qquad \left(i \in \bigcup\nolimits_{j \in R_1^{(1)}} E_j\right). \tag{4.3.7}$$

(ii)

$$x_i^* = +\infty \qquad \left(i \in \bigcup\nolimits_{j \in R_1^{(2)} \cup R_2} E_j\right). \ \square \tag{4.3.8}$$

CHAPTER V
Systems of 1-Bounded Equations

§ 5.1 Introduction

In the last chapter, we studied a method for calculating the minimal nonnegative solutions of systems of nonnegative linear equations. But most systems of nonnegative linear equations encountered are of a special type (called systems of 1-bounded equations). Their minimal nonnegative solutions possess special properties, and calculation of the solutions can be greatly simplified. This chapter is devoted to the study of the minimal nonnegative solutions of the systems of equations of this type.

Definition 5.1.1 *If the system of nonnegative linear equations*

$$x_i = \sum_{k \in E} c_{ik} x_k + b_i \qquad (i \in E) \tag{5.1.1}$$

satisfies the condition

$$\sum_{k \in E} c_{ik} \leqslant 1 \qquad (i \in E), \tag{5.1.2}$$

then it is called a first-type system of 1*-bounded equations; if it satisfies the condition*

$$\sum_{i \in E} c_{ik} \leqslant 1 \qquad (k \in E), \tag{5.1.3}$$

then it is called a second-type system of 1*-bounded equations; the two types of systems of* 1*-bounded equations are known collectively as systems of* 1*-bounded equations.*

Definition 5.1.2 *An ordinary finite-dimensional first- (second-) type system of* 1*-bounded equations is called a first- (second-) type regular system of equations.*

The two types of regular systems of equations are known collectively as regular systems of equations.

We need only present a method for calculating the minimal nonnegative solutions of the regular systems of equations, as was pointed out in §4.3.

§ 5.2 First-type leading-outside systems of equations

Definition 5. 2. 1 *The nonnegative matrix $A = (a_{ik})$ of order n is called a first-type semi-random matrix of order n, or simply, a semi-random matrix of order n if it satisfies the condition*

$$\sum_{k=1}^{n} a_{ik} \leqslant 1 \qquad (i = 1,\ 2,\ \cdots,\ n). \tag{5.2.1}$$

It is called a first-type random matrix of order n, or simply, a random matrix of order n if the equality in (5. 2. 1) *holds for $i = 1,\ 2,\ \cdots,\ n$.*

Suppose $A = (a_{ik})$ is a semi-random matrix of order n. Let

$$s_{ik} = \begin{cases} a_{ik}, & i, \quad k = 1,\ 2,\ \cdots,\ n, \\ 1 - \sum\limits_{j=1}^{n} a_{ij}, & i = 1,\ 2, \cdots,\ n, \quad k = 0, \\ 0, & i = 0, \quad k = 1,\ 2, \cdots,\ n, \\ 1, & i = k = 0. \end{cases} \tag{5.2.2}$$

We define a matrix

$$S = (s_{ij},\ i,\ j \in (0,\ 1,\ \cdots,\ n)). \tag{5.2.3}$$

Definition 5.2.2. *If the semi-random matrix $A = (a_{ik})$ of order n satisfies the condition*

$$i \underset{S}{\rightsquigarrow} 0 \qquad (i = 1,\ 2,\ \cdots,\ n), \tag{5.2.4}$$

then it is called a first-type leading-outside matrix of order n. If A is furthermore irreducible, then it is called a first-type irreducible leading-outside matrix of order n.

Remark 5. 2. 1 Obviously, an irreducible semi-random matrix of order n is a first-type leading-outside matrix of order n iff it is not a random matrix of order n.

Remark 5. 2. 2 It is apparent that a necessary and sufficient condition for a semi-random matrix $A = (a_{ik})$ of order n to be a first-type leading-outside matrix of order n is that it may be transformed to

$$\begin{pmatrix} A_1 & 0 & \cdots & 0 & 0 & \cdots & 0 & 0 \\ 0 & A_2 & \cdots & 0 & 0 & \cdots & 0 & 0 \\ \cdots & \cdots & \cdots & \cdots & \cdots & \cdots & \cdots & \cdots \\ 0 & 0 & \cdots & A_g & 0 & \cdots & 0 & 0 \\ A_{g+1,1} & A_{g+1,2} & \cdots & A_{g+1,g} & A_{g+1} & \cdots & 0 & 0 \\ \cdots & \cdots & \cdots & \cdots & \cdots & \cdots & \cdots & \cdots \\ A_{l,1} & A_{l,2} & \cdots & \cdots & \cdots & \cdots & A_{l,l-1} & A_l \end{pmatrix} \tag{5.2.5}$$

after permuting the rows and columns in the same way if necessary, where A_j ($j = 1, 2, \cdots, l$) are first-type irreducible leading-outside matrices. For each fixed j ($j = g+1, g+2, \cdots, l$), at least one of the matrices A_{jk} ($1 \leqslant k < j$) is nonnull.

Lemma 5.2.1 *If $A = (a_{ik})$ is a first-type leading-outside matrix of order n, then*

$$r < 1, \tag{5.2.6}$$

where r is the maximal eigenvalue *of A.*

Proof. By Remark 5. 2. 2, it is enough to prove the lemma for the case where A is a first-type irreducible leading-outside matrix of order n. This can be obtained immediately by Remark 5. 2. 1, and Remark 2 in [9, Chapter XIII, §2]. □

Definition 5.2.3 *If the coefficient matrix $C = (c_{ik})$ on the right side of the first-type regular system of equations*

$$x_i = \sum_{k=1}^{n} c_{ik}x_k + b_i \qquad (i = 1, 2, \cdots, n) \tag{5.2.7}$$

is a first-type leading-outside matrix, then the system of equations is called a first-type leading-outside system of equations.

Theorem 5.2.1 *If* (5. 2. 7) *is a first-type leading-outside system of equations, then it has a unique (ordinary) solution, i.e., its minimal nonnegative solution.*

Proof. This theorem follows from Remark 5. 2. 2, Lemma 5. 2. 1 and Theorem 4. 3. 1. □

§ 5.3 First-type consistent systems of equations

Definition 5.3.1 *If $P = (p_{ij})$ is a random matrix of order n, then*

$$x_i = \sum_{k=1}^{n} p_{ik}x_k \qquad (i = 1, 2, \cdots, n) \tag{5.3.1}$$

is said to be a first-type random system of homogeneous equations.

It is easy to establish the following theorem.

Theorem 5.3.1 *A first-type random system of homogeneous equations has infinitely many (ordinary) solutions, and its minimal nonnegative solution coincides with the zero solution.* □

Suppose

$$x_i = \sum_{k=1}^{n} c_{ik}x_k + b_i \qquad (i = 1, 2, \cdots, n) \tag{5.3.2}$$

is a first-type regular system of equations. Let

$$s_{ik} = \begin{cases} c_{ik} & i, \ k = 1, 2, \cdots, n, \\ 1 - \sum_{j=1}^{n} c_{ij}, & i = 1, 2, \cdots, n, \quad k = 0, \\ 0, & i = 0, \quad k = 1, 2, \cdots, n, \\ 1, & i = k = 0. \end{cases} \tag{5.3.3}$$

We define a matrix

$$S = (s_{ik}, \ i, \ k \in (0, 1, \cdots, n)), \tag{5.3.4}$$

and let

$$E_1 = \{i: \ i \in (1, 2, \cdots, n), \ i \underset{S}{\not\rightsquigarrow} 0\}, \tag{5.3.5}$$

$$E_2 = E \setminus E_1, \tag{5.3.6}$$

$$E_{11} = \text{the essential subscript set of } (c_{ik}, \ i, \ k \in E_1), \tag{5.3.7}$$

$$E_{12} = E_1 \setminus E_{11}. \tag{5.3.8}$$

Definition 5.3.2 *If the first-type regular system of equations* (5.3.2) *satisfies the condition*

$$b_i = 0 \qquad (i \in E_{11}), \tag{5.3.9}$$

then it is called a first-type consistent system of equations.

Henceforth, (5.3.2) is always assumed to be a first-type consistent system of equations. We can prove the following theorem without difficulty.

Theorem 5.3.2. *The minimal nonnegative solution* x_i^* $(i = 1, 2, \cdots, n)$ *of the first-type consistent system of equations* (5. 3. 2) *is finite, i.e.,*

$$x_i^* < +\infty \qquad (i = 1, 2, \cdots, n); \tag{5.3.10}$$

and

(i) $\{x_i^*, i \in E_{11}\}$ *is the minimal nonnegative solution of the first-type random system of homogeneous equations*

$$x_i = \sum_{k \in E_{11}} c_{ik} x_k \qquad (i \in E_{11}), \tag{5.3.11}$$

hence

$$x_i^* = 0 \qquad (i \in E_{11}). \tag{5.3.12}$$

(ii) $\{x_i^*,\ i\in E_{12}\}$ *is the minimal nonnegative solution, i.e., the unique (ordinary) solution of the first-type leading-outside system of equations*

$$x_i = \sum_{k\in E_{12}} c_{ik}x_k + b_i \qquad (i\in E_{12}). \tag{5.3.13}$$

(iii) $\{x_i^*,\ i\in E_2\}$ *is the minimal nonnegative solution, i.e., the unique (ordinary) solution of the first-type leading-outside system of equations*

$$x_i = \sum_{k\in E_2} c_{ik}x_k + \left(\sum_{k\in E_{12}} c_{ik}x_k^* + b_i\right) \qquad (i\in E_2). \ \square \tag{5.3.14}$$

§ 5.4 Tailed random systems of strictly nonhomogeneous equations

Let

$$x_i = \sum_{k=1}^{n} c_{ik}x_k + b_i \qquad (i = 1,\ 2,\ \cdots,\ n) \tag{5.4.1}$$

be a first-type regular system of equations.

For the definitions of E_1, E_2, E_{11} and E_{12}, see the last section.

Definition 5.4.1 *If*

$$E_2 \bigcup E_{12} = \varnothing, \tag{5.4.2}$$

then the coefficient matrix $C = (c_{ik})$ *on the right side of* (5.4.1) *is called a first-type blockable random matrix of order n or simply, a blockable random matrix of order n. If*

$$E_1 \neq \varnothing \tag{5.4.3}$$

and

$$E_2 \overset{\rightsquigarrow}{C} E_1, \tag{5.4.4}$$

then $C = (c_{ik})$ *is called a tailed random matrix of order n.*

Definition 5.4.2 *If the first-type regular system of equations* (5.4.1) *is strictly nonhomogeneous and the coefficient matrix* $C = (c_{ik})$ *on the right side is a blockable random (tailed random) matrix, then it is called a first-type blockable random (tailed random) system of strictly nonhomogeneous equations, etc.*

Theorem 5.4.1 *If* (5.4.1) *is a tailed random system of strictly nonhomogeneous equations, then its minimal nonnegatiue solution is*

$$x_i^* = +\infty \qquad (i = 1,\ 2,\ \cdots,\ n), \tag{5.4.5}$$

In detail:

(i) $\{x_i^*,\ i \in E_{11}\}$ *is the minimal nonnegative solution of the first-type blockable random system of strictly nonhomogeneous equations*

$$x_i = \sum_{k \in E_{11}} c_{ik} x_k + b_i \qquad (i \in E_{11}). \tag{5.4.6}$$

Hence

$$x_i^* = +\infty \qquad (i \in E_{11}). \tag{5.4.7}$$

(ii) $\{x_i^*,\ i \in E_{12} \bigcup E_2\}$ *is the minimal nonnegative solution of the system of equations with essentially infinite constant terms*

$$x_i = \sum_{k \in E_{12} \bigcup E_2} c_{ik} x_k + \left(\sum_{k \in E_{11}} c_{ik} x_k^* + b_i \right) \qquad (i \in E_{12} \bigcup E_2). \tag{5.4.8}$$

Hence

$$x_i^* = +\infty \qquad (i \in E_{12} \bigcup E_2). \ \square \tag{5.4.9}$$

§ 5.5 Regular systems of equations

Assume

$$x_i = \sum_{k=1}^{n} c_{ik} x_k + b_i \qquad (i = 1, 2, \cdots, n) \tag{5.5.1}$$

is a first-type regular system of equations.

E_1, E_2, E_{11} and E_{12} are the same as in the last section. Let

$$t_{ik} = \begin{cases} c_{ik}, & i, k \in E_{11}, \\ b_i, & i \in E_{11}, \quad k = 0, \\ 0, & i = 0, \quad k \in E_{11}, \\ 1, & i = k = 0. \end{cases} \tag{5.5.2}$$

We define a matrix

$$T_{11} = (t_{ik},\ i,\ k \in \{0\} \bigcup E_{11}), \tag{5.5.3}$$

and let

$$E_{11}^{(1)} = \{i:\ i \in E_{11},\ i \overset{+\!\!\to}{\underset{T11}{}} 0\}, \tag{5.5.4}$$

$$E_{11}^{(2)} = \{i:\ i \in E_{11},\ i \overset{\sim\!\to}{\underset{T11}{}} 0\}, \tag{5.5.5}$$

$$E_{12}^{(1)} = \{i\colon\ i \in E_{12},\ i \overset{\not\sim}{c} E_{11}^{(2)}\},\tag{5.5.6}$$

$$E_{12}^{(2)} = \{i\colon\ i \in E_{12},\ i \tilde{c} E_{11}^{(2)}\},\tag{5.5.7}$$

$$E_{2}^{(1)} = \{i\colon\ i \in E_{2},\ i \overset{\not\sim}{c} E_{11}^{(2)}\},\tag{5.5.8}$$

$$E_{2}^{(2)} = \{i\colon\ i \in E_{2},\ i \tilde{c} E_{11}^{(2)}\}.\tag{5.5.9}$$

From the results of the previous sections, we get the following theorem.

Theorem 5.5.1 *The minimal nonnegative solution* x_i^* $(i = 1, 2, \cdots, n)$ *of the first-type regular system of equations* (5.5.1) *is determined uniquely as follows:* (i) $\{x_i^*,\ i \in E_{11}^{(1)} \bigcup E_{12}^{(1)} \bigcup E_2^{(1)}\}$ *is the minimal nonnegative solution of the first-type consistent system of equations*

$$x_i = \sum_{k \in E_{11}^{(1)} \cup E_{12}^{(1)} \cup E_2^{(1)}} c_{ik} x_k + b_i \qquad (i \in E_{11}^{(1)} \bigcup E_{12}^{(1)} \bigcup E_2^{(1)}).\tag{5.5.10}$$

(ii) $\{x_i^*,\ i \in E_{11}^{(2)} \bigcup E_{12}^{(2)} \bigcup E_2^{(2)}\}$ *is the minimal nonnegative solution of the tailed random system of strictly nonhomogeneous equations*

$$x_i = \sum_{k \in E_{11}^{(2)} \cup E_{12}^{(2)} \cup E_2^{(2)}} c_{ik} x_k + \left(\sum_{k \in E_{11}^{(1)} \cup E_{12}^{(1)} \cup E_2^{(1)}} c_{ik} x_k^* + b_i \right)$$
$$(i \in E_{11}^{(2)} \bigcup E_{12}^{(2)} \bigcup E_2^{(2)}),\tag{5.5.11}$$

hence

$$x_i^* = +\infty \qquad (i \in E_{11}^{(2)} \bigcup E_{12}^{(2)} \bigcup E_2^{(2)}).\ \square\tag{5.5.12}$$

Corollary 5.5.1 *A first-type regular system of equations has a finite minimal nonnegative solution iff the system of equations is a first-type consistent system of equations.* □

Corollary 5.5.2 *A first-type regular system of equations has a unique (ordinary) solution iff the system of equations is a first-type leading-outside system of equations.* □

Corollary 5.5.3 *If a first-type regular system of equations has a unique (ordinary) solution, then this solution coincides with the minimal nonnegative solution.*

Corollary 5.5.4 *The minimal nonnegative solution (i.e., any nonnegative solution) of a first-type regular system of equations is identically infinite iff the system of equations is a tailed random system strictly nonhomogeneous equations.*
□

Definition 5.5.1 *The elements of E_2 are called leading-outside subscripts of C.*

Corollary 5.5.5 *A first-type regular system of equations is a first-type leading-outside system of equations iff every subscript of its coefficient matrix on the right side is a leading-outside subscript.* □

§ 5.6 Pseudo-normal systems of equations

Definition 5.6.1 *If the first-type system of 1-bounded equations*

$$x_i = \sum_{k \in E} c_{ik} x_k + b_i \qquad (i \in E) \tag{5.6.1}$$

satisfies the condition

$$\sum_{k \in E} c_{ik} + b_i \leqslant 1 \qquad (i \in E), \tag{5.6.2}$$

then it is called a pseudo-normal system of equations. If the equality in (5.6.2) *holds for all $i \in E$, then it is said to be a normal system of equations.*

Theorem 5.6.1 *If* (5.6.1) *is a pseudo-normal system of equations, then*

$$0 \leqslant x_i^* \leqslant 1 \qquad (i \in E). \tag{5.6.3}$$

Proof. Put

$$\left.\begin{aligned} x_i^{(0)} &= 0 \qquad (i \in E), \\ x_i^{(n+1)} &= \sum_{k \in E} c_{ik} x_k^{(n)} + b_i \qquad (n \geqslant 1, \quad i \in E). \end{aligned}\right\} \tag{5.6.4}$$

A little reflection shows that

$$0 \leqslant x_i^{(n)} \leqslant 1 \qquad (n \geqslant 1, \ i \in E). \tag{5.6.5}$$

Hence we get (5.6.3) by Theorem 3.2.1. □

Theorem 5.6.2 *Let* (5.6.1) *be a pseudo-normal system of equations. If $i \underset{C}{\to} j$ and $x_i^* = 1$, then $x_j^* = 1$.*

Proof. We know from $i \underset{C}{\to} j$ that there exists a finite subset $\{i, j_1, \cdots, j_s, j\}$ of E such that

$$c_{ij_1} c_{j_1 j_2} \cdots c_{j_s j} > 0. \tag{5.6.6}$$

If $x_{j_1}^* < 1$, then from Theorem 5.6.1,

$$x_i^* = \sum_{k\in E} c_{ik}x_k^* + b_i = \sum_{k\neq j_1} c_{ik}x_k^* + c_{ij_1}x_{j_1}^* + b_i$$

$$< \sum_{k\neq j_1} c_{ik} + c_{ij_1} + b_i = \sum_{k\in E} c_{ik} + b_i \leqslant 1. \tag{5.6.7}$$

This contradicts the hypothesis $x_i^* = 1$, so $x_{j_1}^* = 1$. We can prove $x_{j_2}^* = 1, \cdots, x_{j_s}^* = 1$ in the same way. □

Corollary 5.6.1 *Suppose* (5.6.1) *is a pseudo-normal system of equations. If* $i \overset{\leftrightarrow}{c} j$, *then* x_i^* *and* x_j^* *are simultaneously equal to one or less than one.* □

Theorem 5.6.3 *If* (5.6.1) *is a nonhomogeneous normal system of equations, then the following four statements are equivalent.*

(i) $$x_i^* \equiv 1 \quad (i\in E). \tag{5.6.8}$$

(ii) *The system of equations* (5.6.1) *has no nonnegative nonconstant bounded solution.*

(iii) *The system of equations* (5.6.1) *has no nonconstant bounded solution.*

(iv) *The system of equations* (5.6.1) *has no nonnegative solution* x_i $(i\in E)$ *with* $\inf_{i\in E} x_i = 0$.

Proof. We prove the theorem in the following steps:

(A) To prove (i) $\Longleftrightarrow$ (ii).

Suppose (i) holds, i.e., $x_i^* \equiv 1$ $(i\in E)$. Then (ii) follows from Theorem 3.2.3. Conversely, if (ii) holds, then $x_i^* \equiv c$ $(i\in E)$, owing to Theorem 5.6.1. Since (5.6.1) is a system of nonhomogeneous equations, there exists a subscript $s\in E$ such that $b_s \neq 0$. Substituting $x_i^* \equiv c$ $(i\in E)$ in the sth equation of (5.6.1) we get

$$c = c\sum_{k\in E} c_{sk} + b_s. \tag{5.6.9}$$

Since (5.6.1) is a normal system of equations,

$$1 - \sum_{k\in E} c_{sk} = b_s \neq 0. \tag{5.6.10}$$

It follows from (5.6.9) and (5.6.10) that

$$c = \frac{b_s}{1 - \sum_{k\in E} c_{sk}} = 1. \tag{5.6.11}$$

Consequently (i) holds, so (i) $\Longleftrightarrow$ (ii).

(B) To prove (i) $\Longleftrightarrow$ (iii).

If (i) holds, then by Theorem 3.2.4, we get (iii) immediately; Conversely, it is obvious that if (iii) holds, then (ii) holds. Therefore (i) holds, since (i) $\Longleftrightarrow$ (ii). Consequently (i) $\Longleftrightarrow$ (iii).

(C) To prove (i) $\Longleftrightarrow$ (iv).

If (i) holds, then

$$\inf x_i^* = 1 > 0. \tag{5.6.12}$$

By the minimality of x_i^* $(i \in E)$ we get (iv) immediately. Conversely, if (iv) holds, then

$$\inf_{i \in E} x_i^* = c > 0. \tag{5.6.13}$$

Obviously, $x_i \equiv 1$ $(i \in E)$ is a nonnegative solution of (5.6.1). Now

$$x_i = 1 \leqslant \frac{1}{c}\, x_i^* \leqslant \left(1 + \frac{1}{c}\right) x_i^* \qquad (i \in E). \tag{5.6.14}$$

Hence, by Theorem 3.2.3, we get

$$x_i^* \equiv x_i \equiv 1 \qquad (i \in E). \tag{5.6.15}$$

Consequently (i) holds. So (i) $\Longleftrightarrow$ (iv). □

Corollary 5.6.2 *If $b_i \geqslant c > 0$ $(i \in E)$, then the minimal nonnegative solution of the normal system of equations* (5.6.1) *is*

$$x_i^* \equiv 1 \qquad (i \in E). \ \square \tag{5.6.16}$$

§ 5.7 Pseudo-normal systems of equations of finite dimension

Definition 5.7.1 *If a normal system of equations with dimension n is also a first-type leading-outside system of equations, it is called a normal leading-outside system of equations.*

Obviously, a necessary and sufficient condition for a normal system of equations with dimension n to be a normal leading-outside system of equations is, that it be strictly nonhomogeneous.

Lemma 5.7.1 *The normal leading-outside system of equations*

$$x_i = \sum_{k=1}^{n} c_{ik} x_k + b_i \qquad (i = 1, 2, \cdots, n) \tag{5.7.1}$$

has the unique (ordinary) solution $\bar{x}_i \equiv 1$ $(i = 1, 2, \cdots, n)$, so its minimal nonnegative solution $x_i^* \equiv \bar{x}_i \equiv 1$ $(i \in E)$.

Proof. Evidently $x_i \equiv 1$ $(i = 1, 2, \cdots, n)$ is a (ordinary) solution of (5.7.1).

Hence, we get the lemma immediately by Theorem 5.2.1. □

Suppose

$$x_i = \sum_{k=1}^{n} c_{ik}x_k + b_i \qquad (i = 1, 2, \cdots, n) \tag{5.7.2}$$

is a pseudo-normal system of equations with dimension n.

Let

$$g_{ij} = \begin{cases} c_{ik}, & i, k = 1, 2, \cdots, n, \\ 1 - \left(\sum_{j=1}^{n} c_{ij} + b_i\right), & i = 1, 2, \cdots, n, \quad k = 0, \\ 0, & i = 0, \quad k = 1, 2, \cdots, n, \\ 1, & i = k = 0. \end{cases} \tag{5.7.3}$$

We define a matrix

$$G = (g_{ij}, \ i, j \in (0, 1, 2, \cdots, n)). \tag{5.7.4}$$

Definition 5.7.2 *If the pseudo-normal system of equations* (5.7.2) *with dimension* n *satisfies the condition*

$$i \underset{G}{\rightsquigarrow} 0 \qquad (i = 1, 2, \cdots, n), \tag{5.7.5}$$

then it is called a leading-complement system of equations.

From

$$1 - \sum_{j=1}^{n} c_{ij} \geqslant 1 - \left(\sum_{j=1}^{n} c_{ij} + b_i\right) \qquad (i = 1, 2, \cdots, n), \tag{5.7.6}$$

we see that a leading-complement system of equations must be a first-type leading-outside system of equations.

In what follows, (5.7.2) is assumed to be a leading-complement system of strictly nonhomogeneous equations.

Lemma 5.7.2 *The leading-complement system of strictly nonhomogeneous equations* (5.7.2) *possesses the following properties:*

(i) *It has the unique solution* $\bar{x}_i < \infty$ $(i = 1, 2, \cdots, n)$.

(ii) *Its minimal nonnegative solution* x_i^* $(i = 1, 2, \cdots, n)$ *coincides with its unique* (*ordinary*) *solution* $\bar{x}_i$ $(i = 1, 2, \cdots, n)$.

(iii)

$$0 < x_i^* < 1 \qquad (i = 1, 2, \cdots, n). \tag{5.7.7}$$

Proof. From the fact that a leading-outside system of equations must be a first-type leading-outside system of equations, and Theorem 5.2.1, we see that (i) and (ii) are valid. From the fact that (5.7.2) is a system of strictly nonhomogeneous equations and Corollary 4.2.1, we see the left inequality of (5.7.7) is valid. We now prove the right inequality of (5.7.7).

Suppose, without loss of generality, that

$$C = \begin{pmatrix} C_1 & 0 & \cdots & 0 & 0 & \cdots & 0 & 0 \\ 0 & C_2 & \cdots & 0 & 0 & \cdots & 0 & 0 \\ \cdots & \cdots & \cdots & \cdots & \cdots & \cdots & \cdots & \cdots \\ 0 & 0 & \cdots & C_g & 0 & \cdots & 0 & 0 \\ C_{g+1,1} & C_{g+1,2} & \cdots & C_{g+1,g} & C_{g+1} & \cdots & 0 & 0 \\ \cdots & \cdots & \cdots & \cdots & \cdots & \cdots & \cdots & \cdots \\ C_{l,1} & C_{l,2} & & & & & C_{l,l-1} & C_l \end{pmatrix}, \tag{5.7.8}$$

where C_j ($j = 1, 2, \cdots, l$) are first-type irreducible leading-outside matrices. For every fixed j ($j = g+1, g+2, \cdots, l$) at least one of the matrices C_{jk} ($1 \leqslant k < j$) is nonnull.

Let E_i ($i = 1, 2, \cdots, l$) denote the subscript set of C_i. By Corollary 3.4.1, $\{x_i^*, i \in E_1\}$ is the minimal nonnegative solution of the pseudonormal system of equations

$$x_i = \sum_{k \in E_1} c_{ik} x_k + b_i \qquad (i \in E_1), \tag{5.7.9}$$

and owing to (5.7.5) and (5.7.8), there exists a subscript $j \in E_1$, such that

$$\sum_{k \in E_1} C_{jk} + b_i < 1. \tag{5.7.10}$$

We deduce that

$$0 \leqslant x_i^* \leqslant 1 \qquad (i \in E_1) \tag{5.7.11}$$

from Theorem 5.6.1. Also

$$x_j^* = \sum_{k \in E_1} C_{jk} x_k^* + b_j \leqslant \sum_{k \in E_1} C_{jk} + b_j < 1 \tag{5.7.12}$$

from (5.7.10) and (5.7.11). By the fact that C_i is an irreducible matrix and Corollary 5.6.1, we see that

$$x_i^* < 1 \qquad (i \in E_1). \tag{5.7.13}$$

Similarly, we may prove

$$x_i^* < 1 \qquad (i \in E_s), \tag{5.7.14}$$

where $s = 2, 3, \cdots, g$. Thus

$$x_i^* < 1 \qquad \left(i \in \bigcup_{s=1}^{g} E_s\right). \tag{5.7.15}$$

Obviously,

$$\bigcup_{j=g+1}^{l} E_j \underset{C}{\overset{\approx}{\rightarrow}} \bigcup_{s=1}^{g} E_j. \tag{5.7.16}$$

Hence we know by (5.7.15) and Theorem 5.6.2 that

$$x_i^* < 1 \qquad \left(i \in \bigcup_{j=g+1}^{l} E_j\right). \tag{5.7.17}$$

Combining (5.7.15) and (5.7.17), we obtain the right inequality of (5.7.7), as was to be shown. □

Suppose

$$x_i = \sum_{k=1}^{n} c_{ik} x_k + b_i \qquad (i = 1, 2, \cdots, n) \tag{5.7.18}$$

is a pseudo-normal system of equations with dimension n. Let

$$t_{ik} = \begin{cases} c_{ik}, & i, k = 1, 2, \cdots, n, \\ b_i, & i = 1, 2, \cdots, n, \quad k = 0, \\ 0, & i = 0, \quad k = 1, 2, \cdots, n, \\ 1, & i = k = 0. \end{cases} \tag{5.7.19}$$

We define a matrix

$$T = (t_{ik},\ i,\ k \in (0, 1, \cdots, n)), \tag{5.7.20}$$

and let

$$\hat{E}_1 = \{i:\ i \in (1, 2, \cdots, n),\ i \underset{T}{\overset{\sim}{\not\rightarrow}} 0\}, \tag{5.7.21}$$

$$\hat{E}_2 = \{i:\ i \in (1, 2, \cdots, n),\ i \underset{T}{\overset{\sim}{\rightarrow}} 0\}, \tag{5.7.22}$$

$$\hat{g}_{ik} = \begin{cases} c_{ik}, & i,\ k \in \hat{E}_2, \\ 1 - (\sum\limits_{j \in \hat{E}_2} c_{ij} + b_i), & i \in \hat{E}_2, \quad k = 0, \\ 0, & i = 0, \quad k \in \hat{E}_2, \\ 1, & i = k = 0. \end{cases} \tag{5.7.23}$$

We define another matrix

$$\hat{G} = (\hat{g}_{ik},\ i,\ k \in \{0\} \bigcup \hat{E}_2), \tag{5.7.24}$$

and let

$$\hat{E}_{21} = \{i\colon\ i \in \hat{E}_2 \text{ and } i \underset{\hat{G}}{\not\rightsquigarrow} 0\}, \tag{5.7.25}$$

$$\hat{E}_{22} = \{i\colon\ i \in \hat{E}_2 \text{ and } i \underset{\hat{G}}{\rightsquigarrow} 0\}. \tag{5.7.26}$$

Theorem 5.7.1 *The minimal nonnegative solution* x_i^* $(i = 1, 2, \cdots, n)$ *of the pseudo-normal system of equations* (5.7.18) *with dimension n is determined uniquely in the following way.*

(i) $\{x_i^*,\ i \in \hat{E}_1\}$ *is the zero solution of the system of homogeneous equations*

$$x_i = \sum_{k \in \hat{E}_1} c_{ik} x_k \qquad (i \in E_1). \tag{5.7.27}$$

Hence

$$x_i^* \equiv 0 \qquad (i \in \hat{E}_1). \tag{5.7.28}$$

(ii) $\{x_i^*,\ i \in \hat{E}_{21}\}$ *is the unique (ordinary) solution of the normal leading-outside system of equations*

$$x_i = \sum_{k \in \hat{E}_2} c_{ik} x_k + b_i \qquad (i \in \hat{E}_{21}). \tag{5.7.29}$$

Hence

$$x_i^* \equiv 1 \qquad (i \in \hat{E}_{21}). \tag{5.7.30}$$

(iii) $\{x_i^*,\ i \in \hat{E}_{22}\}$ *is the unique (ordinary) solution of the leading-complement system of strictly nonhomogeneous equations*

$$x_i = \sum_{k \in \hat{E}_{22}} c_{ik} x_k + \left(\sum_{k \in \hat{E}_{21}} c_{ik} x_k^* + b_i \right) \qquad (i \in \hat{E}_{22}). \tag{5.7.31}$$

Hence

$$0 < x_i^* < 1 \qquad (i \in \hat{E}_{22}). \tag{5.7.32}$$

Proof. The proof follows from Lemmas 5.7.1 and 5.7.2. □

Corollary 5.7.1 *We write x_i^* for minimal nonnegative solutions of a pseudo-normal system of equations with dimension n. Necessary and sufficient conditions for $x_i^* \equiv 0$ $(i = 1, 2, \cdots, n)$, $x_i^* \equiv 1$ $(i = 1, 2, \cdots, n)$, and $0 < x_i^* < 1$ $(i = 1, 2, \cdots, n)$ are, respectively, that the system of equations is homogeneous, that it is a normal leading-outside one, and that it is a strictly nonhomogeneous leading-complement one.*

Remark 5.7.1 If (5.7.18) is a normal system of equations with dimension n, then

(i) (5.7.27) is a random system of homogeneous equations;

(ii) $\hat{E}_{21}$, $\hat{E}_{22}$ may be determined easily as follows:

$$\hat{E}_{21} = \{i:\ i \in \hat{E}_2,\ i \underset{C}{\not\rightsquigarrow} \hat{E}_1\}, \tag{5.7.33}$$

$$\hat{E}_{22} = \{i:\ i \in \hat{E}_2,\ i \underset{C}{\rightsquigarrow} \hat{E}_1\}. \tag{5.7.34}$$

§ 5.8 Second-type regular systems of equations

Definition 5.8.1 *If the transpose of the coefficient matrix $C = (c_{ik})$ on the right side of the second-type regular system of equations*

$$x_i = \sum_{k=1}^{n} c_{ik} x_k + b_i \qquad (i = 1,\ 2,\ \cdots,\ n) \tag{5.8.1}$$

is a first-type leading-outside (a first-type blockable random) matrix, then it is called a second-type leading-outside (a second-type blockable random) system of equations.

It is easy to prove the following lemmas.

Lemma 5.8.1 *If (5.8.1) is a second-type leading-outside system of equations, then it has a unique (ordinary) solution, i. e., the minimal nonnegative solution.* □

Lemma 5.8.2 *The second-type blockable random system of homogeneous equations*

$$x_i = \sum_{k=1}^{n} c_{ik} x_k \qquad (i = 1,\ 2,\ \cdots,\ n) \tag{5.8.2}$$

has infinitely many (ordinary) solutions, and its minimal nonnegative solution x_i^* $(i = 1,\ 2,\ \cdots,\ n)$ coincides with the zero solution, i.e.,

$$x_i^* \equiv 0 \qquad (i = 1,\ 2,\ \cdots,\ n).\ \square \tag{5.8.3}$$

Lemma 5.8.3 *If (5.8.1) is a second-type blockable random system of strictly* *nonhomogeneous equations, its minimal nonnegative solution*

$$x^* \equiv +\infty \qquad (i = 1, 2, \cdots, n). \ \square \tag{5.8.4}$$

Suppose (5.8.1) is a second-type regular system of equations. Let

$$t_{ik} = \begin{cases} c_{ik}, & i, k = 1, 2, \cdots, n, \\ b_i, & i = 1, 2, \cdots, n, \quad k = 0, \\ 0, & i = 0, \quad k = 1, 2, \cdots, n, \\ 1, & i = k = 0. \end{cases} \tag{5.8.5}$$

We define a matrix

$$T = (t_{ik}, \ i, \ k \in (0, 1, \cdots, n)), \tag{5.8.6}$$

and put

$$E_1 = \{i: \ i \in (1, 2, \cdots, n), \ i \underset{T}{\not\rightsquigarrow} 0\}, \tag{5.8.7}$$

$$E_2 = \{i: \ i \in (1, 2, \cdots, n), \ i \underset{T}{\rightsquigarrow} 0\}, \tag{5.8.8}$$

$$C_1 = (c_{ik}, \ i, \ k \in E_1), \tag{5.8.9}$$

E_{11} = the leading-outside subscript set of the transpose C_1' of the matrix C_1, (5.8.10)

$$E_{12} = E_1 \setminus E_{11}, \tag{5.8.11}$$

$$C_{12} = (c_{ik}, \ i, \ k \in E_{12}), \tag{5.8.12}$$

$E_{12}^{(1)}$ = the nonessential subscript set of the matrix C_{12}, (5.8.13)

$$E_{12}^{(2)} = E_{12} \setminus E_{12}^{(1)}, \tag{5.8.14}$$

$$C_2 = (c_{ik}, \ i, \ k \in E_2), \tag{5.8.15}$$

E_{21} = the leading-outside subscript set of the matrix C_2', (5.8.16)

$$E_{22} = E_2 \setminus E_{21}, \tag{5.8.17}$$

$$C_{22} = (c_{ik}, \ i, \ k \in E_{22}), \tag{5.8.18}$$

$$E_{22}^{(1)} = \text{the nonessential subscript set of the matrix } C_{22}, \tag{5.8.19}$$

$$E_{22}^{(2)} = E_{22} \setminus E_{22}^{(1)}. \tag{5.8.20}$$

Definition 5.8.2 *If*

$$E_{22}^{(2)} = \varnothing, \tag{5.8.21}$$

then (5.8.1) *is called a second-type consistent system of equations.*

Remark 5.8.1 It is clear that (5.8.1) is a second-type leading-outside system of equations iff

$$E_{12}^{(2)} \bigcup E_{22}^{(2)} = \varnothing, \tag{5.8.22}$$

and will be a second-type blockable random system of strictly nonhomogeneous equations iff

$$E_{22}^{(2)} = \{1, 2, \cdots, n\}. \tag{5.8.23}$$

Theorem 5.8.1 *The minimal nonnegative solution of the second-type regular system of equations* (5.8.1) *is determined as follows:*

(i) $\{x_i^*,\ i \in E_{11} \bigcup E_{12}^{(1)}\}$ *is the unique (ordinary) solution of the second-type leading-outside system of homogeneous equations*

$$x_i = \sum_{k \in E_{11} \bigcup E_{12}^{(1)}} c_{ik} x_k \qquad (i \in E_{11} \bigcup E_{12}^{(1)}), \tag{5.8.24}$$

hence

$$x_i^* = 0 \qquad (i \in E_{11} \bigcup E_{12}^{(1)}). \tag{5.8.25}$$

(ii) $\{x_i^*,\ i \in E_{12}^{(2)}\}$ *is the zero solution of the second-type blockable random system of homogeneous equations*

$$x_i = \sum_{k \in E_{12}^{(2)}} c_{ik} x_k \qquad (i \in E_{12}^{(2)}), \tag{5.8.26}$$

i.e.,

$$x_i^* \equiv 0 \qquad (i \in E_{12}^{(2)}). \tag{5.8.27}$$

(iii) $\{x_i^*,\ i \in E_{21} \bigcup E_{22}^{(1)}\}$ *is the unique (ordinary) solution of the second-type leading-outside system of strictly nonhomogeneous equations*

$$x_i = \sum_{k \in E_{21} \bigcup E_{22}^{(1)}} c_{ik} x_k + b_i \qquad (i \in E_{21} \bigcup E_{22}^{(1)}). \tag{5.8.28}$$

(iv) $\{x_i^*,\ i \in E_{22}^{(2)}\}$ *is the minimal nonnegative solution of the second-type blockable random system of strictly nonhomogeneous equations*

$$x_i = \sum_{k \in E_{22}^{(2)}} c_{ik} x_k + \left(\sum_{k \in E_{21} \cup E_{22}^{(1)}} c_{ik} x_k^* + b_i \right) \qquad (i \in E_{22}^{(2)}), \tag{5.8.29}$$

hence

$$x_i^* = +\infty \qquad (i \in E_{22}^{(2)}). \tag{5.8.30}$$

Proof. These assertions can readily be shown by using Lemmas 5.8.1, 5.8.2, 5.8.3 and Remark 5.8.1. □

Corollary 5.8.1 *A second-type regular system of equations has an ordinary minimal nonnegative solution iff it is a second-type consistent system of equations.*

Corollary 5.8.2 *A second-type regular system of equations has a unique (ordinary) solution iff it is a second-type leading-outside system of equations.*

Corollary 5.8.3 *If a second-type regular system of equations has a unique (ordinary) solution, then this solution coincides with the minimal nonnegative solution.*

Corollary 5.8.4 *The minimal nonnegative solution of a second-type regular system of equations is identically infinite iff the system of equations is a second-type blockable random strictly nonhomogeneous one.* □

PART III

HOMOGENEOUS DENUMERABLE MARKOV CHAINS

CHAPTER VI
General Theory

§ 6.1 Introduction

Let $(\Omega, \mathcal{F}, P)$ be a probability space; $\xi(\omega)$ is a random variable taking values in the set $\{0, 1, 2, \cdots\}$; $x(\omega)$ is a function taking values in the countable set $E = \{1, 2, \cdots\}$ and defined for all $\omega \in \Omega$, $0 \leqslant n < \xi(\omega) + 1$, there n is a nonnegative integer. For every n, let

$$\hat{x}_n(\omega) = \begin{cases} x_n(\omega), & 0 \leqslant n < \xi(\omega) + 1; \\ \Delta, & \xi(\omega) + 1 < n < +\infty. \end{cases}$$

$X(\omega) = \{x_n(\omega), 0 \leqslant n < \xi(\omega) + 1\}$ is called a homogeneous denumerable Markov chain defined on the probability space $(\Omega, \mathcal{F}, p)$ if $\hat{X}(\omega) = \{\hat{x}_n(\omega), n = 0,1,\cdots\}$ is a *discrete parameter Markov chain* (abbreviated to d.p.M.c.) defined in the way of [1, I, §2] with stationary transition probabilities, and $\xi(\omega)$ is called a *living time* of $x(\cdot,\omega)$.

Suppose that $X = \{x_n(\omega),\ 0 \leqslant n < \zeta(\omega) + 1,\ \omega \in \Omega\}$ is a homogeneous denumerable Markov chain (simply called Markov chain) defined on a probability space $(\Omega, \mathcal{F}, P)$, its state space is $E = \{1, 2, \cdots\}$, and the transition probability matrix is $P = (p_{ij}, i, j \in E)$. X is called an *honest* Markov chain, if $P(\zeta(\omega) = +\infty) = 1$. X is turned into the d.p.M.c. defined in [1, I, §2], if it is honest.

We shall take the corresponding concepts and notation about the d.p.M.c. in [1, I, §2] for the homogeneous denumerable Markov chain $X(\omega)$ defined above. Futhermore, by the relation between $X(\omega)$ and $\hat{X}(\omega)$, it is easy to decide that most of the results in [1, I] are true for $X(\omega)$. So we shall frequently use the concepts, notation and some suitable results from [1, I].

In this chapter our main task is to investigate the problems of transition probability, the distribution and moments of the first passage time, the distribution and moments of the times of passage, criteria for recurrence of states, the distribution and moments of additive functionals, Blackwell decomposition of the state space, etc.

§ 6.2 Transition probabilities

Suppose that A and H are subsets of E, and $A \neq \varnothing$. Let

$$ {}_Hp_{iA}^{(n)} = P(x_n \in A,\ x_\nu \notin H, \qquad 0 < \nu < n \mid x_0 = i), \tag{6.2.1}$$

$$ {}_Hp_{iA}^{(\lambda)} = \sum_{n=1}^{\infty} {}_Hp_{iA}^{(n)} \lambda^n \qquad (0 \leqslant \lambda \leqslant 1), \tag{6.2.2}$$

$$ {}_Hp_{iA}^{*} = \sum_{n=1}^{\infty} {}_Hp_{iA}^{(n)}. \tag{6.2.3}$$

Theorem 6.2.1 $\{{}_Hp_{iA}(\lambda),\ i \in E\}$ *is the minimal nonnegative solution of the first-type system of* 1-*bounded equations*

$$x_i = \sum_{j \in E} \lambda p_{ij} x_j + \lambda p_{iA} \qquad (i \in E). \tag{6.2.4}$$

Proof. It is evident that (6.2.4) is a first-type system of 1-bounded equations. Since

$$\left.\begin{aligned} &{}_Hp_{iA}^{(1)} + p_{iA} && (i \in E), \\ &{}_Hp_{iA}^{(n+1)} = \sum_{j \in E \setminus H} p_{ij}\, {}_Hp_{jA}^{(n)} && (i \in E,\ n \geqslant 1)\cdot \end{aligned}\right\} \tag{6.2.5}$$

We have

$$\left.\begin{aligned} &\lambda\, {}_Hp_{iA}^{(1)} = \lambda p_{iA} && (i \in E), \\ &\lambda^{n+1}\, {}_Hp_{iA}^{(n+1)} = \sum_{j \in E \setminus H} \lambda p_{ij} \lambda^n\, {}_Hp_{jA}^{(n)} && (i \in E,\ n \geqslant 1). \end{aligned}\right\} \tag{6.2.6}$$

The theorem follows from (6.2.2), (6.2.6) and Theorem 3.2.3. □

Theorem 6.2.2 $\{{}_Hp_{iA}^{*},\ i \in E\}$ *is the minimal nonnegative solution of the first-type system of* 1-*bounded equations*

$$x_i = \sum_{j \in E \setminus H} p_{ij} x_j + p_{iA} \qquad (i \in E). \tag{6.2.7}$$

Proof. This theorem is a special case of Theorem 6.2.1 for $\lambda = 1$. □

Theorem 6.2.3 $\{{}_Hp_{ij}(\lambda),\ j \in E\}$ *is the minimal nonnegative solution of the second-type system of* 1-*bounded equations*

$$x_j = \sum_{k \in E \setminus H} \lambda p_{kj} x_k + \lambda p_{ij} \qquad (j \in E). \tag{6.2.8}$$

Proof. This theorem follows readily from Theorems 6.2.1 and 3.8.1. □

Theorem 6.2.4 $\{{}_Hp_{ij}^{*}, j \in E\}$ *is the minimal nonnegative solution of the second-type system of* 1-*bounded equations*

$$x_j = \sum_{k \in E \setminus H} p_{kj} x_k + p_{ij} \qquad (j \in E). \tag{6.2.9}$$

Proof. The theorem is a special case of Theorem 6.2.3 for $\lambda = 1$. □

Let

$$e_{hj} = \sum_{n=1}^{\infty} {}_h p_{hj}^{(n)}. \tag{6.2.10}$$

Theorem 6.2.5 *$\{e_{hj}, j \in E\}$ is the minimal nonnegative solution of the second-type system of 1-bounded equations*

$$x_j = \sum_{k \neq h} p_{kj} x_k + p_{hj} \qquad (j \in E). \tag{6.2.11}$$

Proof. The assertion follows from Corollary 3.2.3 and that

$$\left.\begin{aligned} {}_h p_{hj}^{(1)} &= p_{hj} && (j \in E), \\ {}_h p_{hj}^{(n+1)} &= \sum_{k \neq h} {}_h p_{hk}^{(n)} \cdot p_{kj} && (n \geq 1,\ j \in E). \ \square \end{aligned}\right\} \tag{6.2.12}$$

§ 6.3 Distribution and moments of the first passage time

Let

$$ {}_H f_{iA}^{(n)} = P(x_v \notin A \cup H,\ 0 < v < n,\ x_n \in A \mid x_0 = i), \tag{6.3.1}$$

$$ {}_H \Phi_{iA}(\lambda) = \sum_{n=1}^{\infty} {}_H f_{iA}^{(n)} \lambda^n \qquad (0 \leq \lambda \leq 1), \tag{6.3.2}$$

$$ {}_H m_{iA}^{(p)} = \sum_{n=1}^{\infty} n^p \cdot {}_H f_{iA}^{(n)} \qquad (p = 0,\ 1,\ \cdots), \tag{6.3.3}$$

$$ {}_H f_{iA}^{*} = {}_H \Phi_{iA}(1) = {}_H m_{iA}^{(0)} = \sum_{n=1}^{\infty} {}_H f_{iA}^{(n)}. \tag{6.3.4}$$

Theorem 6.3.1 *$\{{}_H \Phi_{iA}(\lambda),\ i \in E\}$ is the minimal nonnegative solution of the pseudo-normal equation system*

$$x_i = \sum_{k \notin A \cup H} \lambda p_{ik} x_k + \lambda p_{iA} \qquad (i \in E). \tag{6.3.5}$$

Proof. Since

$$\left.\begin{aligned} {}_H f_{iA}^{(1)} &= p_{iA} && (i \in E), \\ {}_H f_{iA}^{(n+1)} &= \sum_{k \notin A \cup H} p_{ik} \cdot {}_H f_{kA}^{(n)} && (n \geq 1,\ i \in E), \end{aligned}\right\} \tag{6.3.6}$$

we have

$$\left.\begin{aligned} \lambda_H f_{iA}^{(1)} &= \lambda p_{iA} && (i \in E), \\ \lambda^{n+1}{}_H f_{iA}^{(n+1)} &= \sum_{k \notin A \cup H} \lambda p_{ik} \cdot \lambda^n{}_H f_{kA}^{(n)} && (n \geq 1;\ i \in E). \end{aligned}\right\} \tag{6.3.7}$$

It follows from (6.3.2), (6.3.7) and Corollary 3.2.3 that $\{{}_H\Phi_{iA}(\lambda),\ i \in E\}$ is the minimal nonnegative solution of (6.3.5). Clearly (6.3.5) is a pseudo-normal systems of equations. □

Theorem 6.3.2 $\{{}_Hf_{iA}^*;\ i \in E\}$ *is the minimal nonnegative solution of the pseudo-normal systems of equations*

$$x_i = \sum_{k \notin A \cup H} p_{ik} x_k + p_{iA} \qquad (i \in E). \tag{6.3.8}$$

Proof. This theorem is a special case of Theorem 6.3.1 for $\lambda = 1$. □

When ${}_Hm_{iA}^{(p-1)} = +\infty$, we stipulate that

$$\sum_{l=1}^{p} (-1)^{l-1} C_p^l {}_Hm_{iA}^{(p-l)} = +\infty, \tag{6.3.9}$$

where C_p^l denotes the number of the different combinations of l elements taken from p elements.

Theorem 6.3.3 *For* $p \geq 1$, $\{{}_Hm_{iA}^{(p)};\ i \in E\}$ *is the minimal nonnegative solution of the first-type system of* 1–*bounded equations*

$$x_i = \sum_{k \notin A \cup H} p_{ik} x_k + \sum_{l=1}^{p} (-1)^{l-1} C_p^l {}_Hm_{iA}^{(p-l)} \qquad (i \in E). \tag{6.3.10}$$

Proof. Obviously, (6.3.10) is a first-type systems of 1-bounded equations. It follows from(6.3.6) that

$$\left.\begin{aligned} 1^p{}_H f_{iA}^{(1)} &= p_{iA} \qquad (i \in E), \\ (n+1)^p \cdot {}_H f_{iA}^{(n+1)} &= \sum_{k \notin A \cup H} p_{ik} \cdot n^p \cdot {}_H f_{kA}^{(n)} \\ &\quad + \sum_{l=1}^{p} C_p^l \sum_{k \notin A \cup H} p_{ik} \cdot n^{p-l} \cdot {}_H f_{kA}^{(n)} \qquad (n \geq 1,\ i \in E). \end{aligned}\right\} \tag{6.3.11}$$

We deduce from(6.3.3), (6.3.11) and Corollary 3.2.2 that $\{{}_Hm_{iA}^{(p)};\ i \in E\}$ is the minimal nonnegative solution of the first-type systems of 1–bounded equations

$$x_{i} = \sum_{k\notin A\cup H} p_{ik}x_k + \sum_{l=1}^{p} C_p^l \sum_{k\notin A\cup H} p_{ik}\cdot {}_Hm_{kA}^{(p-l)} + p_{iA} \qquad (i\in E). \tag{6.3.12}$$

Consequently it suffices to prove that

$$\sum_{l=1}^{p} C_p^l \sum_{k\notin A\cup H} p_{ik}\, {}_Hm_{kA}^{(p-l)} + p_{iA} = \sum_{l=1}^{p} (-1)^{l-1} C_p^l\, {}_Hm_{iA}^{(p-l)} \qquad (i\in E). \tag{6.3.13}$$

We are led to the following two cases:

(i) ${}_Hm_{iA}^{(p-1)} = +\infty$.

Since

$$C_p^l = \frac{p!}{l!(p-l)!} = \frac{p}{l}\frac{(p-1)!}{(l-1)!\,((p-1)-(l-1))!}$$

$$= \frac{p}{l} C_{p-1}^{l-1} \geqslant C_{p-1}^{l-1} \qquad (1\leqslant l\leqslant p), \tag{6.3.14}$$

therefore

$$\sum_{l=1}^{p} C_p^l \sum_{k\notin A\cup H} p_{ik}\cdot {}_Hm_{kA}^{(p-1)} + p_{iA}$$

$$\geqslant \sum_{l=1}^{p} C_{p-1}^{l-1} \sum_{k\notin A\cup H} p_{ik}\cdot {}_Hm_{kA}^{((p-1)-(l-1))} + p_{iA}$$

$$= \sum_{l=0}^{p-1} C_{p-1}^{l} \sum_{k\notin A\cup H} p_{ik}\cdot {}_Hm_{kA}^{((p-1)-l)} + p_{iA} = {}_Hm_{A}^{(p-1)} = +\infty. \tag{6.3.15}$$

Hence (6.3.13) follows from (6.3.9).

(ii) ${}_Hm_{iA}^{(p-1)} < +\infty$.

We now have

$${}_Hm_{iA}^{(p-l)} < +\infty \qquad (l = 1, 2, \cdots, p). \tag{6.3.16}$$

Let us first prove

$$\sum_{l=1}^{p} C_p^l \sum_{s=0}^{p-l} C_{p-l}^s (-1)^s\, {}_Hm_{iA}^{(p-l-s)} + \sum_{s=1}^{p} (-1)^s C_p^s\, {}_Hm_{iA}^{(p-s)} = 0. \tag{6.3.17}$$

Let

$$\sum_{l=1}^{p} C_p^l \sum_{s=0}^{p-l} C_{p-l}^s (-1)^s\, {}_Hm_{iA}^{(p-l-s)} + \sum_{s=1}^{p} (-1)^s C_p^s\, {}_Hm_{iA}^{(p-s)} = \sum_{l=1}^{p} \alpha_{p-l}\, {}_Hm_{iA}^{(p-l)}. \tag{6.3.18}$$

But

$$z^p = [1+(z-1)]^p = \sum_{l=0}^{p} C_p^l (z-1)^{p-l}$$

$$= \sum_{l=0}^{p} C_p^l \sum_{s=0}^{p-l} C_{p-l}^s (-1)^s z^{p-l-s} = \sum_{l=1}^{p} \alpha_{p-l} z^{p-l} + z^p. \tag{6.3.19}$$

Thus

$$\alpha_l = 0 \qquad (l=0,\ 1,\ \cdots,\ p-1). \tag{6.3.20}$$

Equation (6.3.17) follows easily from (6.3.18) and (6.3.20).

Now we turn to the proof of (6.3.13). By Theorem 6.3.2, (6.3.13) is valid for $p=1$. Now assume that (6.3.13) is valid for $1,\ 2,\ \cdots,\ p-1$, then

$$\sum_{l=1}^{p} C_p^l \sum_{k \notin A \cup H} p_{ik\,H}m_{kA}^{(p-l)} + p_{iA}$$

$$= \sum_{l=1}^{p-1} C_p^l \left[\sum_{k \notin A \cup H} p_{ik\,H}m_{kA}^{(p-l)} + \sum_{s=1}^{p-l} (-1)^{s-1} C_{p-l\,H}^s m_{iA}^{(p-l-s)} \right.$$

$$\left. + \sum_{s=1}^{p-l} (-1)^s C_{p-l\,H}^s m_{iA}^{(p-l-s)} \right] + \sum_{k \notin A \cup H} p_{ik} \cdot {}_H f_{iA}^* + p_{iA}$$

$$= \sum_{l=1}^{p-1} C_p^l \left[{}_H m_{iA}^{(p-l)} + \sum_{s=1}^{p-l} (-1)^s C_{p-l\,H}^s m_{iA}^{(p-l-s)} \right] + {}_H f_{iA}^*$$

$$= \sum_{l=1}^{p-1} C_p^l \sum_{s=0}^{p-l} (-1)^s C_{p-l\,H}^s m_{iA}^{(p-l-s)} + {}_H m_{iA}^{(0)}$$

$$= \sum_{l=1}^{p} C_p^l \sum_{s=0}^{p-l} (-1)^s C_{p-l\,H}^s m_{iA}^{(p-l-s)} = \sum_{s=1}^{p} (-1)^{s-1} C_{p\,H}^s m_{iA}^{(p-s)}$$

$$= \sum_{l=1}^{p} (-1)^{l-1} C_{p\,H}^l m_{iA}^{(p-l)}. \tag{6.3.21}$$

Therefore (6.3.13) is also valid for p. It follows by induction that (6.3.13) holds. □

Theorem 6.3.4 *If for a certain constant* $0 < C < +\infty$ *we have*

$$ {}_H m_{iA} \leqslant C\,{}_H f_{iA}^* \qquad (i \in E), \tag{6.3.22}$$

then

$$ {}_H m_{iA}^{(p)} \leqslant p!\, C^p\,{}_H f_{iA}^* \qquad (p \geqslant 1,\ i \in E). \tag{6.3.23}$$

Proof. It can be deduced from (6.3.22) that (6.3.23) is valid for the case $p = 1$. Now we assume that (6.3.23) is valid for $p-1$, i.e.,

$$_Hm_{iA}^{(p-1)} \leqslant (p-1)!\, C^{p-1}\, _Hf_{iA}^* \qquad (i \in E). \tag{6.3.24}$$

We wish to show that (6.3.23) is valid for p.

Theorem 6.3.3 implies that $\{_Hm_{iA},\ i \in E\}$ is the minimal nonnegative solution of the first-type systems of 1-bounded equations

$$x_i = \sum_{k \notin A \cup H} p_{ik} x_k + {_Hf_{iA}^*} \qquad (i \in E). \tag{6.3.25}$$

So by Corollary 3.3.3, x_i^* $(i \in E)$ is the minimal nonnegative solution of the first-type systems of 1-bounded equations

$$x_i = \sum_{k \notin A \cup H} p_{ik} x_k + p!\, C^{p-1}\, _Hf_{iA}^* \qquad (i \in E), \tag{6.3.26}$$

hence

$$x_i^* = p!\ C^{p-1}\, _Hm_{iA} \qquad (i \in E). \tag{6.3.27}$$

Then it follows by (6.3.22) that

$$x_i^* \leqslant p!\ C^p\, _Hf_{iA}^* \qquad (i \in E). \tag{6.3.28}$$

It can be seen from the proof of Theorem 6.3.3 that $\{_Hm_{iA}^{(p)},\ i \in E\}$ is the minimal nonnegative solution of the first-type systems of 1-bounded equations

$$x_i = \sum_{k \notin A \cup H} p_{ik} x_k + \sum_{l=1}^{p} C_p^l \sum_{k \notin A \cup H} p_{ik} \cdot {_Hm_{kA}^{(p-l)}} + p_{iA} \qquad (i \in E). \tag{6.3.29}$$

But by virtue of

$$C_p^l = \frac{p}{l}\, C_{p-1}^{l-1} \leqslant p C_{p-1}^{l-1} \qquad (1 \leqslant l \leqslant p) \tag{6.3.30}$$

and (6.3.24), we have

$$\sum_{l=1}^{p} C_p^l \sum_{k \notin A \cup H} p_{ik} \cdot {_Hm_{kA}^{(p-l)}} + p_{iA}$$

$$\leqslant p\left(\sum_{l=1}^{p} C_{p-1}^{l-1} \sum_{k \notin A \cup H} p_{ik} \cdot {_Hm_{kA}^{((p-1)-(l-1))}} + p_{iA}\right)$$

$$= p\left(\sum_{l=0}^{p-1} C_{p-1}^{l} \sum_{k \notin A \cup H} p_{ik} \cdot {}_H m_{kA}^{((p-1)-l)} + p_{iA}\right)$$

$$= p_H m_{iA}^{(p-1)} \leqslant p! C^{p-1} {}_H f_{iA}^{*} \qquad (i \in E). \tag{6.3.31}$$

Thus, by applying Theorem 6.3.1, we get

$${}_H m_{iA}^{(p)} \leqslant x_i^{*} \qquad (i \in E). \tag{6.3.32}$$

Combining (6.3.28) and (6.3.32), we see that (6.3.23) is also valid for p. Therefore, by induction, (6.3.23) is always true. □

§ 6.4 Distribution and moments of the first passage time of a homogeneous finite Markov chain

For homogeneous finite Markov chains, the results in this chapter naturally hold but most of them can still be further strengthened. Since homogeneous finite Markov chains have been rather thoroughly studied, there is no need to devote a whole chapter to them here. Nevertheless we shall devote some space to discussing the distribution and moments of the first passage time, which is very important and has not been adequately studied so far.

Suppose $\{x_n, n \geqslant 0\}$ is a homogeneous finite honest Markov chain, its state space is $E = \{1, 2, \cdots, n\}$.

Let

$$G(j) = \{i\colon\ i \in E,\ i \not\rightsquigarrow j\}, \tag{6.4.1}$$

$$\bar{G}(j) = \{i\colon\ i \in E,\ i \rightsquigarrow j\} = E \setminus G(j), \tag{6.4.2}$$

$$\bar{G}^{(1)}(j) = \{i\colon\ i \in \bar{G}(j),\ i \not\rightsquigarrow G(j)\}, \tag{6.4.3}$$

$$\bar{G}^{(2)}(j) = \{i\colon\ i \in \bar{G}(j),\ i \rightsquigarrow G(j)\} = \bar{G}(j) \setminus \bar{G}^{(1)}(j). \tag{6.4.4}$$

The other notation used in this section is as in §§6.2 and 6.3.

It is easy to prove the following results.

Theorem 6.4.1 *$\{f_{ij}(\lambda), i \in E\}$ is the minimal nonnegative solution of the pseudo-normal systems of equations with dimension n*

$$x_i = \sum_{k \neq j} \lambda p_{ik} x_k + \lambda p_{ij} \qquad (i \in E). \quad □ \tag{6.4.5}$$

Theorem 6.4.2 *$\{f_{ij}^{*}, i \in E\}$ is the minimal nonnegative solution of the normal* *systems of equations with dimension n*

$$x_i = \sum_{k \neq j} p_{ik} x_k + p_{ij} \qquad (i \in E). \tag{6.4.6}$$

Hence

(i)

$$f_{ij}^* = 0 \qquad (i \in G(j)); \tag{6.4.7}$$

(ii)

$$f_{ij}^* = 1 \qquad (i \in \bar{G}^{(1)}(j)); \tag{6.4.8}$$

(iii) $\{f_{ij}^*,\ i \in \bar{G}^{(2)}(j)\}$ *is the unique (ordinary) solution of the leadingcomplement system of strictly nonhomogeneous equations*

$$x_i = \sum_{k \in \bar{G}^{(2)}(j) \setminus \{j\}} p_{ik} x_k + \sum_{k \in \bar{G}^{(1)}(j) \cup \{j\}} p_{ik} \qquad (i \in \bar{G}^{(2)}(j)), \tag{6.4.9}$$

hence

$$0 < f_{ij}^* < 1 \qquad (i \in \bar{G}^{(2)}(j)). \ \square \tag{6.4.10}$$

Theorem 6.4.3 $\{m_{ij};\ i \in E\}$ *is the minimal nonnegative solution of the first-type consistent system of equations*

$$x_i = \sum_{k \neq j} p_{ik} x_j + f_{ij}^* \qquad (i \in E), \tag{6.4.11}$$

hence

$$m_{ij} < +\infty \qquad (i \in E). \tag{6.4.12}$$

In detail,

(i) $\{m_{ij};\ i \in G(j)\}$ *is the zero solution of the random system of homogeneous equations*

$$x_i = \sum_{k \in (j) \setminus \{j\}} p_{ik} x_k \qquad (i \in G(j)), \tag{6.4.13}$$

hence

$$m_{ij} = 0 \qquad (i \in G(j)); \tag{6.4.14}$$

(ii) $\{m_{ij};\ i \in \bar{G}(j)\}$ *is the unique (ordinary) solution of the first-type leading-*

outside system of equations

$$x_i = \sum_{k \in \bar{G}(j) \setminus \{j\}} p_{ik} x_k + f_{ij}^* \qquad (i \in \bar{G}(j)).\square \tag{6.4.15}$$

Theorem 6.4.4 $\{m_{ij}^{(p)}; i \in E\}$ *is the minimal nonnegative solution of the first-type consistent system of equations*

$$x_i = \sum_{k \neq j} p_{ik} x_k + \sum_{l=1}^{p} (-1)^{l-1} C_p^l m_{ij}^{(p-l)} \qquad (i \in E), \tag{6.4.16}$$

hence

$$m_{ij}^{(p)} < +\infty \qquad (i \in E). \tag{6.4.17}$$

We can state the assertion more precisely.

(i) $\{m_{ij}^{(p)}; i \in G(j)\}$ *is the zero solution of the random system of homogeneous equations*

$$x_i = \sum_{k \in G(j) \setminus \{j\}} p_{ik} x_k \qquad (i \in G(j)), \tag{6.4.18}$$

hence

$$m_{ij}^{(p)} = 0 \qquad (i \in G(j)); \tag{6.4.19}$$

(ii) $\{m_{ij}^{(p)}; i \in \bar{G}(j)\}$ *is the unique (ordinary) solution of the first-type leading-outside system of equations*

$$x_i = \sum_{k \in \bar{G}(j) \setminus \{j\}} p_{ik} x_k + \sum_{l=1}^{p} (-1)^{l-1} C_p^l m_{ij}^{(p-l)} \qquad (i \in \bar{G}(j)), \tag{6.4.20}$$

and

$$m_{ij}^{(p)} \leqslant p!\ c^p f_{ij}^* \qquad (j \in \bar{G}(j)), \tag{6.4.21}$$

where

$$0 < c = \max_{i \in \bar{G}(j)} \frac{m_{ij}}{f_{ij}^*} < +\infty. \ \square \tag{6.4.22}$$

Corollary 6.4.1 $f_{ij}(\lambda)$ *is an analytic function of* λ *in the domain* $|\lambda| < 1/c$, *and*

$$f_{ij}(\lambda) = \begin{cases} 0, & i \in G(j); \\ \sum_{p=0}^{\infty} (-1)^p \frac{m_{ij}^{(p)}}{p!} \lambda^p, \quad |\lambda| < \frac{1}{c}, & i \in \bar{G}(j). \ \square \end{cases} \tag{6.4.23}$$

§ 6.5 Distribution and moments of the times of passage

Suppose X is an honest Markov chain. Let

$$k_{iA}^{(n)} = P(x_m \in A \text{ just } n \text{ times, } m > 0 \mid x_0 = i), \tag{6.5.1}$$

$$K_{iA}(\lambda) = \sum_{n=0}^{\infty} k_{iA}^{(n)} \lambda^n \qquad (0 \leqslant \lambda \leqslant 1), \tag{6.5.2}$$

$$K_{iA}^{(p)} = \sum_{n=0}^{\infty} n^p k_{iA}^{(n)} \qquad (p = 0, 1, \cdots), \tag{6.5.3}$$

$$K_{iA}^{*} = K_{iA}(1) = \sum_{n=0}^{\infty} k_{iA}^{(n)}, \tag{6.5.4}$$

$$\bar{K}_{iA}^{*} = P(x_m \in A \text{ infinitely many times, } m \geqslant 0 \mid x_0 = i), \tag{6.5.5}$$

$$\sigma_A = \begin{cases} \inf\{n\colon\ x_n \in A,\ n > 0\}, & \text{if this set is nonempty,} \\ +\infty, & \text{otherwise.} \end{cases} \tag{6.5.6}$$

Evidently, σ_A is an optional random variable.

Theorem 6.5.1 $\{K_{iA}(\lambda),\ i \in E\}$ *is the minimal nonnegative solution of the pseudo-normal system of equations*

$$x_i = \sum_{j \in A} \lambda_A f_{ij}^{*} x_j + (1 - f_{iA}^{*}) \qquad (i \in E). \tag{6.5.7}$$

Proof. From

$$\sum_{j \in A} \lambda_A f_{ij}^{*} + (1 - f_{iA}^{*}) \leqslant \sum_{j \in A} {}_A f_{ij}^{*} + (1 - f_{iA}^{*}) = 1, \tag{6.5.8}$$

we see that (6.5.7) is a pseudo-normal system of equations. Evidently,

$$k_{iA}^{(0)} = 1 - f_{iA}^{*} \qquad (i \in E) \tag{6.5.9}$$

and

$$\begin{aligned} k_{iA}^{(n+1)} &= P(x_m \in A \text{ just } (n+1) \text{ times, } m > 0 \mid x_0 = i) \\ &= P(x_m \in A \text{ just } n \text{ times, } m > \sigma_A \mid x_0 = i) \\ &= \sum_{j \in A} P(x(\sigma_A) = j,\ x_m \in A \text{ just } n \text{ times, } m > \sigma_A \mid x_0 = i) \end{aligned}$$

$$= \sum_{j\in A} P(x(\sigma_A) = j | x_0 = i)$$

$$p(x_m \in A \text{ just } n \text{ times, } m > \sigma_A | x_0 = i,\ x(\sigma_A) = j)$$

$$= \sum_{j\in A} {}_A f_{ij}^* P(x_m \in A \text{ just } n \text{ times, } m > 0 | x_0 = j)$$

$$= \sum_{j\in A} {}_A f_{ij}^* k_{jA}^{(n)}. \tag{6.5.10}$$

It follows from (6.5.9) and (6.5.10) that

$$\left.\begin{aligned} \lambda^0 k_{iA}^{(0)} &= 1 - f_{iA}^* \qquad (i\in E), \\ \lambda^{n+1} k_{iA}^{(n+1)} &= \sum_{j\in A} \lambda_A f_{ij}^* \cdot \lambda^n k_{jA}^{(n)} \qquad (i\in E, n \geqslant 1). \end{aligned}\right\} \tag{6.5.11}$$

Then $\{K_{iA}(\lambda), i\in E\}$ is the minimal nonnegative solution of (6.5.7), owing to (6.5.2), (6.5.11) and Corollary 3.2.3. □

Theorem 6.5.2 *$\{K_{iA}^*; i\in E\}$ is the minimal nonnegative solution of the pseudo-normal system of equations*

$$x_i = \sum_{j\in A} {}_A f_{ij}^* x_j + (1 - f_{iA}^*) \qquad (i\in E). \tag{6.5.12}$$

Proof. This theorem is a special case of Theorem 6.5.1 for $\lambda = 1$. □

Theorem 6.5.3 *For $p \geqslant 1$, $\{K_{iA}^{(p)}; i\in E\}$ is the minimal nonnegative solution of the first-type system of 1-bounded equations*

$$x_i = \sum_{j\in A} {}_A f_{ij}^* x_j + \sum_{l=1}^{p} (-1)^{l-1} K_{iA}^{(p-l)} + (-1)^p (1 - f_{iA}^*) \qquad (i\in E). \tag{6.5.13}$$

Here we stipulate

$$\sum_{l=1}^{p} (-1)^{l-1} K_{iA}^{(p-l)} = +\infty, \tag{6.5.14}$$

if $K_{iA}^{(p-1)} = +\infty$.

Proof. The proof is similar to that of Theorem 6.3.3. □

Theorem 6.5.4

$$\bar{K}_{iA}^* = 1 - K_{iA}^* \qquad (i\in E). \tag{6.5.15}$$

Proof. This theorem follows readily from the definitions of $\bar{K}_{iA}^*$ and K_{iA}^*. □

Corollary 6.5.1 *Suppose the Markov chain is honest. If A is an almost closed set, then $\{\bar{K}^*_{iA}, i \in E\}$ is the minimal nonnegative solution of the normal system of equations*

$$x_i = \sum_{j \in \bar{A}} {}_{\bar{A}}f^*_{ij} x_j + (1 - f^*_{i\bar{A}}) \qquad (i \in E), \tag{6.5.16}$$

where $\bar{A} = E \setminus A$. □

Corollary 6.5.2 *Suppose the Markov chain is honest. A necessary and sufficient condition for a subset A of E to be an almost closed set is that the minimal nonnegative solution x^*_i $(i \in E)$ of the normal system of equations*

$$x_i = \sum_{j \in A} {}_{A}f^*_{ij} x_j + (1 - f^*_{iA}) \qquad (i \in E) \tag{6.5.17}$$

and the minimal nonnegative solution $\bar{x}^*_i$ $(i \in E)$ of the normalsystem of equations

$$x_i = \sum_{j \in \bar{A}} {}_{\bar{A}}f^*_{ij} x_j + (1 - f^*_{i\bar{A}}) \qquad (i \in E) \tag{6.5.18}$$

satisfy the conditions

$$x^*_i + \bar{x}^*_i = 1 \qquad (i \in E), \tag{6.5.19}$$

$$\sum_{i \in E} \bar{x}^*_i > 0. \ \square \tag{6.5.20}$$

§ 6.6 Criteria for recurrence

It is well known that a state s is recurrent if and only if $f^* = 1$ or $p^*_{ss} = +\infty$ or $\bar{K}^*_{ss} = 1$. In principle, we can directly calculate f^*_{ss}, or p^*_{ss}, or $\bar{K}^*_{ss}$ by applying the results in §§6.2, 6.3 and 6.5, so as to determine whether s is recurrent. However this approach is not always practicable. In order to meet the needs in different cases, we proceed to introduce some criteria for recurrence.

Suppose the Markov chain is honest, s is an element of E. Let $D(s) = \{i: s \nrightarrow i\} \cup \{s\}$, $\underline{D}(s) = D(s) \setminus \{s\}$.

Theorem 6.6.1

(I) *If s is nonessential, then s is nonrecurrent.*

(II) *If s is essential, and $D(s)$ is a finite set, then s is recurrent.*

(III) *If s is essential, and $D(s)$ is a denumerable set, then the following statements are equivalent.*

(i) *s is recurrent.*

(ii) *The minimal nonnegative solution of the normal system of equations*

$$x_i = \sum_{k \in \underline{D}(s)} p_{ik} x_k + p_{is} \qquad \left(i \in \underline{D}(s)\right) \tag{6.6.1}$$

is $x_i^* \equiv 1$ $\left(i \in \underline{D}(s)\right)$ *(which can be verified by Theorem* 5.6.3*).*

(iii) *The inequality system*

$$x_i \geqslant \sum_{k \in \underline{D}(s)} p_{kj} x_k + p_{sj} \qquad \left(j \in \underline{D}(s)\right) \tag{6.6.2}$$

has no nonnegative solution x_j $\left(j \in \underline{D}(s)\right)$ *satisfying the condition*

$$\sum_{k \in \underline{D}(s)} p_{ks} x_k + p_{ss} < 1. \tag{6.6.3}$$

(iv) *The inequality system*

$$x_i \geqslant \sum_{k \in D(s)} p_{ik} x_k + p_{is} \qquad \left(i \in D(s)\right) \tag{6.6.4}$$

has no ordinary nonnegative solution.

(v) *The inequality system*

$$x_j \geqslant \sum_{k \in D(s)} p_{kj} x_k + p_{sj} \qquad \left(j \in D(s)\right) \tag{6.6.5}$$

has no ordinary nonnegative solution.

Proof. (I) and (II) are evident. We are going to prove (III).

A little reflection shows that (i) is equivalent to (ii) by Theorem 6.3.2, Corollaries 3.4.1 and 5.6.1; and (i) is equivalent to (iii) by the fact that $e_{ss} = f_{ss}^*$, Theorem 6.2.5, Corollaries 3.4.1 and 3.3.2; moreover, (i) is equivalent to (iv) by Theorem 6.2.2; and (i) is equivalent to (v) by Theorem 6.2.4. □

Let

$$x_i = \sum_{k \in E} c_{ik} x_k + b_i \qquad (i \in E) \tag{6.6.6}$$

be a normal system of equations and $H \subset E$. Let

$$D^{(H)} = \{k\text{: there exists } i \in H, \text{ such that } i \underset{C}{\rightarrow} k\}, \tag{6.6.7}$$

where $C = (c_{ik},\ i, k \in E)$. We construct a Markov chain $\{y_n,\ n \geqslant 0\}$ with phase space

$\hat{E} = \{0\} \cup E$ such that the entries of the transition probability matrix $\hat{P} = (\hat{p}_{ij}, i, j \in E)$ are determined as follows:

$$\hat{p}_{ij} = \begin{cases} r_j, & i = 0, \quad j \in \{0\} \cup D^{(H)}, \\ 0, & i = 0, \quad j \in E \backslash D^{(H)}, \\ c_{ij}, & i, j \in E, \\ b_i, & i \in E, j = 0, \end{cases} \tag{6.6.8}$$

where

$$\sum_{j \in \{0\} \cup D^{(H)}} r_j = 1,$$

and $0 \underset{y_n}{\rightsquigarrow} i$ for all $i \in H$. This is always possible, for example, by putting $r_{n_j} = 1/N$ $(n_j \in H)$ when $H = \{n_1, n_2, \cdots, n_N\}$ is a finite set, or by putting $r_{n_j} = 1/2^j$ $(n_j \in H)$ when $H = \{n_1, n_2, \cdots\}$ is a denumerable set.

Definition 6.6.1 *The Markov chain $\{y_n, n \geqslant 0\}$ defined above is called an H-adjoint Markov chain of the normal system of equations* (6.6.6). *State O is called the additional state of $\{y_n, n \geqslant 0\}$. $\{y_n, n \geqslant 0\}$ is simply referred to as the adjoint Markov chain of* (6.6.6) *if $H = E$.*

Theorem 6.6.2 *If we denote the minimal nonnegative solution of* (6.6.6) *by x_i^* $(i \in E)$ and put*

$$f_{ij}^* = P(\text{there exists } n > 0, \text{ such that } y_n = j \mid x_0 = i) \qquad (i, j \in \hat{E}), \tag{6.6.9}$$

then

$$x_i^* = f_{i0}^* \qquad (i \in E). \tag{6.6.10}$$

Proof. This theorem follows from (6.6.8) by Theorem 6.3.2 and Corollary 3.4.1. □

Theorem 6.6.3 *If we denote by x_i^* $(i \in E)$ the minimal nonnegative solution of the normal system of equations* (6.6.6) *then $x_i^* \equiv 1$ $(i \in H)$ iff the additional state O of the H-adjoint Markov chain of* (6.6.6) *is recurrent.*

Proof. It is easy to complete the proof by applying Theorems 6.6.2 and 6.3.2, Corollary 3.4.1 and Theorem 5.6.2.□

Corollary 6.6.1 *A necessary and sufficient condition for $x_i^* \equiv 1$ $(i \in E)$ is that the adjoint Markov chain of* (6.6.6) *is an irreducible recurrent chain.* □

Let D denote the set of all the nonrecurrent states of an honest Markov chain. Theorem 6.6.4 and Corollary 6.6.2 below are examples of applications of Theorem 6.6.3 and Corollary 6.6.1.

Theorem 6.6.4 *If $H \subset D$, then a necessary and sufficient condition for the system, running out from D with probability* 1 *starting at any state of H is, that* *the additional state of a certain H-adjoint Markov chain of the normal system of equations*

$$x_i = \sum_{j \in D} p_{ij} x_j + p_{i\bar{D}} \qquad (i \in D) \tag{6.6.11}$$

is recurrent.

Proof. Note that the probability of the system running out from D starting at any state i of D is $f^*_{i\bar{D}}$. And by Theorems 6.3.2 and 6.6.3, and Corollary 3.4.1 the theorem is assured. □

Corollary 6.6.2 *A necessary and sufficient condition for the system, running out from D starting at any state of D with probability* 1 *is that the adjoint Markov chain of thenormal system of equations* (6.6.11) *is an irreducible recurrent chain.* □

Subsequently, quite a few problems in our work can be reduced to that of determining if the solution of a normal system of equations is equal to 1. The role of Theorem 6.6.3 and Corollary 6.6.1 is to reduce such problems to those of "criteria for recurrence" which have been rather satisfactorily studied.

§ 6.7 Distribution and moments of additive functionals

Suppose the Markov chain is honest. A and H_N are subsets of E, and $H_N = \{N + 1, N + 2, \cdots\}$. Let

$$\tau_A = \begin{cases} \inf\{n: x_n \in A, n \geqslant 1\}, & \text{if this set is nonempty,} \\ +\infty, & \text{otherwise;} \end{cases} \tag{6.7.1}$$

$$\tau_A^{(n)} = \min(n, \tau_A) \qquad (n \geqslant 1). \tag{6.7.2}$$

Evidently, τ_A, $\tau_A^{(n)}$ are both the optional random variables.

Suppose $V(i)$ $(i \in E)$ is a nonnegative finite-valued function on E. Let

$$\xi_A^{(n)} = \sum_{k=1}^{\tau_A^{(n)}} V(x_{k-1}) \qquad (n \geqslant 1), \tag{6.7.3}$$

$$\xi_A = \sum_{k=1}^{\tau_A} V(x_{k-1}). \tag{6.7.4}$$

It is easily discerned that $\xi_A^{(n)}$ and ξ_A are random variables, and

$$\xi_A = \text{V-}\lim_{n\to\infty} \xi_A^{(n)} = \text{V-}\lim_{N\to\infty} \xi_{A\cup H_N}. \tag{6.7.5}$$

Let

$$F_{iA}^{(n)}(t) = P(\xi_A^{(n)} \leqslant t | x_0 = i), \tag{6.7.6}$$

$$F_{iA}(t) = P(\xi_A \leqslant t | x_0 = i), \tag{6.7.7}$$

$$\varphi_{iA}^{(n)}(\lambda) = \int_0^\infty e^{-\lambda t} dF_{iA}^{(n)}(t) \qquad (\lambda > 0), \tag{6.7.8}$$

$$\varphi_{iA}(\lambda) = \int_0^\infty e^{-\lambda t} dF_{iA}(t) \qquad (\lambda > 0), \tag{6.7.9}$$

$$\psi_{iA}^{(n)}(\lambda) = 1 - \varphi_{iA}^{(n)}(\lambda), \tag{6.7.10}$$

$$\psi_{iA}(\lambda) = 1 - \varphi_{iA}(\lambda), \tag{6.7.11}$$

$$T_{iA}^{(n,p)} = M\{[\xi_A^{(n)}]^p | x_0 = i\}, \tag{6.7.12}$$

$$T_{iA}^{(p)} = M\{[\xi_A]^p | x_0 = i\}, \tag{6.6.13}$$

where $p = 0, 1, \cdots$. Obviously,

$$T_{iA}^{(n,0)} = T_{iA}^{(0)} = 1. \tag{6.7.14}$$

It follows from (6.7.5) and (6.7.6) that

$$\psi_{iA}(\lambda) = \text{V-}\lim_{n\to\infty} \psi_{iA}^{(n)}(\lambda) = \text{V-}\lim_{N\to\infty} \psi_{i,A\cup H_N}(\lambda), \tag{6.7.15}$$

$$T_{iA}^{(p)} = \text{V-}\lim_{n\to\infty} T_{iA}^{(n,p)} = \text{V-}\lim_{N\to\infty} T_{i,A\cup H_N}^{(p)}. \tag{6.7.16}$$

Sometimes we write T_{iA} for $T_{iA}^{(1)}$, and A may be omitted in above notation if $A = \varnothing$.

Theorem 6.7.1 $\psi_{iA}^{(n)}(\lambda)$ *are determined uniquely by the following recursion formulae:*

$$\begin{aligned} &\psi_{iA}^{(1)}(\lambda) = 1 - e^{-\lambda V(i)} \qquad (i \in E), \\ &\psi_{iA}^{(n+1)}(\lambda) = \sum_{j\notin A} p_{ij} e^{-\lambda V(i)} \psi_{jA}^{(n)}(\lambda) + (1 - e^{-\lambda V(i)}) \qquad (n \geqslant 1,\ i \in E). \end{aligned} \tag{6.7.17}$$

Proof. Obviously, the first expression of (6.7.17) is valid. Let us now prove the second.

It is easy to check by the Markov property that

$$F_{iA}^{(n+1)}(t) = \sum_{j\notin A} p_{ij} F_{jA}^{(n)}(t - V(i)) + \sum_{j\in A} p_{ij} P(V(i) \leqslant t \mid x_0 = i) \ (n \geqslant 1,\ i\in E). \tag{6.7.18}$$

Hence we have

$$\varphi_{iA}^{(n+1)}(\lambda) = \sum_{j\notin A} p_{ij} e^{-\lambda V(i)} \varphi_{jA}^{(n)}(\lambda) + \sum_{j\in A} p_{ij} e^{-\lambda V(i)} \quad (n \geqslant 1,\ i\in E). \tag{6.7.19}$$

From (6.7.19) and (6.7.10), we get the second expression of (6.7.17). □

Theorem 6.7.2 *$T_{iA}^{(n,p)}$ is determined uniquely by the following recursion formulae:*

$$\left.\begin{aligned} T_{iA}^{(1,p)} &= V(i)^p \qquad (i\in E), \\ T_{iA}^{(n+1,p)} &= \sum_{l=0}^{p} C_p^l V(i)^l \sum_{j\notin A} p_{ij} T_{jA}^{(n,p-l)} + \sum_{j\in A} p_{ij} V(i)^p \quad (n \geqslant 1,\ i\in E). \end{aligned}\right\} \tag{6.7.20}$$

Proof. It is easy to deduce the theorem from the Markov property.□

Theorem 6.7.3 *$\{\psi_{iA}(\lambda),\ i\in E\}$ is the minimal nonnegative solution of the pseudo-normal system of equations*

$$x_i = \sum_{j\notin A} p_{ij} e^{-\lambda V(i)} x_j + (1 - e^{-\lambda V(i)}) \qquad (i\in E). \tag{6.7.21}$$

Proof. We obtain this theorem by virtue of (6.7.15) and Theorems 6.7.1 and 3.2.1.□

When $V(i)\, T_{iA}^{(p-1)} = +\infty$, we stipulate that

$$\sum_{l=1}^{p} C_p^l (-1)^{l-1} V(i)^l T_{iA}^{(p-l)} = +\infty. \tag{6.7.22}$$

Theorem 6.7.4 *For $p \geqslant 1$, $\{T_{iA}^{(p)},\ i\in E\}$ is the minimal nonnegative solution of the first-type system of 1–bounded equations*

$$x_i = \sum_{j\notin A} p_{ij} x_j + \sum_{l=1}^{p} C_p^l (-1)^{l-1} V(i)^l T_{iA}^{(p-l)} \qquad (i\in E). \tag{6.7.23}$$

Proof. It is easy to complete the proof by using (6.7.16), Theorems 6.7.2 and 3.2.2, and referring to the proof of Theorem 6.3.3.□

Theorem 6.7.5 *Let $G \subset E$. Then a necessary and sufficient condition for*

$$T_{iA}^{(p)} < +\infty \qquad (i \in G) \tag{6.7.24}$$

is that the system of major inequalities

$$x_i \geqslant \sum_{j \notin A} p_{ij} x_j + \sum_{l=1}^{p} C_p^l (-1)^{l-1} V(i)^l T_{iA}^{(p-l)} \qquad (i \in E) \tag{6.7.25}$$

of (6.7.23) *has the nonnegative solution* x_i $(i \in E)$ *satisfying*

$$x_i < +\infty \qquad (i \in G). \tag{6.7.26}$$

Proof. From Theorem 6.7.4 and Corollary 3.3.1, we get the theorem immediately.□

Theorem 6.7.6 *If for a certain constant c,* $0 < c < +\infty$, *we have*

$$T_{iA} \leqslant c \qquad (i \in E), \tag{6.7.27}$$

then

$$T_{iA}^{(p)} < p!\ c^p \qquad (i \in E). \tag{6.7.28}$$

Proof. The proof is similar to that of Theorem 6.3.4.□

Corollary 6.7.1 *If* (6.7.27) *is valid, then* $\varphi_{iA}(\lambda)$ *is analytic in the domain* $|\lambda| < 1/c$, *and*

$$\varphi_{iA}(\lambda) = \sum_{p=0}^{\infty} (-1)^p \frac{T_{iA}^{(p)}}{p!} \lambda^p \qquad \left(|\lambda| < \frac{1}{c},\ i \in E\right). \ \square \tag{6.7.29}$$

Theorem 6.7.7 *Let* $G \subset E$. *Then a necessary and sufficient condition for*

$$F_i(+\infty) = 0 \qquad (i \in G), \tag{6.7.30}$$

$$\psi_i(\lambda) \equiv 1 \qquad (\lambda > 0,\ i \in G), \tag{6.7.31}$$

is that for some λ, $\lambda > 0$, *the additional state of a certain G-adjoint Markov chain of the normal system of equations*

$$x_i = \sum_{j \in E} p_{ij} e^{-\lambda V(i)} x_j + \left(1 - e^{-\lambda V(i)}\right) \qquad (i \in E) \tag{6.7.32}$$

is recurrent.

Proof. This theorem is implied by Theorems 6.7.3 and 6.6.3.□

§ 6.8 Derived Markov chains and criteria for atomic almost closed sets

Starting from a given Markov chain, we will construct on the invariant set a new Markov chain(called the derived chain) in this section. All the results for Markov chains can be carried over to the invariant set. Furthermore, as an example, we obtain the criterion for atomic almost closed sets by using the result of the corollary of Theorem 17.5 in [1, II]. Finally, we will apply this criterion to random walk.

Let Λ be an invariant set of the Markov chain X, and

$$P(\Lambda) \neq 0. \tag{6.8.1}$$

We say Λ is an *invariant set* if

$$\Lambda \in \mathfrak{F}\{x_n,\ n < \zeta + 1\}, \qquad \Lambda \bigcap (\zeta + 1 > m) = \theta_m \Lambda, \tag{6.8.2}$$

where θ_m is the transition operator defined in [6] by E. B. Dynkin.

Theorem 6.8.1 *$\hat{X} = \{x_n(\omega), n < \zeta + 1, \omega \in \Lambda\}$ is a homogeneous denumerable Markov chain on a probability space $(\Omega\Lambda,\ \mathfrak{F}\Lambda,\ P(\cdot|\Lambda))$, its state space is*

$$\hat{E} = \{i\colon\ P_i(\Lambda) \neq 0\} \tag{6.8.3}$$

and the transition probability is

$$\hat{p}_{ij} = p_{ij} \frac{P_j(\Lambda)}{P_i(\Lambda)} \qquad (i,\ j \in \hat{E}), \tag{6.8.4}$$

here

$$P_i(\cdot) = P(\cdot | x_0 = i). \tag{6.8.5}$$

If X is honest, so is $\hat{X}$.

Proof. Let m, l, k be positive integers, $j_l > j_{l-1} > \cdots > j_1$ $(m > j_l)$ be nonnegative integers, i_{m+k}, i_m, i_{j_e}, $i_{j_{e-1}}$, $\cdots$, $i_{j_1} \in \hat{E}$ and

$$P(x_m = i_m,\ x_{j_e} = i_{j_e},\ x_{j_{e-1}} = i_{j_{e-1}},\ \cdots,\ x_{j_1} = i_{j_1},\ \Lambda) \neq 0. \tag{6.8.6}$$

Hence from (6.8.2), the Markov property and

$$(x_m = i_m) = (\zeta + 1 > m,\ x_m = i_m), \tag{6.8.7}$$

we get

$$P(x_{m+k} = i_{m+k} \mid x_m = i_m,\ x_{j_e} = i_{j_e},\ x_{j_{e-1}} = i_{j_{e-1}},\ \cdots,\ x_{j_1} = i_{j_1},\ \Lambda)$$

$$= P(x_{m+k} = i_{m+k} \mid x_m = i_m,\ x_{j_e} = i_{j_e},\ x_{j_{e-1}} = i_{j_{e-1}},\ \cdots,\ x_{j_1} = i_{j_1},\ \Lambda,\ \zeta + 1 > m)$$

$$= \frac{P(x_{m+k} = i_{m+k}, \Lambda, \zeta + 1 > m \mid x_m = i_m, x_{j_e} = i_{j_e}, x_{j_{e-1}} = i_{j_{e-1}}, \cdots,\ x_{j_1} = i_{j_1})}{P(\Lambda, \zeta + 1 > m \mid x_m = i_m, x_{j_e} = i_{j_e}, x_{j_{e-1}} = i_{j_{e-1}}, \cdots, x_{j_1} = i_{j_1})}$$

$$= \frac{P(x_{m+k} = i_{m+k},\ \theta_m \Lambda \mid x_m = i_m,\ x_{j_e} = i_{j_e},\ x_{j_{e-1}} = i_{j_{e-1}},\ \cdots,\ x_{j_1} = i_{j_1})}{P(\theta_m \Lambda \mid x_m = i_m,\ x_{j_e} = i_{j_e},\ x_{j_{e-1}} = i_{j_{e-1}},\ \cdots,\ x_{j_1} = i_{j_1})}$$

$$= \frac{P(x_k = i_{m+k},\ \Lambda \mid x_0 = i_m)}{P(\Lambda \mid x_0 = i_m)}$$

$$= P(x_k = i_{m+k} \mid x_0 = i_m, \Lambda). \tag{6.8.8}$$

For $i,\ j \in \hat{E}$, we have

$$\hat{p}_{ij} = P(x_1 = j \mid x_0 = i,\ \Lambda) = \frac{P(x_1 = j,\ \Lambda \mid x_0 = i)}{P(\Lambda \mid x_0 = i)}$$

$$= \frac{P(x_1 = j \mid x_0 = i)\, P(\Lambda \mid x_0 = i,\ x_1 = j)}{P_i(\Lambda)}$$

$$= p_{ij} \frac{P(\theta_1 \Lambda \mid x_0 = i,\ x_1 = j)}{P_i(\Lambda)}$$

$$= p_{ij} \frac{P(\Lambda \mid x_0 = j)}{P_i(\Lambda)} = p_{ij} \frac{P_j(\Lambda)}{P_i(\Lambda)}. \tag{6.8.9}$$

For $i \in E \setminus \hat{E}$, we have

$$P(x_m = i \mid \Lambda) = \frac{P(\Lambda,\ x_m = i)}{P(\Lambda)} = \frac{P(\theta_m \Lambda,\ x_m = i)}{P(\Lambda)}$$

$$= \frac{P(x_m = i)\, P(\theta_m \Lambda \mid x_m = i)}{P(\Lambda)} = \frac{P(x_m = i)\, P(\Lambda \mid x_0 = i)}{P(\Lambda)}$$

$$= \frac{P(x_m = i)}{P(\Lambda)} \cdot P_i(\Lambda) = 0, \tag{6.8.10}$$

$$p(\zeta < +\infty \mid \Lambda) \leqslant \frac{P(\rho < +\infty)}{P(\Lambda)}. \tag{6.8.11}$$

In view of (6.8.8)—(6.8.11), the theorem is established. □

Definition 6.8.1 $\hat{X}$ *is called the derived Markov chain of the Markov chain* X *on* Λ.

Theorem 6.8.2 *Let* C *be an almost closed set of the honest Markov chain* X, *and*

$$\mathscr{L}(C) \doteq \Lambda. \tag{6.8.12}$$

Then a necessary and sufficient condition for C *to be an atomic almost closed set (perfectly nonatomic almost closed set) of* X *is that* $\hat{E}$ *is an atomic almost closed set (perfectly nonatomic almost closed set) of the derived Markov chain* $\hat{X}$ *of* X *on* Λ. *For the meaning of the notation* $\mathscr{L}(C)$, *see* [1,II,§17].

For ease of exposition, the proof will be presented as the outcome of a series of lemmas.

Lemma 6.8.1 *Let*

$$A = \{i\colon\ P_i(\Lambda) > a\}, \tag{6.8.13}$$

where $0 < a < 1$ *is a constant. We are going to prove that* $C\Delta A$ *is an irreturnable set, i.e.,* $P(\bar{\mathscr{L}}(C\Delta A)) = 0$. *Hence* C *is an atomic almost closed set (perfectly nonatomic almost closed set) of* X *iff* A *is an atomic almost closed set (perfectly nonatomic almost closed set) of* X.

Proof. By the corollary of Theorem 17.1 in [1, I] we know

$$\mathscr{L}(A) \doteq \Lambda. \tag{6.8.14}$$

We get this lemma immediately by (6.8.12) and (6.8.14). □

Lemma 6.8.2 *A necessary and sufficient condition for any subset* A^0 *of* A *to be an almost closed set of* X *is that* A^0 *is an almost closed set of* $\hat{X}$. *Therefore,* A *is an atomic almost closed set (perfectly nonatomic almost closed set) of* X *iff* A *is an atomic almost closed set (perfectly nonatomic almost closed set) of* $\hat{X}$.

Proof. By $A^0 \subseteq A$ and (6.8.14) we know

$$P(\bar{\mathscr{L}}(A^0) \cap \bar{\Lambda}) \leqslant P(\mathscr{L}(A) \cap \bar{\Lambda}) = 0, \tag{6.8.15}$$

$$P(\bar{\mathscr{L}}(A^0) \cap \bar{\Lambda}) \leqslant P(\mathscr{L}(A \cap \bar{\Lambda}) = 0, \tag{6.8.16}$$

where $\bar{\Lambda} = \Omega \setminus \Lambda$, then

$$P(\bar{\mathscr{L}}(A^0) \mid \Lambda) = \frac{P(\bar{\mathscr{L}}(A^0) \cap \Lambda)}{P(\Lambda)}$$

$$= \frac{P(\bar{\mathfrak{L}}(A^0)) - P(\bar{\mathfrak{L}}(A^0) \bigcap \bar{\Lambda})}{P(\Lambda)} = \frac{P(\bar{\mathfrak{L}}(A^0))}{P(\Lambda)} \tag{6.8.17}$$

and

$$P(\underline{\mathfrak{L}}(A^0) \mid \Lambda) = \frac{P(\underline{\mathfrak{L}}(A^0))}{P(\Lambda)}. \tag{6.8.18}$$

Hence necessary and sufficient conditions for both

$$P(\bar{\mathfrak{L}}(A^0)) = P(\underline{\mathfrak{L}}(A^0)) \tag{6.8.19}$$

and

$$P(\bar{\mathfrak{L}}(A^0)) > 0 \tag{6.8.20}$$

to hold are that both

$$P(\bar{\mathfrak{L}}(A^0) \mid \Lambda) = P(\underline{\mathfrak{L}}(A^0) \mid \Lambda) \tag{6.8.21}$$

and

$$P(\bar{\mathfrak{L}}(A^0)) > 0 \tag{6.8.22}$$

hold. □

Lemma 6.8.3 *Let*

$$B = \{i:\ 0 < P_i(\Lambda) \leqslant a\}. \tag{6.8.23}$$

Then

$$A \bigcap B = \varnothing, \qquad A \bigcup B = \hat{E}, \tag{6.8.24}$$

$$P(\bar{\mathfrak{L}}(B) \mid \Lambda) = 0. \tag{6.8.25}$$

Hence, A is an almost closed set (perfectly nonatomic almost closed set) of $\hat{X}$ iff $\hat{E}$ is an atomic almost closed set (perfectly nonatomic almost closed set) of $\hat{X}$.

Proof. Based on (6.8.13), (6.8.23) and the definition of $\hat{E}$, we get (6.8.24). By (6.8.14) and (6.8.24), it can be seen that

$$P(\bar{\mathfrak{L}}(B) \mid \Lambda) = P(x_n \in B \text{ infinitely many times} \mid \Lambda)$$

$$= P(x_n \notin A \text{ infinitely many times} \mid \Lambda) = 0, \tag{6.8.26}$$

and the lemma is proved. □

Summing up the above three lemmas, the Theorem 6.8.2 is thereby established.

Theorem 6.8.3 *An almost closed set A of E is an atomic almost closed set iff the system of equations*

$$x_i = \sum_{j\in\hat{E}} p_{ij} \frac{P_j(\mathfrak{L}(A))}{P_i(\mathfrak{L}(A))} x_j \qquad (i\in\hat{E}) \tag{6.8.27}$$

has no nonconstant bounded solution, where

$$\hat{E} = \{i:\ P_i(\mathfrak{L}(A) > 0)\}. \tag{6.8.28}$$

Proof. By Theorems 6.8.1 and 6.8.2 and the corollary of Theorem 17.5 in [1, I], we immediately get this theorem.□

Let the state space E of the Markov chain X be the set of all integers. Assume the transition probabilities are

$$p_{ij} = \begin{cases} a_i, & j = i-1, \\ b_i, & j = i+1, \\ 0 & |j-i| > 1, \end{cases} \tag{6.8.29}$$

where

$$a_i > 0, \qquad b_i > 0, \qquad a_i + b_i = 1 \qquad (i\in E). \tag{6.8.30}$$

Then the Markov chain is also called a random walk.

Let

$$A_a = \begin{cases} (\cdots,\ n-2,\ n-1), & \text{if } a = 1, \\ (n,\ n+1,\ \cdots), & \text{if } a = 2, \end{cases} \tag{6.8.31}$$

$$\left.\begin{aligned} x_i &= -b_0\left(1 + \frac{b_{-1}}{a_{-1}} + \cdots + \frac{b_{-1}b_{-2}\cdots b_{i+1}}{a_{-1}a_{-2}\cdots a_{i+1}}\right), \quad \text{if } i < -1, \\ x_{-1} &= -b_0 \qquad x_0 = 0, \qquad x_1 = a_0, \\ x_i &= a_0\left(1 + \frac{a_1}{b_1} + \cdots + \frac{a_1a_2\cdots a_{i-1}}{b_1b_2\cdots b_{i-1}}\right), \quad \text{if } i > 1, \\ r_1 &= \lim_{i\to-\infty} x_i, \qquad r_2 = \lim_{i\to+\infty} x_i. \end{aligned}\right\} \tag{6.8.32}$$

The Blackwell decomposition of E has been presented in [11, Theorem 5.1]. Now, by using the method presented in Theorem 6.8.3, we are going to derive this result.

Theorem 6.8.4

(A) *If at least one of r_1, r_2 is infnite, then E is an atomic almost closed set.*

(B) *If both* r_1, r_2 *are finite, then* E *can be partitioned into two atomic almost closed sets:*

$$E = A_1 + A_2. \tag{6.8.33}$$

Proof. The proof of Conclusion (A) is the same as that of [1, Theorem 5.1]. So it is omitted.

We wish to show Conclusion (B). To this end, we need a lemma.

Lemma 6.8.4 *Suppose*

$$s_i > 0 \quad (i = 1, 2, \cdots), \qquad \sum_{i=1}^{\infty} s_i < +\infty.$$

If we put

$$\bar{s}_i = \sum_{k=1}^{\infty} s_k \qquad (i = 1, 2, \cdots), \tag{6.8.34}$$

then

$$\sum_{i=1}^{\infty} \frac{s_i}{\bar{s}_i \bar{s}_{i+1}} = +\infty. \tag{6.8.35}$$

Proof. Since $1/\bar{s}_{i+1} \uparrow +\infty$ $(i \uparrow +\infty)$, it suffices to prove

$$\sum_{i=1}^{\infty} \frac{s_i}{\bar{s}_i} = +\infty. \tag{6.8.36}$$

Since we have

$$\sum_{i=n}^{\infty} \frac{s_i}{\bar{s}_i} \geqslant \sum_{i=n}^{\infty} \frac{s_i}{\bar{s}_n} = \frac{1}{\bar{s}_n} \sum_{i=n}^{\infty} s_i = 1 \tag{6.8.37}$$

for any n, (6.8.36) holds. The desired conclusion follows. □

Now, we turn to the proof of Conclusion (B) of Theorem 6.8.4. If both r_1, r_2 are finite, then by [11, §5] A_1 is an almost closed set, and

$$P(\mathcal{L}(A_1)) = \frac{r_2 - x_i}{r_2 - r_1} \qquad (i \in E). \tag{6.8.38}$$

By (6.8.38) and Theorem 6.8.3, for A_1 to be an atomic almost closed set, we need only to prove that the system of equations

$$u_i = a_i \frac{r_2 - x_{i-1}}{r_2 - x_i} u_{i-1} + b_i \frac{r_2 - x_{i+1}}{r_2 - x_i} u_{i+1} \qquad (i \in E) \tag{6.8.39}$$

has no nonconstant bounded solution. It follows by (6.8.39) that

$$u_{i+1} - u_i = \frac{a_i(r_2 - x_{i-1})}{b_i(r_2 - x_{i+1})}(u_i - u_{i-1})$$

$$= (r_2 - x_1)(r_2 - x_0)\frac{a_1 a_2 \cdots a_i}{b_1 b_2 \cdots b_i}\frac{1}{(r_2 - x_{i+1})(r_2 - x_i)}(u_1 - u_0)$$

$$(i > 0). \tag{6.8.40}$$

Hence

$$u_{i+1} - u_0 = \frac{(r_2 - x_1)(r_2 - x_0)}{a_0}\left(\frac{a_0}{(r_2 - x_1)(r_2 - x_0)}\right.$$

$$\left. + a_0 \sum_{k=1}^{i} \frac{a_1 a_2 \cdots a_k}{b_1 b_2 \cdots b_k}\frac{1}{(r_2 - x_{k+1})(r_2 - x_k)}\right)(u_1 - u_0) \qquad (i > 0). \tag{6.8.41}$$

Since $r_2 < +\infty$, by Lemma 6.8.4,

$$\frac{a_0}{(r_2 - x_1)(r_2 - x_0)} + a_0 \sum_{k=1}^{\infty} \frac{a_1 a_2 \cdots a_k}{b_1 b_2 \cdots b_k}\frac{1}{(r_2 - x_{k+1})(r_2 - x_k)} = +\infty. \tag{6.8.42}$$

Therefore, if u_i $(i \in E)$ is a bounded solution of (6.8.39), then we have $u_1 - u_0 = 0$ from (6.8.41). Hence $u_i \equiv$ constant. Consequently, A_1 is an atomic almost closed set. In the same way we can prove that A_2 is an atomic almost closed set. □

CHAPTER VII
Martin Exit Boundary Theory

§ 7.1 Introduction

The theory of the Martin exit boundary for a Markov chain was first established by Doob [12] and then generalized by Hunt. Many works appeared thereafter, but in all those works there are always some restrictions imposed on the Markov chain. For example, in [13] all the states of the Markov chain are assumed to be nonrecurrent, and in [14] it is assumed that there exists at least one state which can reach all the other states. So the Martin exit boundary theory for general Markov chains has not been established. Moreover, some subjects involved in the exit boundary theory have not been studied in detail. For example, only definitions of the atomic exit point and the nonatomic exit point are given, but no effective criteria. The aim of the present chapter is:

(i) to establish the Martin exit boundary theory for general Markov chains;

(ii) to present the criteria for the excessive functions, potential functions, minimal excessive functions, minimal potential functions, and minimal harmonic functions, atomic exit points and nonatomic exit points, and for the existence of atomic exit space and nonatomic exit space;

(iii) to present the criteria for Blackwell decompositions of atomic almost closed sets, completely nonatomic almost closed sets and state spaces.

Hunt's paper [13] is no doubt one of the few classical works of boundary theory of Markov processes, but it is not readily accessible. So Dynkin [15] systematized Hunt's works making the arguments comparatively rigorous and clear. Therefore, in this chapter, all the passages which are similar to that in [15] are intentionally written very briefly in order to avoid repetition.

§ 7.2 Decomposition for Markov chains

Suppose $X(\omega)=\{x_n(\omega),\ 0\leqslant n<\zeta(\omega)+1,\ \omega\in\Omega\}$ is a Markov chain defined on a probability space $(\Omega, \mathscr{F}, P)$, with state space $E=\{0, 1, 2, \cdots\}$ and transition probability matrix $P=(p_{ij},\ i, j\in E)$. It is well known that E may be decomposed as

$$E = E_0 \bigcup \bigcup_{a\in\mathfrak{A}} E_a, \tag{7.2.1}$$

where $\mathfrak{A}$ may be empty, finite or denumerable, and $0 \notin \mathfrak{A}$; E_0 is the set of all nonrecurrent states, so it may be empty; $E_a(a \in \mathfrak{A})$ is an irreducible recurrent set. Thus

$$E_a \bigcap E_{a'} = \varnothing \qquad (a \neq a';\ a, a' \in \{0\} \bigcup \mathfrak{A}). \tag{7.2.2}$$

Let

$$E^A = \bigcup_{a \in A} E_a. \tag{7.2.3}$$

Namely, E^A is the set of all recurrent states.

Definition 7.2.1 *If $E^{\prime} = \varnothing$, then $X(\omega)$ is called a transient chain. If for all $\omega \in \Omega$, there exists $N \geqslant 0$ such that*

$$x_n(\omega) \in E^A, \qquad N \leqslant n < \zeta(\omega) + 1, \tag{7.2.4}$$

then $X(\omega)$ is called a pseudo-recurreut chain. If E^A is in addition an irreducible set, then $X(\omega)$ is called a pseudo-irreducible recurrent chain.

Lemma 7.2.1 *If Λ is an invariant set, then*

$$P_i(\Lambda) \geqslant \sum_{j \in E} p_{ij} p_j(\Lambda) \qquad (i \in E). \tag{7.2.5}$$

Proof.

$$P_i(\Lambda) \geqslant \sum_{j \in E} P_i(x_1 = j, \zeta(\omega) + 1 > 1, \Lambda)$$

$$= \sum_{j \in E} P_i(x_1 = j, \theta_1 \Lambda) = \sum_{j \in E} P_i(x_1 = j) P_i(\theta_1 \Lambda | x_1 = j)$$

$$= \sum_{j \in E} P_{ij} P_j(\Lambda). \ \square \tag{7.2.6}$$

Let

$$\Lambda_0 = \{\omega\colon x_n \in E_0, 0 \leqslant n < \zeta(\omega) + 1\}, \tag{7.2.7}$$

$$\Lambda_a = \{\omega\colon \text{there exists } N \geqslant 0 \text{ such that } x_n \in E_a \text{ when}$$

$$N \leqslant n < \zeta(\omega) + 1\} \qquad (a \in A). \tag{7.2.8}$$

Lemma 7.2.2 *For any fixed $a \in A$, Λ_a is an invariant set, and $P_i(\Lambda_a)\,(i \in E)$ is determined uniquely as follows:*

$$P_i(\Lambda_a) = 1 \qquad (i \in E_a), \tag{7.2.9}$$

$$P_i(\Lambda_a) = 0 \qquad (i \in E^{A} \setminus E_a), \tag{7.2.10}$$

and $P_i(\Lambda_a)$ $(i \in E_0)$ *is the minimal nonnegative solution of the pseudoregular equation system*

$$x_i = \sum_{j \in E_0} p_{ij} x_j + \sum_{j \in E_a} p_{ij} \qquad (i \in E_0). \tag{7.2.11}$$

Proof. Obviously, Λ_a is an invariant set. Equations (7.2.9) and (7.2.10) are trivial.

Let

$$f_{iE_a}^{(n)} = P_i(x_\nu \notin E_a,\ 0 < \nu < n,\ x_n \in E_a) \qquad (i \in E_0). \tag{7.2.12}$$

It is clear that

$$\left.\begin{aligned} f_{iE_a}^{(1)} &= \sum_{j \in E_a} p_{ij} \qquad (i \in E_0), \\ f_{iE_a}^{(n+1)} &= \sum_{k \in E_0} p_{ik} f_{kE_a}^{(n)} \qquad (n \geqslant 1), \end{aligned}\right\} \tag{7.2.13}$$

$$P_i(\Lambda_a) = \sum_{n=1}^{\infty} f_{iE_a}^{(n)} \qquad (i \in E_0). \tag{7.2.14}$$

Equations (7.2.13), (7.2.14) and Corollary 3.2.3 imply immediately that $P_i(\Lambda_a)$ $(i \in E_0)$ is the minimal nonnegative solution of (7.2.11), as desired.□

Lemma 7.2.3 Λ_0 *is an invariant set, and*

$$P_i(\Lambda_0) = \begin{cases} 1 - \sum_{a \in A} P_i(\Lambda_a) & (i \in E_0), \\ 0 & (i \in E^{A}). \ \square \end{cases} \tag{7.2.15}$$

Theorem 7.2.1 (Decomposition Theorem of Markov Chains)

(i)

$$\Lambda_a \cap \Lambda_{a'} = \varnothing \qquad (a \neq a';\ a, a' \in \{0\} \cup A), \tag{7.2.16}$$

$$P\left(\Omega \setminus \bigcup_{a \in \{0\} \cup A} \Lambda_a\right) = 0; \tag{7.2.17}$$

(ii) *If* $P(\Lambda_0) > 0$, *then*

$$X^0(\omega) = \{x_n(\omega),\ 0 \leqslant n < \zeta(\omega) + 1,\ \omega \in \Lambda_0\}$$

is a transient chain on the probability space $(\Omega\Lambda_0, \mathcal{F}\Lambda_0, P(\cdot|\Lambda_0))$, *its state space is*

$$E_0^+ = \{i: P_i(\Lambda_0) > 0\} \subset E_0 \tag{7.2.18}$$

and the transition probabilities of E_0^+ *are*

$$\overset{0}{p}_{ij} = p_{ij} \frac{P_j(\Lambda_0)}{p_i(\Lambda_0)} \qquad (i, j \in E_0^+); \tag{7.2.19}$$

(iii) *If* $\mathfrak{A} \neq \varnothing$, $a \in \mathfrak{A}$, *then*

$$X^a = \{x_n(\omega), 0 \leqslant n < \zeta(\omega) + 1, \omega \in \Lambda_a\}$$

is a pseudo-irreducible recurrent chain on the probability space $(\Omega\Lambda_a, \mathcal{F}\Lambda_a, P(\cdot|\Lambda_a))$, *its state space is*

$$E_a^+ = \{i: P_i(\Lambda_a) > 0\} \supset E_a, \tag{7.2.20}$$

and the transition probabilities of E_a^+ *are*

$$\overset{a}{p}_{ij} = p_{ij} \frac{P_j(\Lambda_a)}{P_i(\Lambda_a)} \qquad (i, j \in E_a^+). \tag{7.2.21}$$

Proof. Since Λ_a $(a \in \{0\} \cup \mathfrak{A})$ is an invariant set, and by Theorem 6.8.1, it is easy to complete the proof of this theorem. □

§ 7.3 Limit behaviour of excessive functions

Definition 7.3.1 *A nonnegative function* $(+\infty$ *is accessible*$)$ f_i $(i \in E)$ *is called an excessive function with respect to* P *(or briefly called excessive function) if*

$$f_i \geqslant \sum_{j \in E} p_{ij} f_j \qquad (i \in E), \tag{7.3.1}$$

and it is called a harmonic function if the equalities hold in (7.3.1).

Let

$$\Omega_\infty = \{\omega: \zeta(\omega) = +\infty\}, \tag{7.3.2}$$

namely, Ω_∞ is the set of the honest samples.

Theorem 7.3.1 *Let* f_i $(i \in E)$ *be an excessive function. If* $f_s < +\infty$, *then the ordinary limits*

$$\lim_{n\to\infty} f_{x_n(\omega)} \tag{7.3.3}$$

exist P_s-almost on Ω_∞. If $f_s = +\infty$, then on Ω_∞, "P_s-almost" has only two possible cases: that for all n, $f_{x_n(\omega)} = +\infty$, or that $f_{x_n(\omega)}$ tends to ordinary limit.

Proof. Though what is discussed in [15] are the transient chains, the arguments in [15, §5] are in fact also valid for the general case. So we can get the theorem by repeating the arguments there word for word. □

§ 7.4 Green functions and Martin kernels

Let

$$f_{ij}^{(n)} = P_i(x_v \neq j,\ 0 \leqslant v < n,\ x_n = j), \tag{7.4.1}$$

$$f_{ij}^* = P_i(\text{there exists } n \geqslant 0, \text{ such that } x_n = j), \tag{7.4.2}$$

$$F^* = (f_{ij}^*,\ i, j \in E). \tag{7.4.3}$$

Obviously, we have

$$f_{ij}^* = \sum_{n=0}^{\infty} f_{ij}^{(n)}. \tag{7.4.4}$$

Theorem 7.4.1 *For a fixed j, f_{ij}^* $(i \in E)$ is determined uniquely as follows:*
(i) $f_{jj}^* = 1.$ (7.4.5)
(ii) *f_{ij}^* $(i \neq j)$ is the minimal nonnegative solution of the pseudoregular system of equations*

$$x_i = \sum_{k \neq j} p_{ik} x_k + p_{ij} \qquad (i \neq j). \tag{7.4.6}$$

Proof. From (7.4.1) we know $f_{jj}^{(0)} = 1$ and $f_{jj}^{(n)} = 0$ $(n \geqslant 1)$. Hence, (7.4.4) implies (7.4.5) immediately. On the other hand, by virtue of (7.4.1),

$$f_{ij}^{(0)} = 0 \qquad (i \neq j). \tag{7.4.7}$$

Thus from (7.4.4) we deduce

$$f_{ij}^* = \sum_{n=1}^{\infty} f_{ij}^{(n)} \qquad (i \neq j). \tag{7.4.8}$$

By using the Markov property and (7.4.1) we get

$$\left.\begin{aligned} & f_{ij}^{(1)} = p_{ij} \qquad (i \neq j), \\ & f_{ij}^{(n+1)} = \sum_{k \neq j} p_{ik} f_{kj}^{(n)} \qquad (n \geqslant 1,\ i \neq j). \end{aligned}\right\} \tag{7.4.9}$$

Hence, (7.4.8) and (7.4.9) and Theorem 3.2.3 can be applied to obtain (ii). □

Theorem 7.4.2 *For a fixed j, f_{ij}^* $(i\in E)$ is an excessive function.*

Proof. From

$$0\leqslant f_{ij}^*\leqslant 1 \qquad (i\in E) \tag{7.4.10}$$

and Theorem 7.4.1 we know

$$\left.\begin{aligned}&\sum_{k\in E} p_{jk}f_{kj}^*\leqslant\sum_{k\in E} p_{jk}\leqslant 1=f_{jj}^*,\\&\sum_{k\in E} p_{ik}f_{kj}^*=\sum_{k\neq j} p_{ik}f_{kj}^*+p_{ij}f_{jj}^*,\\&\qquad=\sum_{k\neq j} p_{ik}f_{kj}^*+p_{ij}=f_{ij}^* \qquad (i\neq j).\end{aligned}\right\} \tag{7.4.11}$$

So f_{ij}^* $(i\in E)$ is an excessive function. □

Theorem 7.4.3 *Suppose $f_i\geqslant 0$ $(i\in E)$. Let*

$$f=\begin{pmatrix}f_0\\f_1\\\vdots\end{pmatrix}. \tag{7.4.12}$$

Then F^*f is an excessive function.

Proof. The assertion follows readily from Theorem 7.4.2 and

$$F^*f=\sum_{k\in E} f_k\begin{pmatrix}f_{0k}^*\\f_{1k}^*\\\vdots\end{pmatrix}. \tag{7.4.13}$$

Definition 7.4.1 *$f_{ij}^*(i, j\in E)$ is called a Green function with respect to P, or briefly, a Green function.*

Theorem 7.4.4 *Suppose $p_{ij}^{(n)}$ is an element of P^n. Let*

$$g_{ij}=\sum_{n=0}^{\infty} p_{ij}^{(n)} \qquad (i, j\in E). \tag{7.4.14}$$

Then

$$g_{ij}=f_{ij}^*g_{jj} \qquad (i, j\in E). \tag{7.4.15}$$

If j is a nonrecurrent state, then

$$g_{ij} = f_{ij}^* g_{jj} < +\infty \qquad (i, j \in E), \tag{7.4.16}$$

here we stipulate $0 \cdot \infty = 0$.

Proof. The assertion follows readily from [1, Theorem 5.2]. □

Definition 7.4.2 *The measure* γ *on* E *is called a standard measure with respect to* P *(or briefly, a standard measure), if it satisfies the following conditions*:

(i) γ *is finite, i.e.,*

$$\sum_{i \in E} \gamma_i < +\infty; \tag{7.4.17}$$

(ii)

$$n_j = \sum_{i \in E} \gamma_i f_{ij}^* > 0 \qquad (j \in E). \tag{7.4.18}$$

Theorem 7.4.5 *There always exists a standard measure. If* γ *is a standard measure, then*

$$0 < n_j < +\infty \qquad (j \in E). \tag{7.4.19}$$

Proof. For example, $\gamma_i = 2^{-(i+1)}$ $(i \in E)$ is a standard measure, so there always exists a standard measure. If γ is a standard measure, then

$$0 < n_j = \sum_{i \in E} \gamma_i f_{ij}^* \leqslant \sum_{i \in E} \gamma_i < +\infty \qquad (j \in E). \tag{7.4.20}$$

Hence (7.4.19) is valid. □

Henceforth γ always denotes a standard measure. We say that a nonnegative function f_i $(i \in E)$ is γ-integrable if

$$\sum_{i \in E} f_i \gamma_i < +\infty.$$

Definition 7.4.3

$$K(i, j) = \frac{f_{ij}^*}{n_j} \qquad (i, j \in E) \tag{7.4.21}$$

is called a Martin kernel with respect to P, or briefly, a Martin kernel.

§ 7.5 *h*-chains

Definition 7.5.1 *The function* $a_{ij}(i, j \in E)$ *defined on* $E \times E$ *is called a transition function on E, if*

$$a_{ij} \geqslant 0 \qquad (i, j \in E), \tag{7.5.1}$$

$$\sum_{j \in E} a_{ij} \leqslant 1 \qquad (i \in E). \tag{7.5.2}$$

If $a_{ij}(i, j \in E)$ is a transition function on E, then $A = (a_{ij}, i, j \in E)$ is called a transition matrix on E.

By the definition, transition probabilities $p_{ij}(i, j \in E)$ of a Markov chain are transition functions on E, and $P = (p_{ij}, i, j \in E)$ is a transition matrix on E.

Assume $h_i(i \in E)$ is a γ-integrable excessive function. It is easy to prove that $0 \leqslant h_i < +\infty$ $(i \in E)$. Let

$$h = \begin{pmatrix} h_0 \\ h_1 \\ \vdots \end{pmatrix}, \tag{7.5.3}$$

$$E^h = \{i: h_i > 0\}, \tag{7.5.4}$$

$$p_{ij}^h = p_{ij} \frac{h_j}{h_i}. \tag{7.5.5}$$

Since

$$p_{ij}^h = p_{ij} \frac{h_j}{h_i} \geqslant 0 \qquad (i, j \in E^h) \tag{7.5.6}$$

and

$$\sum_{j \in E^h} p_{ij}^h = \frac{1}{h_i} \sum_{j \in E^h} p_{ij} h_j \leqslant \frac{h_i}{h_i} \leqslant 1, \tag{7.5.7}$$

$P^h = (p_{ij}^h, i, j \in E)$ is a transition matrix on E^h.

Lemma 7.5.1 *If h_i $(i \in E)$ is an excessive function, then*

$$p_{ik} = 0 \qquad (i \notin E^h,\ k \in E^h). \tag{7.5.8}$$

Proof. If $i \notin E^h$ and $k \in E^h$, then

$$0 = h_i \geqslant \sum_{j \in E} p_{ij} h_j \geqslant p_{ik} h_k. \tag{7.5.9}$$

Hence, $h_k > 0$ implies $p_{ik} = 0$. □

Lemma 7.5.2 *If $h_i(i \in E)$ is an excessive function, then for every fixed $j \in E^h$, f_{ij}^* $(i \in E)$ is determined uniquely as follows:*

$$f_{jj}^* = 1, \tag{7.5.10}$$

$$f_{ij}^* = 0 \qquad (i \notin E^h); \tag{7.5.11}$$

and $f_{ij}^*(i \in E^h \setminus \{j\})$ *is the minimal nonnegative solution of the pseudoregular system of equations*

$$x_i = \sum_{k \neq j} p_{ik} x_k + p_{ij} \qquad (i \in E^h \setminus \{j\}). \tag{7.5.12}$$

Proof. We deduce our lemma from Theorem 7.4.1, Lemma 7.5.1 and Corollary 3.4.1.□

Theorem 7.5.1 *Suppose* $X^h(\omega) = \{x_n^h(\omega)\}$ *is a Markov chain with transition probability matrix* P^h. *Let*

$$f_{ij}^{*h} = P(\text{there exists an } n \geqslant 0 \text{ such that } x_n^h = j | x_0^h = i) \qquad (i, j \in E^h). \tag{7.5.13}$$

Then

$$f_{ij}^{*h} = f_{ij}^* \frac{h_j}{h_i} \qquad (i, j \in E^h). \tag{7.5.14}$$

Proof. We suppose that $i, j \in E^h$. Since

$$f_{jj}^{*h} = 1 = f_{jj}^* = f_{jj}^* \frac{h_j}{h_j},$$

it suffices to prove the theorem for the case $i \neq j$. From Theorem 7.4.1, we see that $f_{ij}^{*h}(i \in E^h \setminus \{j\})$ is the minimal nonnegative solution of the pseudo-normal system of equations

$$x_i = \sum_{k \in E^h \setminus \{j\}} p_{ik} \frac{h_k}{h_i} x_k + p_{ij} \frac{h_j}{h_i} \qquad (i \in E^h \setminus \{j\}). \tag{7.5.15}$$

Thus, from Theorem 3.3.3 we know that $h_i f_{ij}^{*h}$ $(i \in E^h \setminus \{j\})$ is the minimal nonnegative solution of the pseudo-regular system of equations

$$x_i = \sum_{k \in E^h \setminus \{j\}} p_{ik} x_k + p_{ij} h_j \qquad (i \in E^h \setminus \{j\}). \tag{7.5.16}$$

Hence, from Lemma 7.5.2 and Corollary 3.3.3, we obtain

$$h_i f_{ij}^{*h} = f_{ij}^* h_j \qquad (i \in E^h \setminus \{j\}), \tag{7.5.17}$$

i.e.,

$$f_{ij}^{*h} = f_{ij}^* \frac{h_j}{h_i} \qquad (i \in E^h \setminus \{j\}). \ \square \tag{7.5.18}$$

Let

$$\gamma_i^h = h_i \gamma_i \qquad (i \in E^h), \tag{7.5.19}$$

$$n_j^h = \sum_{i \in E^h} \gamma_i^h f_{ij}^{*h} \qquad (j \in E^h). \tag{7.5.20}$$

Theorem 7.5.2

$$n_j^h = \left(\sum_{i \in E^h} \gamma_i f_{ij}^* \right) h_j = n_j h_j \tag{7.5.21}$$

and γ_i^h $(i \in E^h)$ *is a standard measure with respect to* P^h.

Proof. From (7.5.14), (7.5.19) and (7.5.20) we obtain the first equality of (7.5.21). From Lemma 7.5.2, (7.4.18) and the first equality of (7.5.21), we obtain the second equality of (7.5.21). From (7.5.21) and the fact that $\gamma_i (i \in E)$ is a standard measure, we deduce that γ_i^h $(i \in E^h)$ is a standard measure with respect to P^h. □

Let

$$K^h(i, j) = \frac{f_{ij}^{*h}}{n_j^h} \qquad (i, j \in E^h). \tag{7.5.22}$$

Definition 7.5.2 P^h *and the Markov chain with* P^h *as its transition probability matrix are called an h-chain; the excessive functions, the standard measures, Green functions* f_{ij}^{*h} $(i, j \in E^h)$ *and Martin kernels* $K^h(i, j)$ $(i, j \in E^h)$ *with respect to* P^h *are called h-excessive functions, h-standard measures, h-excessive functions* and *h-Martin kernels, respectively.*

Theorem 7.5.3

$$K^h(i, j) = \frac{K(i, j)}{h_i} \qquad (i, j \in E^h) \tag{7.5.23}$$

and

$$\sum_{i \in E^h} \gamma_i^h K^h(i, j) = 1 \qquad (j \in E^h). \tag{7.5.24}$$

Proof. Equations (7.5.14), (7.5.21), (7.5.22) and (7.4.20) imply that

$$K^h(i, j) = \frac{f_{ij}^{*h}}{n_j^h} = \frac{f_{ij}^*(h_j/h_i)}{n_j h_j} = \frac{1}{h_i} \frac{f_{ij}^*}{n_j} = \frac{K(i, j)}{h_i} \qquad (i, j \in E^h). \tag{7.5.25}$$

By (7.4.18), (7.4.21), (7.5.11), (7.5.20) and (7.5.23) we deduce

$$\sum_{i \in E^h} \gamma_i^h K^h(i, j) = \sum_{i \in E^h} h_i \gamma_i \frac{K(i, j)}{h_i} = \sum_{i \in E^h} \gamma_i K(i, j)$$

$$= \sum_{i\in E^h} \gamma_i \frac{f_{ij}^*}{n_j} = \frac{1}{n_j} \sum_{i\in E^h} \gamma_i f_{ij}^* = \frac{1}{n_j} \sum_{i\in E} \gamma_i f_{ij}^*$$

$$= \frac{n_j}{n_j} = 1 \qquad (j\in E^h). \ \square \tag{7.5.26}$$

Lemma 7.5.3

$$f_{ij}^* \geqslant f_{ik}^* f_{kj}^* \qquad (i, j, k\in E). \tag{7.5.27}$$

Proof. For every $s\in E$, let

$$\sigma^{(s)} = \begin{cases} \inf(n: x_n = s, n \geqslant 0), \\ +\infty. \end{cases}$$

Then

$$f_{ij}^* = \mathrm{P}(\sigma^{(j)} < +\infty | x_0 = i)$$

$$\geqslant P(\sigma^{(k)} < +\infty, \sigma^{(k)} \leqslant \sigma^{(j)} < +\infty | x_0 = i)$$

$$= P(\sigma^{(k)} < +\infty | x_0 = i) P(\sigma^{(k)} \leqslant \sigma^{(j)} < +\infty | x_0 = i, \sigma^{(k)} < +\infty)$$

$$= f_{ik}^* P(\sigma^{(k)} \leqslant \sigma^{(j)} < +\infty | x_0 = i, x(\sigma^{(k)}) = k)$$

$$= f_{ik}^* P(\sigma^{(j)} < +\infty | x(0) = k)$$

$$= f_{ik}^* f_{kj}^*. \ \square$$

Lemma 7.5.4

(i) *For a fixed* $i\in E$ *and* $a\in A$,

$$f_{ij}^* = \text{const.} \qquad (j\in E_a). \tag{7.5.28}$$

(ii)

$$f_{ij}^* = 1 \qquad (i, j\in E_a), \tag{7.5.29}$$

and

$$f_{ij}^* = 0 \qquad (i\in E_a, j\notin E_a). \tag{7.5.30}$$

Proof. Since all $E_a (a\in A)$ are disjoint and irreducible recurrent classes, we obtain (7.5.29) and (7.5.30) readily. Suppose $j, k\in E_a$. Owing to (7.5.29) and Lemma 7.5.3, we have

$$f_{ij}^* \geqslant f_{ik}^* f_{kj}^* = f_{ik}^*. \tag{7.5.31}$$

Similarly,

$$f_{ik}^* \geqslant f_{ij}^*. \tag{7.5.32}$$

Hence

$$f_{ij}^{*} = f_{ik}^{*}. \tag{7.5.33}$$

So (7.5.28) holds, as desired. □

Theorem 7.5.4 *For each fixed* $i \in E^h$,

$$K^h(i, j) \leqslant \frac{1}{h_i n_i} \qquad (j \in E^h). \tag{7.5.34}$$

Proof. Using Lemma 7.5.3 we get

$$K^h(i, j) = \frac{K(i, j)}{h_i} = \frac{1}{h_i}\frac{f_{ij}^{*}}{n_j} = \frac{1}{h_i}\frac{f_{ij}^{*}}{\sum\limits_{k \in E} \gamma_k f_{kj}^{*}}$$

$$\leqslant \frac{1}{h_i}\frac{f_{ij}^{*}}{\sum\limits_{k \in E} \gamma_k f_{ki}^{*} f_{ij}^{*}} = \frac{1}{h_i}\frac{1}{\sum\limits_{k \in E} \gamma_k f_{ki}^{*}} = \frac{1}{h_i n_i} \qquad (i, j \in E^h). \square \tag{7.5.35}$$

Theorem 7.5.5 *Let*

$$E_a^h = E_a \cap E^h. \tag{7.5.36}$$

Then

(i) *For each fixed* $i \in E^h$, $a \in \mathscr{A}$,

$$K^h(i, j) = \text{const.} \qquad (j \in E_a^h) \tag{7.5.37}$$

provided $E_a^h \neq \varnothing$;

(ii) *for each* $a \in \mathscr{A}$

$$K^h(i, j) \neq \text{const.} \qquad (i, j \in E_a^h), \tag{7.5.38}$$

and

$$K^h(i, j) = 0 \qquad (i \in E_a^h, j \notin E_a^h), \tag{7.5.39}$$

provided $E_a^h \neq \varnothing$.

Proof. From Theorem 7.5.3 and Lemma 7.5.4, we get the theorem readily. □

Lemma 7.5.5 *If* $h_i (i \in E)$ *is an excessive function, then*

$$K(i, j) = 0 \qquad (i \notin E^h, j \in E^h). \tag{7.5.40}$$

Proof. Suppose $i \notin E^h$, $j \in E^h$. Since

$$h_i \geqslant \sum_{k \in E} p_{ik} h_k, \tag{7.5.41}$$

it is easy to obtain

$$h_i \geqslant \sum_{k \in E} p_{ik}^{(n)} h_k \qquad (n = 1, 2, \cdots). \tag{7.5.42}$$

Thus from $i \notin E^h$, $j \in E^h$ we deduce

$$p_{ij}^{(n)} = 0, \tag{7.5.43}$$

equivalently,

$$i \nrightarrow j. \tag{7.5.44}$$

Therefore,

$$f_{ij}^* = 0. \tag{7.5.45}$$

Then by (7.4.21) we obtain (7.5.40) immediately. □

§ 7.6 Limit theorem for Martin kernels

Theorem 7.6.1 *For any finite measure μ, the ordinary limit*

$$\lim_{n \to \infty} \sum_{i \in E} K(i, x_n) \mu_i \tag{7.6.1}$$

exists on Ω_∞ (P_r-a.s.). Particularly, for any $i \in E$, the ordinary limit

$$\lim_{n \to \infty} K(i, x_n) \mu_i \tag{7.6.2}$$

exists P_r-almost on Ω_∞ (P_r-a.s.).

Proof. In view of Lemmas 7.2.1, 7.2.2 and 7.2.3, and Theorem 7.2.1 we see that for any $a \in \{0\} \bigcup \mathfrak{A}$, the Markov chain $X^a = \{x_n(\omega), 0 \leqslant n < \zeta(\omega) + 1, \omega \in \Lambda_a\}$ on the probability space $(\Omega\Lambda_a, \mathfrak{F}\Lambda_a, P(\cdot|\Lambda_a))$ is a $P(\Lambda_a)$-chain.

Let us prove the theorem in the following steps:

(1) Suppose $i \notin E_a^+$. Then the definition of Λ_a and Lemma 7.5.5 imply that

$$K(i, x_n(\omega)) = 0 \qquad (n \geqslant 0) \tag{7.6.3}$$

on $(\Lambda_a \bigcap \Omega_\infty) \backslash \Delta_1$, where

$$\Delta_1 \subset \Lambda_a \bigcap \Omega_\infty, \tag{7.6.4}$$

$$P_{\gamma^{P(\Lambda a)}}(\Delta_1) = 0. \tag{7.6.5}$$

Hence we have

$$\lim_{n\to\infty} \sum_{i\notin E_a^+} K(i, x_n(\omega))\mu_i = 0 \tag{7.6.6}$$

on $(\Lambda_a \bigcap \Omega_\infty)\backslash\Delta_1$.

(2) Suppose $i\in E_a^+$, $a\in\mathfrak{A}$. Then by the definition of Λ_a we see

$$f_{ij}^{*P(\Lambda_a)} = 1 \qquad (i\in E_a^+, j\in E_a), \tag{7.6.7}$$

so

$$K^{P(\Lambda_a)}(i, j) = \frac{1}{\sum_{k\in E_a^+} \gamma_k P_k(\Lambda_a)} \qquad (i\in E_a^+, j\in E_a). \tag{7.6.8}$$

If $\omega\in\Lambda_a$, then by the definition of Λ_a, we know that there exists $N\geqslant 0$, such that

$$x_n(\omega)\in E_a \tag{7.6.9}$$

when $n\geqslant N$. Hence from (7.6.8) we deduce that

$$K(i, x_n(\omega)) = \frac{1}{\sum_{k\in E_a^+} \gamma_k P_k(\Lambda_a)} \qquad (i\in E_a^+) \tag{7.6.10}$$

when $n\geqslant N$. So we have

$$\lim_{n\to\infty} \sum_{i\in E_a^+} K^{P(\Lambda_a)}(i, x_n(\omega))P_i(\Lambda_a)\mu_i = \frac{\sum_{i\in E_a^+} P_i(\Lambda_a)\mu_i}{\sum_{k\in E_a^+} \gamma_k P_k(\Lambda_a)} < +\infty \tag{7.6.11}$$

on $(\Delta_a \bigcap \Omega_\infty)$.

(3) Suppose $i\in E_a^+$, $a=0$. Using Theorem 7.2.1, we see that a $P(\Lambda_a)$-chain is a transient chain. When the chain is a transient chain, our definition of Martin kernel coincides to that of Dynkin in [15]. So by [15, Theorem 3], we see that there exists the ordinary limit

$$\lim_{n\to\infty} \sum_{i\in E_0^+} K^{P(\Lambda_0)}(i, x_n(\omega))P_i(\Lambda_0)\mu_i \tag{7.6.12}$$

on $(\Lambda_a \bigcap \Omega_\infty)\backslash\Delta_2$, where

$$\Delta_2 \subset \Lambda_0 \bigcap \Omega_\infty, \tag{7.6.13}$$

$$P_{\gamma^{P(\Lambda_0)}}(\Delta_2) = 0. \tag{7.6.14}$$

(4) From Theorem 7.5.3, we deduce that

$$\sum_{i\in E_a^+} K(i, x_n)\mu_i = \sum_{i\in E_a^+} \frac{K(i, x_n)}{P_i(\Lambda_a)} P_i(\Lambda_a)\mu_i$$

$$= \sum_{i\in E_a^+} K^{P(\Lambda_a)}(i, x_n) P_i(\Lambda_a)\mu_i \qquad (a\in\{0\}\bigcup\mathfrak{A}) \tag{7.6.15}$$

$\left(P_{\gamma^{P(\Lambda_a)}}\text{-a.s.}\right)$ on Λ_a.

(5) If

$$\Delta \subset \Lambda_a \tag{7.6.16}$$

and

$$P_{\gamma^{P(\Lambda_a)}}(\Delta) = 0, \tag{7.6.17}$$

then

$$P_\gamma(\Delta) = \sum_{i\in E} \gamma_i P_i(\Delta) = \sum_{i\in E} \gamma_i P_i(\Lambda_a\Delta)$$

$$= \sum_{i\in E_a^+} \gamma_i P(\Lambda_a\Delta) = \sum_{i\in E_a^+} \gamma_i P_i(\Lambda_a) P_i(\Delta|\Lambda_a)$$

$$= \sum_{i\in E_a^+} \gamma_i^{P(\Lambda_a)} P_i(\Delta|\Lambda_a) = P_{\gamma^{P(\Lambda_a)}}(\Delta) = 0. \tag{7.6.18}$$

Based on steps (1)—(5) and Theorem 7.2.1, we establish the theorem. □

§ 7.7 Martin boundaries

Let

$$N(i) = \begin{cases} i, & \text{if } i\in E_0, \\ \inf\limits_{k\in E_a} k, & \text{if } i\in E_a, \quad a\in\mathfrak{A}, \end{cases} \tag{7.7.1}$$

$$d^h(i, j) = |2^{-N(i)} - 2^{-N(j)}| + \sum_{s\in E^h} |K^h(s, i) - K^h(s, j)| n_s^h 2^{-N(s)} \quad (i, j\in E^h). \tag{7.7.2}$$

Lemma 7.7.1[1]

$$d^h(i, j) = d(i, j) \qquad (i, j \in E^h). \tag{7.7.3}$$

Proof. The assertion follows from (7.5.23), (7.5.21) and (7.5.40).□

Lemma 7.7.2 *If there exist* $a, a' \in \mathcal{A}$ *such that* $i_1, i_2 \in E_a \cap E^h$, $j_1, j_2 \in E_{a'} \cap E^h$, *then*

$$d^h(i_1, j_1) = d^h(i_2, j_2). \tag{7.7.4}$$

Proof. We get our lemma immediately from Theorem 7.5.5.□

Lemma 7.7.3

(i)
$$d^h(i, j) \geqslant 0 \qquad (i, j \in E^h). \tag{7.7.5}$$

(ii) $d^h(i, j) = 0 \Longleftrightarrow$ *There exists* $a \in \mathcal{A}$, *such that* $i, j \in E_a \cap E^h (i = j$ *or* $i \neq j)$.

(7.7.6)

(iii)
$$d^h(i, j) = d^h(j, i) \qquad (i, j \in E^h). \tag{7.7.7}$$

(iv)
$$d^h(i, j) + d^h(j, k) \geqslant d^h(i, k) \qquad (i, j, k \in E^h). \tag{7.7.8}$$

Proof. This lemma follows from (7.7.1), (7.7.2) and Theorem 7.5.5.□

Theorem 7.7.1 *The state space* E^h *of an h-chain forms a metric space by introducing the metric* d^h *defined by* (7.7.2) *and identifying the states in a recurrent class* $E_a (a \in \mathcal{A})$; *and*

$$d^h(i, j) \leqslant 3 \qquad (i, j \in E^h). \tag{7.7.9}$$

Proof. The assertion follows from Lemmas 7.7.2 and 7.7.3 and Theorem 7.5.4.□

We deduce from Lemma 7.1.1 that the metric $d^h(i, j)$ is just the metric $d(i, j)$. So we will henceforth talk about the metric d only.

Without difficulty, we can prove the next theorem.

Theorem 7.7.2 *An infinite sequence* $\{j_n\}$ *in* E^h *is a Cauchy sequence in the metric space* E^h *if and only if both the following two conditions are satisfied:*

(i) $\lim_{n\to\infty} N(j_n)$ *exists* (*fnite or* $+\infty$);

(ii) *for any* $i \in E^h$, $K^h(i, j_n)$ *is a real Cauchy sequence.* □

Based on (7.5.23) and (7.5.40), it is easy to prove the following.

---

1 We get the definitions of $d(i, j)$ $(i, j \in E)$ by deleting all h in (7.7.2).

Theorem 7.7.3 *An infinite sequence $\{j_n\}$ in E^h is a Cauchy sequence in the metric space E^h if and only if it is a Cauchy sequence in the metric space E.* □

Let us now take the completion of the metric space E^h for the metric d^h, then we have a complete metric space E^{h*} (the metric is still denoted by d^h). So E^h is a dense subset of E^{h*}. It is evident that if $E_a^h = E_a \bigcap E^h \neq \varnothing$ $(a \in \mathfrak{A})$, then $E_a^h = E_a$. $\partial E^h = E^{h*} \setminus E_0^h$ is called the Martin boundary[1], where $E_0^h = E_0 \bigcap E^h$.

Suppose $i \in E^h$, $\xi \in \partial E^h$. Clearly,

$$K^h(i, \xi) = \lim_{\substack{j_n \underset{d}{\to} \xi \\ j_n \in E^h}} K^h(i, j_n) \tag{7.7.10}$$

($j_n \underset{d}{\to} \xi$ denotes that the limit of the sequence $j_n (n = 1, 2, \cdots)$ with respect to the topology induced by the metric d is ξ) defines an h-excessive function satisfying

$$\sum_{i \in E^h} \gamma_i^h K^h(i, \xi) \leqslant 1. \tag{7.7.11}$$

We know by the definition that the function $K^h(\cdot, \cdot)$ is continuous on the product space $E^h \times E^{h*}$, and

$$K^h(i, \xi) = \frac{1}{h_i} K(i, \xi) \qquad (i \in E^h,\ \xi \in E^{h*}). \tag{7.7.12}$$

Theorem 7.7.4 *E^{h*} is a sequentially compact complete metric space, also a compact complete metric space, and*

$$d^h(\xi_1, \xi_2) = |2^{-N(\xi_1)} - 2^{-N(\xi_2)}| + \sum_{s \in E^h} |K^h(s,\xi_1) - K^h(s,\xi_2)| n_s^h 2^{-N(s)} \leqslant 3 \qquad (\xi_1, \xi_2 \in E^{h*}), \tag{7.7.13}$$

where

$$N(\xi) = +\infty, \qquad \xi \in E^{h*} \setminus E. \tag{7.7.14}$$

Hence, the infinite sequence $\{\xi_n)$ in E^{h*} is a Cauchy sequence if and only if both the following two conditions are satisfied:

(i) $\lim_{n \to \infty} N(\xi_n)$ *exists (finite or* $+\infty$);

(ii) *for any $i \in E^h$, $K^h(i, \xi_n)$ is a Cauchy sequence in the real number field.*

Proof. By [16, I, §3, Theorem 2], we need only to prove that E^{h*} is sequentially compact and that (7.7.13) hold.

---

1 Notice that all the points identified in E^{h*} and/or in ∂E^h will from now on be considered as one point.

We get (7.7.13) from Lemma 7.5.4, Theorem 7.5.2, (7.7.2) and (7.7.9), and the fact that $d^h(\xi_1, \xi_2)$ is a continuous function of ξ_1, ξ_2.

From Lemma 7.5.4 and (7.7.10), we know that for any $i \in E^h$,

$$K^h(i, \xi) \leqslant \frac{1}{h_i n_i} < +\infty \qquad (\xi \in E^{h*}). \tag{7.7.15}$$

Suppose $\{\xi\}$ is an infinite subset of E^{h*}. We choose from it an infinite sequence ξ_n $(n \geqslant 0)$ such that

$$\lim_{n\to\infty} N(\xi_n) = +\infty. \tag{7.7.16}$$

And from (7.7.15) we know that $K^h(0, \xi_n)$ $(n \geqslant 0)$ is a real bounded sequence. So we may choose from it an infinite subsequence ξ_{0n} $(n \geqslant 0)$ such that $K^h(0, \xi_{0n})$ $(n \geqslant 0)$ is a Cauchy sequence in the real number field. In the same way, we may choose from ξ_{0n} $(n \geqslant 0)$ an infinite subsequence ξ_{1n} $(n \geqslant 0)$ such that $K^h(1, \xi_{1n})$ $(n \geqslant 0)$ is a Cauchy sequence of real. Proceeding in this way, we will get a denumerable collection of sequences as follows:

$$\begin{array}{ccccc}
\xi_{00}, & \xi_{01}, & \cdots, & \xi_{0n}, & \cdots \\
\xi_{10}, & \xi_{11}, & \cdots, & \xi_{1n}, & \cdots \\
\cdots & \cdots & \cdots & \cdots & \cdots \\
\xi_{i0}, & \xi_{i1}, & \cdots, & \xi_{in}, & \cdots \\
\xi_{i+1,0}, & \xi_{i+1,1}, & \cdots, & \xi_{i+1,n}, & \cdots \\
\cdots & \cdots & \cdots & \cdots & \cdots
\end{array}$$

where $\xi_{i+1,\,n}(n \geqslant 0)$ is an infinite subsequence of ξ_{in} $(n \geqslant 0)$, and for each $i \in E^h$, $K^h(i, \xi_{in})(n \geqslant 0)$ is a Cauchy sequence in the real number field. Hence ξ_{nn} $(n \geqslant 0)$ is an infinite subsequence of $\{\xi\}$ with properties: (1) $\lim_{n\to\infty} N(\xi_{nn}) = +\infty$; (2) for each $i \in E^h$, $K^h(i, \xi_{nn})(n \geqslant 0)$ is a Cauchy sequence in the real number field. So by the completeness of E^{h*} we see that ξ_{nn} $(n \geqslant 0)$ is a convergent sequence of $\{\xi\}$. Hence E^{h*} is sequentially compact. This completes the proof. □

It is easy to prove the following.

Theorem 7.7.5 *An infinite sequence $\{\xi_n\}$ is a Cauchy sequence in E^{h*} if and only if it is a Cauchy sequence in E^*.* □

By Theorem 7.6.1 we obtain

Theorem 7.7.6 *For any initial state $i \in E$ and almost all the honest samples, there exists*

$$\text{d-}\lim_{n\to\infty} x_n = x_\infty \in \partial E, \tag{7.7.17}$$

where d-$\lim_{n\to\infty} x_n$ *denotes the limit of the sequence* x_n $(n=0, 1, 2, \cdots)$ *with respect to the topology induced by the metric d.* □

§ 7.8 Distribution of x_ζ

If $\zeta < +\infty$, then x_ζ is a point in the space E. If $\zeta = +\infty$, then $x_\zeta = x_\infty$ and $x_\infty \in \partial E$.

Suppose D is a finite subset of E. Let

$$\tau = \sup\{n\colon x_n \in D\}. \tag{7.8.1}$$

The random variable τ may take values: 0, 1, $\cdots$ and $+\infty$. If for all $n \geqslant 0$, $x_n \notin D$, then τ is undefined. Let

$$L_0(i) = P_i(\tau = 0) = P_i(x_0 \in D,\ x_n \notin D,\ n \geqslant 1). \tag{7.8.2}$$

Thus

$$L_0(i) = 0, \quad \text{if } i \text{ is a recurrent state}, \tag{7.8.3}$$

and

$$P_i(x_\tau = j) = \sum_{m=0}^{\infty} P_i(\tau = m,\ x_m = j)$$

$$= \sum_{m=0}^{\infty} P_i(x_m = j) L_0(j). \tag{7.8.4}$$

From (7.4.16), (7.8.3) and (7.8.4), we deduce

$$P_i(x_\tau = j) = \begin{cases} 0, & \text{if } j \text{ is a recurrent state}, \\ f_{ij}^* g_{jj} L_0(j), & \text{if } j \text{ is a nonrecurrent state}. \end{cases} \tag{7.8.5}$$

Hence

$$P_\gamma(x_\tau = j) = \begin{cases} 0, & \text{if } j \text{ is a recurrent state}, \\ n_j g_{jj} L_0(j), & \text{if } j \text{ is a nonrecurrent state}. \end{cases} \tag{7.8.6}$$

From (7.4.21), (7.8.5) and (7.8.6), we obtain

$$P_i(x_\tau = j) = K(i, j) P_\gamma(x_\tau = j). \tag{7.8.7}$$

From the above relations and by referring to [15, §10], it is easy to prove that, if f is a continuous function or a nonnegative Borel measurable function on E^*, then

$$E_i f(x_\zeta) = \int_{E^*} K(i, \xi) f(\xi) \mu_1(d\xi), \tag{7.8.8}$$

where

$$\mu_1(\Gamma) = P_\gamma(x_\zeta \in \Gamma) \tag{7.8.9}$$

and Γ is a Borel set in E^*. Consequently, if we let $f = \chi_\Gamma$, it follows that

$$P_i(x_\zeta \in \Gamma) = \int_\Gamma K(i, \xi)\mu_1(d\xi), \tag{7.8.10}$$

where

$$\chi_\Gamma(\xi) = \begin{cases} 0, & \xi \notin \Gamma, \\ 1, & \xi \in \Gamma. \end{cases} \tag{7.8.11}$$

It is also easy to show that

$$\mu_1(j) = n_j g_{jj}\left(1 - \sum_{k \in E} p_{jk}\right) \qquad (j \in E). \tag{7.8.12}$$

§ 7.9 Martin expressions of excessive functions

Suppose h is a γ-integrable excessive function. Let

$$\mu_h(\Gamma) P^h_{\gamma_h}(x_\zeta \in \Gamma), \tag{7.9.1}$$

where Γ is a Borel set in E^{h*}. Applying (7.8.8) to a h-chain, we find

$$E^h_i f(x_\zeta) = \frac{1}{h_i}\int_{E*} K(i, \xi) f(\xi)\mu_h(d\xi) \qquad (i \in E). \tag{7.9.2}$$

Putting $f = 1$ in (7.9.2), we get

$$h_i = \int_{E*} K(i, \xi)\mu_h(d\xi) \qquad (i \in E). \tag{7.9.3}$$

Equation (7.9.3) is called the *Martin expression of excessive function* h, μ_h is called the *spectral measure of* h.

Applying (7.8.12) to an h-chain, we obtain

$$\mu_h(j) = n_j g_{jj}\left(h_j - \sum_{k \in E} p_{jk} h_k\right) \qquad (j \in E_0 \cap E^h). \tag{7.9.4}$$

It is clear that (7.9.4) also hold for $j \in E_0 \setminus E_0 \cap E^h$. Then we have

$$\mu_h(j) = n_j g_{jj}\left(h_j - \sum_{k\in E} p_{jk} h_k\right) \qquad (j\in E). \tag{7.9.5}$$

§ 7.10 Exit space

Let

$$\bar{f}_{ij}^* = P_i(\text{there exists } n>0, \text{ such that } x_n = j) \qquad (i, j\in E). \tag{7.10.1}$$

By the definitions of f_{ij}^* and $\bar{f}_{ij}^*$, it is apparent that

$$\bar{f}_{ij}^* = \begin{cases} f_{ij}^* & (i\neq j), \\ \sum_{k\in E} p_{ik} f_{kj}^* & (i=j). \end{cases} \tag{7.10.2}$$

From [1, I, §10] we have

$$g_{jj}(1-\bar{f}_{jj}^*) = 1 \qquad (j\in E_0). \tag{7.10.3}$$

Theorem 7.10.1 *For any $k\in E_0$, we have*

$$\mu_{K(\cdot,k)} = \delta_k, \tag{7.10.4}$$

where δ_k is the unit measure at k.

Proof. From (7.4.21) and (7.10.2) we deduce

$$\sum_{s\in E} p_{ks} K(s, k) = \sum_{s\in E} p_{ks} \frac{f_{sk}^*}{n_k} = \frac{1}{n_k}\sum_{k\in E} p_{ks} f_{sk}^*$$

$$= \frac{1}{n_k}\bar{f}_{kk}^* = \frac{1}{n_k}(1-(1-\bar{f}_{kk}^*)) = \frac{1}{n_k}(f_{kk}^* - (1-\bar{f}_{kk}^*))$$

$$= K(k, k) - \frac{1-\bar{f}_{kk}^*}{n_k}. \tag{7.10.5}$$

From (7.9.5), (7.10.3) and (7.10.5) it follows that

$$\mu_{K(\cdot,k)}(k) = n_k g_{kk}\left[K(k, k) - \sum_{s\in E} p_{ks} K(s, k)\right]$$

$$= n_k g_{kk}\frac{1-\bar{f}_{kk}^*}{n_k} = g_{kk}(1-\bar{f}_{kk}^*) = 1. \tag{7.10.6}$$

By (7.5.25) and (7.9.1), we find

$$\mu_{K(\cdot,k)}(E^*) = \sum_{i\in E} \gamma_i K(i, k) = 1. \tag{7.10.7}$$

Then we obtain our theorem readily from (7.10.6) and (7.10.7). □

Definition 7.10.1 *Let*

$$B = \{\xi:\ \xi\in\partial E,\quad \mu_{K(\cdot,\xi)} = \delta_\xi\}. \tag{7.10.8}$$

We say that B is the exit space, and each point of B is called an exit point.

In the same way as the proof of [15, Theorem 5], we obtain

Theorem 7.10.2 *The exit space B is a Borel set of ∂E. For any γ-integrable excessive function h, we have $\mu_h(\partial E\backslash B)=0$. If $\xi\in B$, then $K(\cdot, \xi)$ is a harmonic function, and*

$$\sum_{i\in E} \gamma_i K(i, \xi) = 1. \tag{7.10.9}$$

§ 7.11 Uniqueness theorem

Following the proof of [15, Theorem 6], we establish

Theorem 7.11.1 *Every γ-integrable excessive function h can be expressed uniquely as follows:*

$$h_i = \int_{E_0\cup B} K(i, \xi)\mu(d\xi) \qquad (i\in E), \tag{7.11.1}$$

where μ is a finite measure on the Borel set $E_0\bigcup B$; conversely, for any finite measure μ on $E_0\bigcup B$, (7.11.1) defines a γ-integrable excessive function. This function is harmonic if and only if $\mu(E_0)=0$. □

§ 7.12 Minimal excessive functions

Definition 7.12.1 *A nonzero excessive function h is said to be minimal, if $h = h_1+h_2$ (where h_1 and h_2 are also excessive functions) implies $h_1=c_1h$, $h_2 = c_2h$ (c_1 and c_2 are constants).*

As in the proof of [15, Theorem 7], it is easy to get

Theorem 7.12.1 *The general form of γ-integrable minimal excessive functions is $cK(\cdot, \xi)$, where $\xi\in E_0\bigcup B$, and c is an arbitary positive constant.* □

§ 7.13 Terminal random variables

We shall write briefly P for P_y. In this section, the point ω in the probability space Ω is identified with the point $(x_0(\omega), x_1(\omega), \cdots, x_{\zeta(\omega)}(\omega))$ (if $\zeta(\omega) < +\infty$) or $(x_0(\omega), x_1(\omega), \cdots)$ (if $\zeta(\omega) = +\infty$) in the sample space.

Define a transformation T on $\Omega_1 = \{\omega: \zeta(\omega) \geqslant 1\}$: T maps $\{i_0, i_1, \cdots, i_n\}$ and $\{i_0, i_1, \cdots\}$ to $\{i_1, i_2, \cdots, i_n\}$ and $\{i_1, i_2, \cdots\}$, respectively.

For any random variable $\varphi(\omega)$, let

$$\hat{T}\varphi(\omega) = \begin{cases} \varphi(T\omega), & \omega \in \Omega_1, \\ 0, & \omega \notin \Omega_1. \end{cases} \tag{7.13.1}$$

Definition 7.13.1 *If for a random variable φ we have*

$$\hat{T}\varphi = \varphi, \tag{7.13.2}$$

then φ is called a terminal random variable. For any measurable subset Λ of Ω, if x_Λ is a terminal random variable, then Λ is called a terminal set.

Obviously, any terminal random variable is zero on $\Omega \backslash \Omega_\infty$, hence terminal sets belong to Ω_∞.

If $\Lambda_1, \Lambda_2 \in \mathcal{F}$, and $P((\Lambda_1 \backslash \Lambda_1\Lambda_2) \bigcup (\Lambda_2 \backslash \Lambda_2\Lambda_1)) = 0$, then we say that Λ_1 and Λ_2 are equivalent, and write $\Lambda_1 \doteq \Lambda_2$.

Definition 7.13.2 *A positive probability terminal set which cannot be partitioned into two positive probability terminal sets is called an atomic terminal set.*

Obviously, on an atomic terminal set, any terminal random variable is constant.

Let[1]

$$\bar{\mathfrak{L}}(A) = \{\omega: \omega \in \Omega_\infty, \text{ there are infinitely many } n, \\ \text{such that } x_n(\omega) \in A\}, \tag{7.13.3}$$

$$\underline{\mathfrak{L}}(A) = \{\omega: \omega \in \Omega_\infty, \text{ there exists } N \geqslant 0, \text{ such that} \\ x_n(\omega) \in A \text{ provided } n \geqslant N\}. \tag{7.13.4}$$

Definition 7.13.3 *Suppose A is a subset of E. If*

$$P(\bar{\mathfrak{L}}(A)) = P(\underline{\mathfrak{L}}(A)) > 0, \tag{7.13.5}$$

1 We give the definitions for the following notations and terms without directly using [1, I, §17], as the Markov chain here allows $P(\Omega_\infty) < 1$, in contrast to that in [1].

then A is called an almost closed set. If

$$P(\bar{\mathfrak{L}}(A)) = 0, \tag{7.13.6}$$

then A is called an irreturnable set.

Obviously, if A is an *irreturnable set*, or an almost closed set, then $\bar{\mathfrak{L}}(A) \doteq \underline{\mathfrak{L}}(A)$. In this case, we denote by $\mathfrak{L}(A)$ any terminal set equivalent to $\bar{\mathfrak{L}}(A)$ or $\underline{\mathfrak{L}}(A)$.

Definition 7.13.4 *An almost closed set which can not be partitioned into two disjoint almost closed sets is called an atomic almost closed set; an almost closed set which does not contain any atomic almost closed set is called a completely nonatomic almost closed set.*

Definition 7.13.5 *Assume $\xi \in B$. If $\mu_1(\xi) > 0$, then ξ is called an atomic exit point. Otherwise, it is called a nonatomic exit point.*

It is easy to prove the following two theorems:

Theorem 7.13.1 *A one-to-one correspondence can be set up between each two of the following three sets, the set of atomic terminal sets Λ, the set of atomic almost closed sets A and the set of atomic exit points ξ, subject to the following condition:*

$$\Lambda \doteq (\mathfrak{L}(A)) \doteq \{\omega\colon x_\infty(\omega) = \xi\}. \ \square \tag{7.13.7}$$

Theorem 7.13.2 *Every bounded harmonic function h corresponds to a nonnegative bounded terminal random variable φ, such that*

$$h_i = E_i \varphi, \tag{7.13.8}$$

and vice versa. Further, the equality

$$\lim_{n \to \infty} h_{x_n(\omega)} = \varphi(\omega) \tag{7.13.9}$$

holds almost everywhere on Ω_∞. $\square$

§ 7.14 Criteria for potentials and excessive functions, Riesz decomposition

The results of this section will be applied in §7.15, but they are also of independent interest.

Definition 7.14.1 *A function f_i $(i \in E)$ is called a potential of a nonnegative function v_i $(i \in E)$, if*

$$f_i = \sum_{j \in E} p_{ij} v_i. \tag{7.14.1}$$

Theorem 7.14.1 *A necessary and sufficient condition for the function f_i $(i \in E)$ to be a potential of a nonnegative function is, that there exists a nonnegative function v_i $(i \in E)$, such that f_i $(i \in E)$ is the minimal nonnegative solution of the first-type system of 1-bounded equations*

$$x_i = \sum_{j \in E} p_{ij} x_j + v_i \qquad (i \in E). \tag{7.14.2}$$

Proof. The assertion follows immediately from Theorem 3.8.1 and the definition of potentials. □

Theorem 7.14.1 demonstrates that our "Theory of the Minimal Nonnegative Solution" is nothing other than the potential theory which is most active today in the study of Markov processes.

Theorem 7.14.2 *h_i $(i \in E)$ is an excessive function if and only if there exists a nonnegative function v_i $(i \in E)$, such that h_i $(i \in E)$ is the nonnegative solution of the first-type system of 1-bounded equations*

$$x_i = \sum_{j \in E} p_{ij} x_j + v_i \qquad (i \in E). \ \square \tag{7.14.3}$$

Definition 7.14.2 *Suppose h_i $(i \in E)$ is an excessive function. Let*

$$v_i = h_i - \sum_{j \in E} p_{ij} h_j \qquad (i \in E). \tag{7.14.4}$$

Then v_i $(i \in E)$ is called the excess of h_i $(i \in E)$.

Evidently, the excess v_i $(i \in E)$ of the excessive function h_i $(i \in E)$ is determined uniquely by $h_i (i \in E)$.

From Theorems 7.14.1 and 7.14.2 it is easy to prove

Theorem 7.14.3 (Riesz Decomposition). *Suppose $h_i (i \in E)$ is an excessive function. Then h_i may be decomposed uniquely into the following form:*

$$h_i = b_i + f_i \qquad (i \in E), \tag{7.14.5}$$

where b_i $(i \in E)$ is a harmonic function and f_i $(i \in E)$ is the potential of the excess v_i $(i \in E)$ of h_i $(i \in E)$. □

§ 7.15 Criteria for minimal harmonic functions, minimal potentials and minimal excessive functions

Definition 7.15.1 *If a minimal excessive function is also a harmonic function (potential), then it is called a minimal harmonic function (minimal potential).*

In the papers [12], [13], [15], etc., the definition of the minimal harmonic function is given without any effective criteria. This is inadequate for applications. For example, if we want to distinguish the exit space B from the boundary ∂E, we should first be able to decide whether or not $K(\cdot, \xi)$ is a minimal harmonic function for any point ξ in ∂E. The aim of this section is to present the criteria for minimal harmonic functions, as well as minimal potentials and minimal excessive functions.

It is easy to prove that if h_i $(i\in E)$ is a minimal excessive function, then $0 \leqslant h_i < +\infty$ $(i\in E)$. So in what follows, we only consider finitevalued excessive functions.

Theorem 7.15.1 *A finite-valued harmonic function b_i $(i\in E)$ is minimal if and only if the system of equations*

$$x_i = \sum_{j\in E^b} p_{ij} \frac{b_j}{b_i} x_j \qquad (i\in E^b) \tag{7.15.1}$$

has no nonconstant nonnegative bounded solution, where $E^b = \{i\colon b_i > 0\}$.

Proof. (1) Necessity.

Suppose $b_i (i\in E)$ is a finite-valued minimal harmonic function.

If (7.15.1) has a nonconstant nonnegative bounded solution $b_i^{(1)}$ $(i\in E^b)$, then $b_i^{(1)}$ $(i\in E^b)$ is bounded. We may, without loss of generality, assume $b_i^{(1)} \leqslant 1$ $(i\in E^b)$. Let

$$b_i^{(2)} = 1 - b_i^{(1)} \qquad (i\in E^b). \tag{7.15.2}$$

Thus

$$0 \leqslant b_i^{(2)} \leqslant 1 \qquad (i\in E^b). \tag{7.15.3}$$

It is easy to show that $b_i^{(2)}$ $(i\in E^b)$ is also a nonconstant nonnegative bounded solution of (7.15.1). And by (7.15.2), $b_i^{(1)}$ $(i\in E^b)$ is not proportional to $b_i^{(2)} (i\in E^b)$. Let

$$b_i^{(1)} = b_i^{(2)} = 0 \qquad (i\notin E^b). \tag{7.15.4}$$

Hence from Lemma 7.5.1 we deduce

$$\sum_{j\in E} p_{ij} b_j b_j^{(1)} = \sum_{j\in E^b} p_{ij} b_j b_j^{(1)} = b_i \sum_{j\in E^b} p_{ij} \frac{b_j}{b_i} b_j^{(1)} = b_i b_i^{(1)} \qquad (i\in E^b), \tag{7.15.5}$$

$$\sum_{j\in E} p_{ij} b_j b_j^{(1)} = \sum_{j\in E^b} p_{ij} b_j b_j^{(1)} = \sum_{j\in E^b} 0 b_j b_j^{(1)} = 0 = b_i b_j^{(1)} \qquad (i\notin E^b). \tag{7.15.6}$$

It can be seen from (7.15.5) and (7.15.6) that $b_i b_i^{(1)} (i\in E)$ is a harmonic function. Similarly, $b_i b_i^{(2)}$ is also a harmonic function. It is evident that

$$b_i = b_i b_i^{(1)} + b_i b_i^{(2)} \qquad (i\in E). \tag{7.15.7}$$

And $b_i b_i^{(1)}\ (i\in E)$ is not proportional to $b_i b_i^{(2)}$. Hence $b_i\ (i\in E)$ and $b_i b_i^{(1)}\ (i\in E)$ are not proportional. But this contradicts the minimality of $b_i\ (i\in E)$. Therefore, it is impossible that (7.15.1) has a nonconstant nónnegative bounded solution.

(2) Sufficiency.

Suppose (7.15.1) has no nonconstant nonnegative bounded solution.

If $b_i (i\in E)$ is not minimal, then we have

$$b_i = b_i^{(1)} + b_i^{(2)} \qquad (i\in E), \tag{7.15.8}$$

and $b_i^{(1)}\ (i\in E)$ is a harmonic function which is not proportional to $b_i\ (i\in E)$. Consequently,

$$0 \leqslant \frac{b_i^{(1)}}{b_i} \leqslant 1, \qquad \frac{b_i^{(1)}}{b_i} \neq \text{const.} \qquad (i\in E^b). \tag{7.15.9}$$

Noting that $b_i^{(1)} = 0 (i\notin E^b)$, we get

$$\sum_{j\in E^b} p_{ij} \frac{b_j}{b_i} \cdot \frac{b_j^{(1)}}{b_j} = \frac{1}{b_i} \sum_{j\in E^b} p_{ij} b_j^{(1)} = \frac{1}{b_i} \sum_{j\in E} p_{ij} b_j^{(1)} = \frac{b_i^{(1)}}{b_i}, \qquad (i\in E^b). \tag{7.15.10}$$

So $b_i^{(1)}/b_i (i\in E^b)$ is a nonconstant nonnegative bounded solution of (7.15.1). But this contradicts the hypothesis. Thus $b_i\ (i\in E)$ must be minimal. □

It is easy to proof the following three theorems:

Theorem 7.15.2 *The finite-valued potential $f_i (i\in E)$ is minimal if and only if its excess (take $f_i (i\in E)$ as an excessive function) $v_i (i\in E)$ possesses the following property: there exists $i_0 \in E_0$, such that*

$$\left.\begin{aligned} 0 < v_{i_0} < +\infty, \\ v_i = 0 \qquad (i \neq i_0), \end{aligned}\right\} \tag{7.15.11}$$

In other words, the potential of the nonnegative function $v_i\ (i\in E)$ is minimal if and only if it possesses property (7.15.11). □

Theorem 7.15.3 *The finite-valued potential f_i $(i \in E)$ is minimal if and only if there exists a constant $c > 0$ and $j \in E_0$, such that*

$$f_i = c g_{ij} \qquad (i \in E). \ \square \tag{7.15.12}$$

Theorem 7.15.4 *The finite-valued excessive function h_i $(i \in E)$ is minimal if and only if one of the following two conditions is satisfied.*

(i) *h_i $(i \in E)$ is a minimal harmonic function;*

(ii) *h_i $(i \in E)$ is a minimal potential.* $\square$

§ 7.16 Atomic exit spaces and nonatomic exit spaces

Definition 7.16.1 *The set of all the atomic exit points, denoted by B_1, is called an atomic exit space; the set of all the nonatomic exit points, denoted by B_2, is called a nonatomic exit space.*

Theorem 7.16.1 *Suppose $\xi \in B$. Then $\xi \in B_1$ if and only if $K(\cdot, \xi)$ is bounded.*

Proof. (1) Necessity.

Suppose ξ is an atomic exit boundary point, namely

$$\mu_1(\xi) > 0. \tag{7.16.1}$$

From (7.8.10) we know

$$P_i(x_\zeta = \xi) = K(i, \xi)\mu_1(\xi) \qquad (i \in E). \tag{7.16.2}$$

Hence

$$K(i, \xi) = \frac{P_i(x_3 = \xi)}{\mu_1(\xi)} \leqslant \frac{1}{\mu_1(\xi)} < +\infty \qquad (i \in E). \tag{7.16.3}$$

Thus $K(\cdot, \xi)$ is bounded.

(2) Sufficiency.

Suppose $\xi \in B$, and $K(\cdot, \xi)$ is bounded. $K(\cdot, \xi)$ is then a bounded harmonic function. So from Theorem 7.13.2 it is seen that there exists a nonnegative bounded terminal random variable φ, such that

$$K(\cdot, \xi) = E_{i\varphi} \qquad (i \in E). \tag{7.16.4}$$

Put

$$\Lambda = \{\omega \colon \varphi(\omega) \neq 0\}. \tag{7.16.5}$$

Hence from $\xi \in B$ and (7.16.4) we see that

$$P(\Lambda) \neq 0. \tag{7.16.6}$$

It is evident that Λ is a terminal set. We are now going to show that Λ is an atomic terminal set. In fact, if there exist terminal sets Λ_1 and Λ_2, such that

$$\Lambda = \Lambda_1 \bigcup \Lambda_2, \quad \Lambda_1 \bigcap \Lambda_2 = \varnothing, \quad P(\Lambda_1) > 0, \quad P(\Lambda_2) > 0, \tag{7.16.7}$$

then the function

$$\varphi_j(\omega) = \begin{cases} \varphi(\omega), & \omega \in \Lambda_j \\ 0, & \omega \notin \Lambda_j \end{cases} \qquad (j = 1, 2) \tag{7.16.8}$$

is still a bounded terminal random variable. Accordingly, by virtue of Theorem 7.13.2 we know that for any $j \in (1, 2)$,

$$f_i^{(j)} = E_i \varphi_j \qquad (i \in E) \tag{7.16.9}$$

is a bounded harmonic function which is not identically zero, and

$$\lim_{n \to \infty} f_{x_n(\omega)}^{(j)} \neq 0 \qquad \left(\omega \in \Lambda_j \bigcap \Omega_\infty\right) \tag{7.16.10}$$

and

$$\lim_{n \to \infty} f_{x_n(\omega)}^{(j)} = 0 \qquad \left(\omega \notin \Lambda_j \bigcap \Omega_\infty\right) \tag{7.16.11}$$

exist almost everywhere on Ω_∞. From (7.16.7), (7.16.10) and (7.16.11) we affirm that $f_i^{(1)}$ $(i \in E)$ and $f_i^{(2)}$ $(i \in E)$ are not proportional. From (7.16.4), (7.16.6), (7.16.8) and (7.16.9), we see

$$K(i, \xi) = f_i^{(1)} + f_i^{(2)} \qquad (i \in E). \tag{7.16.12}$$

Hence, from the fact that $f_i^{(1)}$ $(i \in E)$ and $f_i^{(2)}$ $(i \in E)$ are not proportional we know $K(i, \xi)$ $(i \in E)$ is not proportional to $f_i^{(1)}$ $(i \in E)$ or $f_i^{(2)}$ $(i \in E)$. But this contradicts the minimality of $K(\cdot, \xi)$. So Λ must be an atomic terminal set. Consequently, since an atomic terminal set, a terminal random variable is constant, as mentioned in §7.13, we obtain

$$\varphi(\omega) = \begin{cases} c, & \omega \in \Lambda, \\ 0, & \omega \notin \Lambda, \end{cases} \tag{7.16.13}$$

where $c > 0$ is a constant. Combining (7.16.4) and (7.16.13), we obtain

$$K(i, \xi) = c P_i(\Lambda) \qquad (i \in E). \tag{7.16.14}$$

From the fact that Λ is an atomic terminal set and Theorem 7.13.1 we know there exists an atomic exit boundary point ξ^0 such that

$$P_i(\Lambda) = P_i(x_\zeta = \xi^0) = K(i, \xi^0)\mu_1(\xi^0). \tag{7.16.15}$$

From (7.16.14) and (7.16.15), we get

$$K(i, \xi) = K(i, \xi^0)c\mu_1(\xi^0). \tag{7.16.16}$$

From (7.16.16) and (7.10.9), we see

$$K(i, \xi) = K(i, \xi^0) \qquad (i \in E). \tag{7.16.17}$$

Hence we have $\xi = \xi^0$, i.e., ξ is an atomic exit boundary point. □

Definition 7.16.2 *Every finite minimal harmonic function is called a minimal solution of the system of equations*

$$x_i = \sum_{j \in E} p_{ij} x_j \qquad (i \in E). \tag{7.16.18}$$

Theorem 7.16.2 $B_1 = \varnothing$ *if and only if the system of equations* (7.16.18) *has no bounded minimal solution. And* $B_2 = \varnothing$ *if and only if the equation system* (7.16.18) *has no γ-integrable unbounded minimal solution.*

Proof. The theorem follows immediately from Theorems 7.12.1 and 7.16.1 and the simple fact that for any standard measure γ a bounded nonnegative function on E is always γ-integrable. □

§ 7.17 Blackwell decomposition of the state space

In this section we suppose

$$\sum_{j \in E} p_{ij} = 1 \qquad (i \in E). \tag{7.17.1}$$

Theorem 7.17.1 *Suppose A is an almost closed set. Then it is a complete nonatomic almost closed set if and only if the system of equations*

$$x_i = \sum_{j \in E} p_{ij} \frac{P_j(\mathcal{L}(A))}{P_i(\mathcal{L}(A))} x_j \qquad (i \in E^{P(\mathcal{L}(A))}) \tag{7.17.2}$$

has no bounded minimal solution, where $E^{P(\mathcal{L}(A))} = \{i\colon P_i(\mathcal{L}(A)) > 0\}$.

Proof. From Theorem 6.8.2 we know that A is a complete nonatomic almost closed set if and only if E is a complete nonatomic almost closed set of a $P(\mathcal{L}(A))$-

chain. We get our theorem readily by applying Theorems 7.13.1 and 7.16.2 to the $P(\mathfrak{L}(A))$-chain. □

From Theorems 6.8.3 and 7.17.2 we obtain at once the criterion for Blackwell decomposition of the state space E. But the statements are omitted since they are only just repetitions of these two theorems.

CHAPTER VIII
Martin Entrance Boundary Theory

§ 8.1 Introduction

For simplicity, we shall identify the Markov chain $X(\omega)=\{x_n(\omega),\ n<\zeta(\omega)+1\}$ with its transition probability matrix $P=(p_{ij};\ i,\ j\in E)$, so P is also called a Markov chain.

Definition 8.1.1 *A nonnegative function v_i $(j\in E)$ defined on E is called an excessive measure $($of $P)$, if*

$$v_j \geqslant \sum_{k\in E} v_k p_{kj} \qquad (j\in E). \tag{8.1.1}$$

The excessive measure v_j $(j\in E)$ is called a harmonic measure, if for all $j\in E$, $v_j=\sum_{k\in E} v_k p_{kj}$, i. e., the equalities hold in (8.1.1). *The excessive measure v_j $(j\in E)$ is said to be finite positive, if*

$$0<v_j<+\infty \qquad (j\in E). \tag{8.1.2}$$

The theory of the Martin entrance boundary of a Markov chain has been developed in [12], [13], [15] along with the theory of the Martin exit boundary. The methods used in these works consist essentially in taking advantage of a certain finite positive excessive measure to reduce the problem of the entrance boundary of a Markov chain into that of exit boundary of another Markov chain (called the adjoint Markov chain). Hence, the Martin entrance boundary theory, set up so far, is only for those Markov chains satisfying the following two conditions.

(i) There exists a finite positive excessive measure α_j $(j\in E)$.

(ii) An exit boundary can be established for the adjoint Markov chain $\tilde{P}=(\tilde{p}_{ij}, i,j\in E)$, where

$$\tilde{p}_{ij}=\frac{\alpha_j}{\alpha_i}p_{ji} \qquad (i,j\in E). \tag{8.1.3}$$

Since we have established the exit boundary theory for general Markov chains in Chapter VII, the condition (ii) is already disposed of.

The aim of this chapter is to get rid of the condition (i) as well and establish the Martin entrance boundary theory for general Markov chains.

This chapter can be divided into two parts. The aim of the first part (§§8.2—8.3) is: (A) to give necessary and sufficient conditions for (i), thus enabling us to see clearly the range of application for the entrance boundary theory so far developed by our predecessors. (B) to make clear why condition (i) does not hold in the general case; thereby revealing the way of developing the entrance boundary theory for general Markov chains. This is the connecting link to the second part (§§8.4—8.6) which is devoted to the establishment of the entrance boundary theory for general Markov chains.

§ 8.2 The first group of lemmas

Definition 8.2.1 *If E is an irreducible recurrent class, then P is called a irreducible recurrent chain.*

Lemma 8.2.1 *If P is an irreducible recurrent Markov chain, then there exists a finite positive excessive measure.*

Proof. The lemma follows readily from [1, I, Corollary 1 of Theorem 9.5 and Theorem 9.7].□

Lemma 8.2.2 *If the Markov chain P is a transient chain, then there exists a finite positive excessive measure.*

Proof. Let

$$v_j = \sum_{i \in E} 2^{-(i+1)} g_{ij} \qquad (j \in E). \tag{8.2.1}$$

Since P is a transient chain, we know

$$0 \leqslant g_{ij} < +\infty, \qquad g_{jj} \geqslant 1 \qquad (i,j \in E). \tag{8.2.2}$$

It follows by Theorem 7.4.4 that

$$g_{ij} = f^*_{ij} g_{jj} \qquad (i,j \in E). \tag{8.2.3}$$

Equations (8.2.1) and (8.2.2) yield

$$v_j = \sum_{i \in E} 2^{-(i+1)} g_{ij} \geqslant 2^{-(j+1)} g_{jj} > 0 \qquad (j \in E). \tag{8.2.4}$$

By (8.2.1), (8.2.2) and (8.2.3) we obtain

$$v_j = \sum_{i \in E} 2^{-(i+1)} g_{ij} = \left(\sum_{i \in E} 2^{-(i+1)} f_{ij} \right) g_{ij}$$

$$\leqslant \left(\sum_{i\in E} 2^{-(i+1)} \right) g_{jj} = g_{jj} < +\infty \qquad (j \in E). \tag{8.2.5}$$

Invoking the Markov property, we see

$$\begin{aligned} p_{ij}^{(0)} &= \delta_{ij}, \\ p_{ij}^{(n+1)} &= \sum_{k\in E} p_{ik}^{(n)} p_{kj} \qquad (n = 0, 1, \cdots; i, j \in E), \end{aligned} \tag{8.2.6}$$

where $\delta_{ii} = 1$, $\delta_{ij} = 0 \quad (i \neq j)$. Hence by virtue of (8.2.2) and Corollary 3.2.3, we affirm that for any fixed $i \quad (i \in E)$, $\{g_{ij}, j \in E\}$ is the minimal nonnegative solution of the system of nonnegative linear equations

$$x_j = \sum_{k\in E} x_k p_{ki} + \delta_{ij} \qquad (j \in E). \tag{8.2.7}$$

From (8.2.1) and Theorem 3.3.2, we see that $v_j \quad (j \in E)$ is the minimal nonnegative solution of the system of nonnegative linear equations

$$x_j = \sum_{k\in E} x_k p_{kj} + 2^{-(j+1)} \qquad (j \in E). \tag{8.2.8}$$

Hence

$$v_j \geqslant \sum_{k\in E} v_k p_{kj} \qquad (j \in E). \tag{8.2.9}$$

We know by (8.2.4), (8.2.5) and (8.2.9) that $v_j \quad (j \in E)$ is a finite positive excessive measure. □

Lemma 8.2.3 *Suppose Markov chain P is irreducible recurrent. If* $v_j \quad (j \in E)$ *is an excessive measure, but not a harmonic measure, then*

$$v_j = +\infty \qquad (j \in E). \tag{8.2.10}$$

Proof. The hypothesis implies that there exist $i_0 \in E$ and a constant $a > 0$ such that

$$v_j \geqslant 0, \qquad v_j \geqslant \sum_{k\in E} v_k p_{kj} + \delta_{i_0 j} a \qquad (j \in E), \tag{8.2.11}$$

i.e., v_j is the nonnegative solution of the system of nonnegative linear inequalities

$$x_j \geqslant \sum_{k\in E} x_k p_{kj} + \delta_{i_0 j} a \qquad (j \in E). \tag{8.2.12}$$

But from the proofs of Corollary 3.3.3 and Lemma 8.2.2, we see that $\{ag_{i_0j}, j\in E\}$ is the minimal nonnegative solution of the system of nonnegative linear equations

$$x_j = \sum_{k\in E} x_k p_{kj} + \delta_{i_0j} a \qquad (j\in E). \tag{8.2.13}$$

Hence, by Theorem 3.3.1,

$$v_j \geqslant a g_{i_0j} \qquad (j\in E). \tag{8.2.14}$$

But the fact that P is an irreducible recurrent chain yields that

$$g_{i_0j} = +\infty \qquad (j\in E). \tag{8.2.15}$$

Consequently, (8.2.10) follows by $a>0$ and (8.2.14), and the lemma is thereby established. □

Let

$$E_{00} = \{i: i\in E_0, \text{ and there exists } j\in E\backslash E_0 \text{ such that } i\sim j\}, \tag{8.2.16}$$

$$E_1 = E\backslash E_{00}, \tag{8.2.17}$$

$$P_1 = (p_{ij}, \quad i, \quad j\in E_1). \tag{8.2.18}$$

Definition 8.2.2 *If $E_{00}=\varnothing$, then P is called a bridgingless chain.*

Obviously we have

Lemma 8.2.4 *$P_1 = (p_{ij}, \quad i, \quad j\in E_1)$ is a bridgingless chain.* □

§ 8.3 Properties of finite excessive measures

Theorem 8.3.1 *$v_j \quad (j\in E)$ is a finite excessive measure of P if and only if both the following two conditions are satisfied:*

(i) $v_j = 0 \quad (j\in E_{00})$, (8.3.1)

(ii) $v_j \quad (j\in E_1)$ *is a finite excessive measure of P_1.*

Proof. Obviously, it suffices to prove the necessity.

Suppose $v_j \quad (j\in E)$ is a finite excessive measure of P. In what follows we only prove that (i) holds since (ii) is the immediate consequence of (i).

If $E_{00}=\varnothing$, then (i) is trivial, so suppose $E_{00}\neq\varnothing$. Hence, if we arbitrarily choose an element i_0 from E_{00}, then there exist $a_0\in\mathcal{A}$ and $j\in E_{a_0}$, such that

$$p_{i_0j_0} > 0. \tag{8.3.2}$$

Since $v_j \quad (j\in E)$ is an excessive measure, we see

$$v_j \geqslant 0, \qquad v_j \geqslant \sum_{k \in E_{a_0}} v_k p_{kj} + \delta_{j_0 j} v_{i_0} p_{i_0 j_0} \qquad (j \in E_{a_0}). \tag{8.3.3}$$

If $v_{i_0} \neq 0$, then from (8.3.2) and (8.3.3) it can be deduced that $v_j \ (j \in E_{a_0})$ is an excessive measure but not a harmonic measure of $(p_{ij}, \ i, \ j \in E_{a_0})$. Therefore,

$$v_j = +\infty \qquad (j \in E_{a_0}) \tag{8.3.4}$$

by $a_0 \in \mathcal{A}$ and Lemma 8.2.3. But this contradicts the finiteness of $v_j \ (j \in E)$. So $v_{i_0} \neq 0$ is impossible. Since $i_0 \in E_{00}$ is arbitrary, we get (i), as desired. □

Theorem 8.3.2 *The finite positive excessive measure of P exists if and only if P is a bridgingless chain.*

Proof. By Theorem 8.3.1, we need only prove the sufficiency of the condition. Suppose P is a bridgingless chain, so we have

$$p_{ij} = 0 \qquad (i \in E_a, \ \ j \in E_{a'}, \ \ a, \ \ a' \in \{0\} \bigcup \mathcal{A}, \ \ a \neq a'). \tag{8.3.5}$$

From Lemmas 8.2.1 and 8.2.2 we see that for any $a \in \{0\} \bigcup \mathcal{A}$, there exists $v_j \ (j \in E_a)$ such that

$$0 < v_j < +\infty, \qquad v_j \geqslant \sum_{k \in E_a} v_k p_{kj} \qquad (j \in E_a). \tag{8.3.6}$$

It follows by (8.3.5) and (8.3.6) that

$$0 < v_j < +\infty, \qquad v_j \geqslant \sum_{k \in E} v_k p_{kj} \qquad (j \in E). \tag{8.3.7}$$

Therefore, $v_j \ (j \in E)$ is a finite positive excessive measure of P. □

Theorem 8.3.3 *There exists a finite excessive measure* $v_j \ (j \in E)$ *of P, with the following properties:*

(8.3.8)

(i) $v_j = 0 \ \ (j \in E_{00})$;

(ii) $v_j \ \ (j \in E_1)$ *is a finite positive excessive measure of* P_1.

Proof. The theorem follows readily from Lemma 8.2.4, Theorems 8.3.1 and 8.3.2. □

§ 8.4 The second group of lemmas

Since P_1 is a bridgingless chain, we see by Theorem 8.3.2 that there exists a finite positive excessive measure of P_1. Suppose $\alpha_j \ (j \in E_1)$ is a finite positive excessive measure of P_1. Let

$$\tilde{p}_{ij} = \frac{\alpha_j}{\alpha_i} p_{ji} \qquad (i, \ j \in E_1), \tag{8.4.1}$$

$$\tilde{P} = (\tilde{p}_{ij}, \ i, \ j \in E_1). \tag{8.4.2}$$

Lemma 8.4.1 *v_j $(j \in E)$ is a finite excessive (finite harmonic) measure of P if and only if both the following two conditions are satisfied:*

(i) $v_j = 0$ $(j \in E_{00})$; and (8.4.3)

(ii) v_j/α_j $(j \in E_1)$ is a finite excessive (finite harmonic) function of $\tilde{P}$.

Proof. The theorem follows immediately from Theorem 8.3.3 and the definition of $\tilde{P}$. □

Let

$$\left.\begin{aligned} {}_i p_{ij}^{(1)} &= p_{ij} \qquad (i, \ j \in E_1, \ i \neq j), \\ {}_i p_{ij}^{(n+1)} &= \sum_{k \in E_1 \setminus (i)} {}_i p_{ik}^{(n)} p_{ki} \qquad (n = 1, \ 2, \ \cdots; \ i, \ j \in E_1, \ i \neq j), \end{aligned}\right\} \tag{8.4.4}$$

$$e_{ij} = \begin{cases} 1 & (i, \ j \in E_1, \ i = j), \\ \sum_{n=1}^{\infty} {}_i p_{ij}^{(n)} & (i, \ j \in E_1, \ i \neq j), \end{cases} \tag{8.4.5}$$

$$\left.\begin{aligned} {}_j \tilde{p}_{ij}^{(1)} &= \tilde{p}_{ij} & (i, \ j \in E_1, \ i \neq j), \\ {}_j \tilde{p}_{ij}^{(n+1)} &= \sum_{k \in E_1 \setminus (j)} \tilde{p}_{ik}\, {}_j \tilde{p}_{kj}^{(n)} & (n = 1, \ 2, \ \cdots; \ i, \ j \in E_1, \ i \neq j), \end{aligned}\right\} \tag{8.4.6}$$

$$\tilde{f}_{ij}^{*} = \begin{cases} 1 & (i, \ j \in E, \ i = j), \\ \sum_{n=1}^{\infty} {}_j \tilde{p}_{ij}^{(n)} & (n = 1, 2, \cdots; i, j \in E_1, \ i \neq j). \end{cases} \tag{8.4.7}$$

It is easy to verify that

$$\tilde{f}_{ij}^{*} = \frac{\alpha_j}{\alpha_i} e_{ji} \qquad (i, \ j \in E_1). \tag{8.4.8}$$

Definition 8.4.1 *The nonnegative functions L_i $(i \in E_1)$ defined on E_1 are said to be standard with respect to $(\alpha, \ P_1)$, if*

$$0 < \sum_{j \in E_1} e_{ij} L_j < +\infty \qquad (i \in E_1), \tag{8.4.9}$$

and

$$\sum_{j\in E_1} \alpha_j L_j < +\infty. \tag{8.4.10}$$

Lemma 8.4.2 *There exists a function L_i $(i \in E_1)$ standard with respect to (α, P_1).*

Proof. By Theorem 7.4.5, there exists a standard measure γ_i $(i \in E_1)$ with respect to $\tilde{P}$, hence

$$0 < \sum_{j\in E_1} \gamma_j \tilde{f}^*_{ji} < +\infty \qquad (i \in E_1). \tag{8.4.11}$$

Let

$$L_j = \frac{\gamma_j}{\alpha_j} \qquad (j \in E_1). \tag{8.4.12}$$

It is easy to verify that L_j $(j \in E_1)$ is a standard function with respect to (α, P_1), proving the lemma. □

Definition 8.4.2 *Suppose v_j $(j \in E)$ is a finite excessive measure of P, β_j $(j \in E_1)$ is a nonnegative function on E_1. If*

$$\sum_{j\in E_1} v_j \beta_j < +\infty, \tag{8.4.13}$$

then we say v_j $(j \in E)$ is β-finite.

Lemma 8.4.3 *Suppose v_j $(j \in E)$ is a finite excessive function of P, then there exists a function L_i $(i \in E_1)$ standard with respect to (α, P_1) such that v_j $(j \in E)$ is L-finite.*

Proof. From Lemma 8.4.2, we see that there exists a function standard with respect to (α, P_1). Suppose $L_j^{(1)}$ $(j \in E_1)$ is a function standard with respect to (α, P_1).

Let

$$L_i = \begin{cases} L_j^{(1)}, & \text{if } v_j = 0, \\ \min(2^{-(j+1)} v_j^{-1}, L_j^{(1)}), & \text{if } v_j > 0. \end{cases} \tag{8.4.14}$$

It is easy to verify that L_j $(j \in E_1)$ is a function standard with respect to (α, P_1) and v_j $(j \in E)$is L-finite. □

Lemma 8.4.4 *If L_j $(j \in E_1)$ is a function standard with respect to (α, P_1), then $\alpha_j L_j (j \in E_1)$ is a measure standard with respect to $\tilde{P}$.*

Proof. By (8.4.8), we get this lemma readily. □

§ 8.5 Entrance boundary

Suppose L_i $(i \in E_1)$ is a function standard with respect to (α, P). From Lemma 8.4.4, we know that $\alpha_j L_j$ $(j \in E_1)$ is a measure standard with respect to $\tilde{P}$. The notations $K(i, j)$, $d(i, j)$, B, etc., concerning the exit boundary of $\tilde{P}$ of §7.7 (the measure standard with respect to P is selected to be $\alpha_i L_i$ $(i \in E_1)$) are replaced now by $K_{\tilde{p}}(i, j)$, $d_{\tilde{p}}(i, j)$, B_{P_1}, etc., respectively.

Let

$$E_{01} = E_0 \backslash E_{00}, \tag{8.5.1}$$

$$\hat{d}(i, j) = \begin{cases} d_{\tilde{p}}(i, j), & i, j \in E_1, \\ 0, & i \in E_{00}, \quad j \in E_{00}, \\ 1, & i \in E_{00}, \quad j \in E_1 \quad \text{if} \quad i \in E_1, \quad j \in E_{00}. \end{cases} \tag{8.5.2}$$

In view of (8.5.2) and §7.7, it is easy to show that the state space E is a metric space with metric $\hat{d}$. By identifying every state in E_∞, and also in each recurrent class E_a $(a \in A)$, and taking the completion of the space with respect to the metric $\hat{d}$, we get a complete metric space $\hat{E}$. $\partial \hat{E} = \hat{E} \backslash E_{01}$ is called the Martin entrance boundary of P. Obviously $\hat{E} \supset E_{\tilde{p}}^*$, $\partial \hat{E} \supset \partial E_{\tilde{p}}$.

§ 8.6 Entrance space and the expressions of excessive measures

Let $\hat{B} = B_{\tilde{p}}$, and call $\hat{B}$ the entrance space of P. Let

$$\hat{K}(i, \xi) = \begin{cases} \alpha_i K_{\tilde{p}}(i, \xi), & i \in E_1, \quad \xi \in E_{01} \bigcup B_{\tilde{p}}, \\ 0, & i \in E_{00}, \quad \xi \in E_{01} \bigcup B_{\tilde{p}}. \end{cases} \tag{8.6.1}$$

Theorem 8.6.1 *Any L-finite excessive measure v_j $(j \in E)$ of P may be uniquely expressed in the following form*

$$v_j = \int_{E_{01} \bigcup \hat{B}} \hat{K}(j, \xi) \mu(d\xi) \qquad (j \in E), \tag{8.6.2}$$

where μ is a finite measure on $E_{01} \bigcup \hat{B}$, and vice versa. The measure becomes harmonic if and only if $\mu(E_{01}) = 0$.

Proof. The theorem follows from Theorem 7.11.1, (8.6.1) and Lemma 7.4.1. □

PART IV

HOMOGENEOUS DENUMERABLE MARKOV PROCESSES

CHAPTER IX
Minimal Q-Processes

§ 9.1 Introduction

Assume that $X(\omega)=\{x(t,\ \omega),\ t<\sigma(\omega)\}$ is a homogeneous denumerable Markov process defined on a complete probability space $(\Omega,\ \mathscr{F},\ P)$, with the denumerable set $E=\{1,2,\cdots\}$ as its minimal state space, $(p_{ij}(t), t\geqslant 0, i,j\in E)$ as its standard transition probability matrix, and assume its Q matrix satisfies the relation:

$$0<q_i\equiv -q_{ii}<+\infty, \qquad \sum_{j\in E} q_{ij}=0 \qquad (i\in E). \tag{9.1.1}$$

Under the above assumptions, we may, without effecting the transition probability matrix, further assume that $X(\omega)$ has the property (D) mentioned in §1.1.

It is well known that the Markov processes with Q as densitymatrix are not necessarily unique, we generally called them Q-processes, and call the part $\{x(t,\ \omega),\ t<\tau(\omega)\}$ before the first infinity τ the minimal Q-process.

The purpose of this chapter is to investigate the problems of transition probability, the distribution and moments of the first passage time, the distribution and moments of functionals of integral-type and classification of states, etc., of the minimal Q-processes.

We assume $q_i>0$ and $\sum_{j\in E} q_{ij}=0$ $(i\in E)$ merely for convenience of discussions. In fact, we can still obtain the corresponding results for the general case of $q_i\geqslant 0$ $(i\in E)$ and $\sum_{j\in E} q_{ij}\leqslant 0$.

§ 9.2 Transition probabilities

The result in this section is nothing new, but the form of statement is different from the usual one. For convenience of reference, we include it.

Suppose A is a nonempty subset of E. Let

$$p_{iA}^{\min}(t)=P(x(t)\in A,\quad t<\tau|x_0=i), \tag{9.2.1}$$

and

$$p_{iA}^{\min}(\lambda) = \int_0^\infty e^{-\lambda t} p_{iA}^{\min}(t)dt, \qquad \lambda \geqslant 0. \tag{9.2.2}$$

Theorem 9.2.1 *$\{p_{iA}^{\min}(\lambda),\ i \in E\}$ is the minimal nonnegative solution of the first-type system of* 1*-bounded equations*

$$x_i = \sum_{k \neq i} \frac{q_{ik}}{\lambda + q_i} x_k + \frac{\delta_{iA}}{\lambda + q_i} \qquad (i \in E). \tag{9.2.3}$$

Proof. It is easy to complete the proof by [1, II, §17]. □

§ 9.3 Distribution and moments of the first passage time

Suppose $\tau^{(n)}(\omega)$ is the time of the nth jump of $x(\cdot,\ \omega)$, i.e., (9.3.1)

$$\tau^{(0)}(\omega) \equiv 0,$$

$$\tau^{(n)}(\omega) = \inf\{t: \quad t > \tau^{(n-1)}(\omega),$$

$$x(t,\ \omega) = x(\tau^{(n-1)}(\omega),\ \omega\}. \tag{9.3.2}$$

Let

$$\tau(\omega) = \mathrm{V} - \lim_{n \to \infty} \tau^{(n)}(\omega) \tag{9.3.3}$$

be the first infinity of $x(\cdot, \omega)$.

Lemma 9.3.1 *$\tau^{(1)}$ and $x_\tau^{(1)}$ are conditionally independent under $x_0 = i$ $(i \in E)$, hence for $j \neq i$ and $q_{ij} \neq 0$, we have*

$$P(\tau^{(1)} \leqslant t | x_0 = i, \quad x_\tau^{(1)} = j) = P(\tau^{(1)} \leqslant t | x_0 = i) = \begin{cases} 0, & t < 0, \\ 1 - e^{-q_i t}, & t \geqslant 0. \end{cases} \tag{9.3.4}$$

For the proof of this lemma, see [11, §2]. □

Suppose $\mathcal{F}_1$, $\mathcal{F}_2$ are two sub-σ-algebras of $\mathcal{F}$, $\Delta \in \mathcal{F}$, $P(\Delta) \neq 0$; $\mathcal{F}_1$ and $\mathcal{F}_2$ are conditionally independent with respect to Δ, i.e., if $\Lambda_1 \in \mathcal{F}_1$ and $\Lambda_2 \in \mathcal{F}_2$, then

$$P(\Lambda_1 \Lambda_2 | \Delta) = Q(\Lambda_1 | \Delta) P(\Lambda_2 | \Delta). \tag{9.3.5}$$

Lemma 9.3.2 *Suppose nonnegative functions X and Y are measurable with respect to $\mathcal{F}_1$ and $\mathcal{F}_2$, respectively, Λ_1, $\Lambda_1' \in \mathcal{F}_1$, Λ_2, $\Lambda_2' \in \mathcal{F}_2$, $P(\Lambda_1' \Delta \Lambda_2') \neq 0$. Then*

$$P(X + Y \leqslant t, \quad \Lambda_1, \quad \Lambda_2 | \Lambda_1' \Delta \Lambda_2')$$

$$= \int_{-0}^{t} P(Y \leqslant t-u, \quad \Lambda_2 | \Lambda_1' \Delta \Lambda_2') dP(X \leqslant u, \quad \Lambda_1 | \Lambda_1' \Delta \Lambda_2'). \tag{9.3.6}$$

Proof. The proof proceeds in the following three steps:

(1) To prove: Λ_1 and Λ_2 are conditionally independent with respect to $\Lambda_1' \Delta$, i.e.,

$$P(\Lambda_1 \Lambda_2 | \Lambda_1' \Delta) = P(\Lambda_1 | \Lambda_1' \Delta) P(\Lambda_2 | \Lambda_1' \Delta). \tag{9.3.7}$$

In fact, since

$$P(\Lambda_1 \Lambda_2 | \Lambda_1' \Delta) = \frac{P(\Lambda_1 \Lambda_2 \Lambda_1' \Delta)}{P(\Lambda_1' \Delta)}$$

$$= \frac{P(\Lambda_1 \Lambda_1' | \Delta \Lambda_2) P(\Delta \Lambda_2)}{P(\Delta) P(\Lambda_1' | \Delta)}$$

$$= \frac{P(\Lambda_1 \Lambda_1' | \Delta \Lambda_2)}{P(\Lambda_1' | \Delta)} \cdot \frac{P(\Delta \Lambda_2)}{P(\Delta)}$$

$$= P(\Lambda_1 | \Lambda_1' \Delta) P(\Lambda_2 | \Delta) = P(\Lambda_1 | \Lambda_1' \Delta) P(\Lambda_2 | \Lambda_1' \Delta), \tag{9.3.8}$$

consequently (9.3.7) holds.

(2) By using the result of (1) twice, we may obtain that Λ_1 and Λ_2 are conditionally independent with respect to $\Lambda_1' \Delta \Lambda_2'$, that is,

$$P(\Lambda_1 \Lambda_2 | \Lambda_1' \Delta \Lambda_2') = P(\Lambda_1 | \Lambda_1' \Delta \Lambda_2') P(\Lambda_2 | \Lambda_1' \Delta \Lambda_2'). \tag{9.3.9}$$

(3) We can now prove (9.3.6).

(A) Suppose

$$P(\Lambda_1 \Lambda_1' \Delta \Lambda_2 \Lambda_2') = 0. \tag{9.3.10}$$

Then

$$P(X + Y \leqslant t, \quad \Lambda_1, \quad \Lambda_2 | \Lambda_1' \Delta \Lambda_2') = 0. \tag{9.3.11}$$

It follows from (2) and (9.3.10) that

$$P(\Lambda_1 | \Lambda_1' \Delta \Lambda_2') P(\Lambda_2 | \Lambda_1' \Delta \Lambda_2') = P(\Lambda_1 \Lambda_2 | \Lambda_1' \Delta \Lambda_2') = 0. \tag{9.3.12}$$

Consequently,

$$P(\Lambda_1 | \Lambda_1' \Delta \Lambda_2') = 0, \tag{9.3.13}$$

or

$$P(\Lambda_2|\Lambda_1'\Delta\Lambda_2') = 0. \tag{9.3.14}$$

If (9.3.13) holds, then for any $u \in (-\infty, +\infty)$ we have

$$P(X \leqslant u, \quad \Lambda_1|\Lambda_1'\Delta\Lambda_2') = 0. \tag{9.3.15}$$

Thus

$$\int_{-0}^{t} P(Y \leqslant t-u, \quad \Lambda_2|\Lambda_1'\Delta\Lambda_2')dP(X \leqslant u, \quad \Lambda_1|\Lambda_1'\Delta\Lambda_2') = 0. \tag{9.3.16}$$

If (9.3.14) holds, then for arbitrary t, $u \in (-\infty, \ +\infty)$ we have

$$P(Y \leqslant t-u, \quad \Lambda_2|\Lambda_1'\Delta\Lambda_2') = 0. \tag{9.3.17}$$

Thus (9.3.16) also holds. It can be seen from (9.3.11) and (9.3.16) that (9.3.6) holds.

(B) Suppose

$$P(\Lambda_1\Lambda_2\Delta\Lambda_1'\Lambda_2') \neq 0. \tag{9.3.18}$$

It follows from (2) that

$$\begin{aligned}
&P(X+Y \leqslant t, \quad \Lambda_1, \quad \Lambda_2|\Lambda_1'\Delta\Lambda_2')\\
&= P(\Lambda_1\Lambda_2|\Lambda_1'\Delta\Lambda_2')P(X+Y \leqslant t|\Lambda_1\Lambda_1'\Delta\Lambda_2\Lambda_2')\\
&= P(\Lambda_1|\Lambda_1'\Delta\Lambda_2')P(\Lambda_2|\Lambda_1'\Delta\Lambda_2')\\
&\quad \times \int_{-0}^{t} p(Y \leqslant t-u|\Lambda_1\Lambda_1'\Delta\Lambda_2\Lambda_2')dp(\Lambda_1\Lambda_1'\Delta\Lambda_2\Lambda_2')\\
&= \int_{-0}^{t} P(\Lambda_2|\Lambda_1'\Delta\Lambda_2')P(Y \leqslant t-u|\Lambda_1'\Delta\Lambda_2\Lambda_2')\\
&\quad \times dP(\Lambda_1|\Lambda_1'\Delta\Lambda_2')P(X \leqslant u|\Lambda_1\Lambda_1'\Delta\Lambda_2')\\
&= \int_{-0}^{t} P(Y \leqslant t-u, \quad \Lambda_2|\Lambda_1'\Delta\Lambda_2')dP(X \leqslant u, \quad \Lambda_1|\Lambda_1'\Delta\Lambda_2').
\end{aligned} \tag{9.3.19}$$

Therefore (9.3.6) holds. □

Suppose $F_i(t)$ $(i = 1, \ 2)$ is a nonnegative nondecreasing right continuous function defined on $(-\infty, \ +\infty)$, $F_i(t) = 0$ $(i = 1, \ 2)$ for $t < 0$, and

$$F_i(+\infty) = \lim_{t\to+\infty} F_i(t) \leqslant 1 \qquad (i = 1,\ 2). \tag{9.3.20}$$

Let

$$F(t) = \int_{-0}^{t} F_1(t-u)dF_2(u) \qquad (t \geqslant 0), \tag{9.3.21}$$

$$\varphi_i(\lambda) = \int_{-0}^{\infty} e^{-\lambda t} dF_i(t) \qquad (i = 1,\ 2,\quad \lambda \geqslant 0), \tag{9.3.22}$$

$$\varphi(\lambda) = \int_{-0}^{\infty} e^{-\lambda t} dF(t) \qquad (\lambda \geqslant 0), \tag{9.3.23}$$

$$m_i^{(p)} = \int_{-0}^{\infty} t^p dF_i(t) \qquad (i = 1,\ 2,\quad p = 0,\ 1,\ \cdots), \tag{9.3.24}$$

$$m^{(p)} = \int_{-0}^{\infty} t^p dF(t) \qquad (p = 0,\ 1,\ \cdots). \tag{9.3.25}$$

It is easy to prove the following two lemmas.

Lemma 9.3.3

$$\varphi(\lambda) = \varphi_1(\lambda)\varphi_2(\lambda). \quad \square \tag{9.3.26}$$

Lemma 9.3.4

$$m^{(p)} = \sum_{l=0}^{p} C_p^l m_1^{(l)} m_2^{(p-l)} \qquad (p = 0,\ 1,\ \cdots). \quad \square \tag{9.3.27}$$

Suppose A and H are subsets of E, and $A \neq \emptyset$. Let

$$\sigma_A(\omega) = \begin{cases} \inf\{t\colon \tau^{(1)}(\omega) \leqslant t < \tau(\omega),\ x_t(\omega) \in A\}, & \text{if this set is nonempty,} \\ +\infty, & \text{otherwise,} \end{cases} \tag{9.3.28}$$

$${}_H\sigma_A(\omega) = \begin{cases} \sigma_A(\omega), & \text{if } \sigma_A(\omega) \leqslant \sigma_H(\omega), \\ +\infty, & \text{otherwise,} \end{cases} \tag{9.3.29}$$

$${}_Hf_{iA}^{(n)}(t) = P({}_H\sigma_A(\omega) \leqslant t,\ {}_H\sigma_A(\omega) = \tau^{(n)}(\omega) | x_0 = i) \qquad (n = 1,\ 2,\ \cdots), \tag{9.3.30}$$

$${}_Hf_{iA}(t) = P({}_H\sigma_A(\omega) \leqslant t | x_0 = i), \tag{9.3.31}$$

$${}_H\Phi_{iA}^{(n)}(\lambda) = \int_{-0}^{\infty} e^{-\lambda t} d_H f_{iA}^{(n)}(t) \qquad (\lambda \geqslant 0), \tag{9.3.32}$$

$$ {}_H\Phi_{iA}(\lambda) = \int_{-0}^{\infty} e^{-\lambda t} d_H f_{iA}(t) \qquad (\lambda \geqslant 0), \tag{9.3.33}$$

$$ {}_H m_{iA}^{(n,p)} = \int_{-0}^{\infty} t^p d_H f_{iA}^{(n)}(t) \qquad (p = 0, \quad 1, \quad \cdots), \tag{9.3.34}$$

$$ {}_H m_{iA}^{(p)} = \int_{-0}^{\infty} t^p d_H f_{iA}(t) \qquad (p = 0, \quad 1, \quad \cdots), \tag{9.3.35}$$

$$ {}_H f_{iA}^{*} = {}_H\Phi_{iA}(0) = {}_H m_{iA}^{(0)} = P({}_H\sigma_A(\omega) < +\infty | x_0 = i). \tag{9.3.36}$$

Lemma 9.3.5

$$P(x_\tau^{(1)} = j, \tau^{(1)} \leqslant t | x_0 = i) = \frac{q_{ij}}{q_i} P(\tau^{(1)} \leqslant t | x_0 = i) \qquad (j \neq i). \tag{9.3.37}$$

Proof. Obviously (9.3.37) holds provided $q_{ij} = 0$, since in this case both sides of (9.3.37) are equal to zero. (9.3.37) follows readily from Lemma 9.3.1 when $q_{ij} \neq 0$, proving the lemma. □

Lemma 9.3.6

$$P(x_{\tau^{(1)}} = j, \ x_{\tau^{(\nu)}} \notin A \bigcup H, \ 1 < \nu < n+1, \ x_{\tau^{(n+1)}} \in A, \ \tau^{(n+1)} \leqslant t | x_0 = i)$$

$$= \frac{q_{ij}}{q_i} \int_{-0}^{t} {}_H f_{jA}^{(n)}(t-u) dp(\tau^{(1)} \leqslant u | x_0 = i) \qquad (j \neq i). \tag{9.3.38}$$

Proof. Obviously (9.3.38) holds provided $q_{ij} = 0$, since in this case both sides of (9.3.38) equal zero. When $q_{ij} \neq 0$, it follows from Lemmas 9.3.1 and 9.3.2, the homogeneity and strong Markov property that

$$P(x_{\tau^{(1)}} = j, \ x_{\tau^{(\nu)}} \notin A \bigcup H, \quad 1 < \nu < n+1, \quad x_{\tau^{(n+1)}} \in A, \quad \tau^{(n+1)} \leqslant t | x_0 = i)$$

$$= \frac{q_{ij}}{q_i} P(x_{\tau^{(\nu)}} \notin A \bigcup H, \quad 1 < \nu < n+1, \quad x_{\tau^{(n+1)}} \in A,$$

$$(\tau^{(n+1)} - \tau^{(1)}) + \tau^{(1)} \leqslant t | x_0 = i, \quad x_{\tau^{(1)}} = j)$$

$$= \frac{q_{ij}}{q_i} \int_{-0}^{t} P(x_{\tau^{(\nu)}} \notin A \bigcup H, \quad 1 < \nu < n+1, \quad x_{\tau^{(n+1)}} \in A,$$

$$\tau^{(n+1)} - \tau^{(1)} \leqslant t - u | x_0 = i, \quad x_{\tau^{(1)}} = j)$$

$$\times \, dP(\tau^{(1)} \leqslant u | x_0 = i, \quad x_{\tau^{(1)}} = j)$$

$$= \frac{q_{ij}}{q_i} \int_{-0}^{t} P\big(x_{\tau^{(\nu)}} \notin A \cup H, \quad 0 < \nu < n, \quad x_{\tau^{(n)}} \in A,$$

$$\tau^{(n)} \leqslant t - u \,|\, x_0 = j\big) \cdot dP\big(\tau^{(1)} \leqslant u \,|\, x_0 = i\big)$$

$$= \frac{q_{ij}}{q_i} \int_{-0}^{t} P\big({}_H\sigma_A \leqslant t - u, \quad {}_H\sigma_A = \tau^{(n)} \,|\, x_0 = j\big)$$

$$\times \, dP\big(\tau^{(1)} \leqslant u \,|\, x_0 = i\big)$$

$$= \frac{q_{ij}}{q_i} \int_{-0}^{t} {}_H f_{jA}^{(n)}(t - u) dP\big(\tau^{(1)} \leqslant u \,|\, x_0 = i\big). \tag{9.3.39}$$

Therefore this lemma is proved. □

Lemma 9.3.7 ${}_H f_{iA}^{(n)}(t)$ *is determined uniquely by the following recursion formulae*

$$\left. \begin{aligned} {}_H f_{iA}^{(1)}(t) &= \sum_{j \in A \setminus \{i\}} \frac{q_{ij}}{q_i} P\big(\tau^{(1)} \leqslant t \,|\, x_0 = i\big) \qquad (i \in E), \\ {}_H f_{iA}^{(n+1)}(t) &= \sum_{j \notin \{i\} \cup A \cup H} \frac{q_{ij}}{q_i} \int_{-0}^{t} {}_H f_{jA}^{(n)}(t - u) \\ &\quad \times dP\big(\tau^{(1)} \leqslant u \,|\, x_0 = i\big) \qquad (n \geqslant 1, \quad i \in E). \end{aligned} \right\} \tag{9.3.40}$$

Proof. It follows immediately from Lemma 9.3.5 that

$${}_H f_{iA}^{(1)}(t) = P\big({}_H\sigma_A \leqslant t, \quad {}_H\sigma_A = \tau^{(1)} \,|\, x_0 = i\big)$$

$$= P\big(x_{\tau^{(1)}} \in A, \quad \tau^{(1)} \leqslant t \,|\, x_0 = i\big)$$

$$= \sum_{j \in A \setminus \{i\}} P\big(x_{\tau^{(1)}} = j, \quad \tau^{(1)} \leqslant t \,|\, x_0 = i\big)$$

$$= \sum_{j \in A \setminus \{i\}} \frac{q_{ij}}{q_i} P\big(\tau^{(1)} \leqslant t \,|\, x_0 = i\big). \tag{9.3.41}$$

Owing to Lemma 9.3.6, it can be seen that

$${}_H f_{iA}^{(n+1)}(t) = P\big({}_H\sigma_A \leqslant t, \quad {}_H\sigma_A = \tau^{(n+1)} \,|\, x_0 = i\big)$$

$$= P\big(x_{\tau^{(\nu)}} \notin A \cup H, \; 0 < \nu < n+1, \; x_{\tau^{(n+1)}} \in A, \; \tau^{(n+1)} \leqslant t \,|\, x_0 = i\big)$$

$$= \sum_{j\notin\{i\}\cup A\cup H} P(x_{\tau^{(1)}} = j,\quad x_{\tau^{(v)}} \notin A \cup H,$$

$$1 < v < n+1,\quad x_{\tau^{(n+1)}} \in A,\quad \tau^{(n+1)} \leqslant t | x_0 = i)$$

$$= \sum_{j\notin\{i\}\cup A\cup H} \frac{q_{ij}}{q_i} \int_{-0}^{t} {}_H f_{jA}^{(n)}(t-u) dP(\tau^{(1)} \leqslant u | x_0 = i). \ \square \qquad (9.3.42)$$

Lemma 9.3.8 ${}_H\Phi_{iA}^{(n)}(\lambda)$ *is determined uniquely by the following recursion formulae*

$$\left.\begin{aligned} {}_H\Phi_{iA}^{(1)}(\lambda) &= \sum_{j\in A\setminus\{i\}} \frac{q_{ij}}{\lambda + q_i} \qquad (i \in E), \\ {}_H\Phi_{iA}^{(n+1)}(\lambda) &= \sum_{j\notin\{i\}\cup A\cup H} \frac{q_{ij}}{\lambda + q_i}\, {}_H\Phi_{jA}^{(n)}(\lambda) \qquad (n \geqslant 1,\quad i \in E). \end{aligned}\right\} \qquad (9.3.43)$$

Proof. Taking Laplace-Stieltjes transformations on both sides of (9.3.40) and taking account of Lemmas 9.3.1 and 9.3.3, we obtain the lemma immediately. □

Lemma 9.3.9 ${}_H m_{iA}^{(n,p)}$ *is determined uniquely by the following recursion formulae*

$$\left.\begin{aligned} {}_H m_{iA}^{(1,p)} &= p!\left(\frac{1}{q_i}\right)^p \sum_{j\in A\setminus\{i\}} \frac{q_{ij}}{q_i} \qquad (i \in E), \\ {}_H m_{iA}^{(n+1,p)} &= \sum_{l=0}^{p} C_p^l\, l! \left(\frac{1}{q_i}\right)^l \sum_{j\notin\{i\}\cup A\cup H} \frac{q_{ij}}{q_i}\, {}_H m_{jA}^{(n,p-l)} \qquad (n \geqslant 1, i \in E). \end{aligned}\right\} \qquad (9.3.44)$$

Proof. By virtue of Lemmas 9.3.1, 9.3.4 and 9.3.7, we obtain this lemma. □

Lemma 9.3.10

$${}_H f_{iA}(t) = \sum_{n=1}^{\infty} {}_H f_{iA}^{(n)}(t), \qquad (9.3.45)$$

$${}_H\Phi_{iA}(\lambda) = \sum_{n=1}^{\infty} {}_H\Phi_{iA}^{(n)}(\lambda), \qquad (9.3.46)$$

$${}_H m_{iA}^{(p)} = \sum_{n=1}^{\infty} {}_H m_{iA}^{(n,p)}. \qquad (9.3.47)$$

Proof. This lemma follows from (9.3.30) ~ (9.3.35). □

Theorem 9.3.1 $\{{}_H\Phi_{iA}(\lambda),\ i \in E\}$ *is the minimal nonnegative solution of the pseudo-normal system of equations*

$$x_i = \sum_{j \notin \{i\} \cup A \cup H} \frac{q_{ij}}{\lambda + q_i} x_j + \sum_{j \in A \setminus \{i\}} \frac{q_{ij}}{\lambda + q_i} \qquad (i \in E). \tag{9.3.48}$$

Proof. The assertion follows readily from (9.3.46), Lemma 9.3.8 and Corollary 3.2.3. □

Theorem 9.3.2 *$\{_H f_{iA}^*, \ i \in E\}$ is the minimal nonnegative solution of the pseudo-normal system of equations*

$$x_i = \sum_{j \notin \{i\} \cup A \cup H} \frac{q_{ij}}{q_i} x_j + \sum_{j \in A \setminus \{i\}} \frac{q_{ij}}{q_i} \qquad (i \in E). \tag{9.3.49}$$

Proof. This theorem is a special case of Theorem 9.3.1 for $\lambda = 0$. □

Theorem 9.3.3 *For $p \geqslant 1$, $\{_H m_{iA}^{(p)}, \ i \in E\}$ is the minimal nonnegative solution of the first-type system of 1-bounded equations*

$$x_i = \sum_{j \notin \{i\} \cup A \cup H} \frac{q_{ij}}{q_i} x_j + \frac{p}{q_i} {}_H m_{iA}^{(p-1)} \qquad (i \in E). \tag{9.3.50}$$

Proof. We know from (9.3.47), Lemma 9.3.9 and Corollary 3.2.2 that $\{_H m_{iA}^{(p)}, \ i \in E\}$ is the minimal nonnegative solution of the first-type system of 1-bounded equations

$$x_i = \sum_{j \notin \{i\} \cup A \cup H} \frac{q_{ij}}{q_i} x_j + \sum_{l=1}^{p} C_p^l l! \left(\frac{1}{q_i}\right)^l \sum_{j \notin \{i\} \cup A \cup H} \frac{q_{ij}}{q_i} {}_H m_{jA}^{(p-l)}$$
$$+ p! \left(\frac{1}{q_i}\right)^p \sum_{j \in A \setminus \{i\}} \frac{q_{ij}}{q_i} \qquad (i \in E). \tag{9.3.51}$$

Hence

$${}_H m_{iA}^{(p)} = \sum_{l=0}^{p} C_p^l l! \left(\frac{1}{q_i}\right)^l \sum_{j \notin \{i\} \cup A \cup H} \frac{q_{ij}}{q_i} {}_H m_{jA}^{(p-l)}$$
$$+ p! \left(\frac{1}{q_i}\right)^p \sum_{j \in A \setminus \{i\}} \frac{q_{ij}}{q_i} \qquad (p = 1, 2, \cdots, \ i \in E). \tag{9.3.52}$$

Consequently

$$\sum_{l=1}^{p} C_p^l l! \left(\frac{1}{q_i}\right)^l \sum_{j \notin \{i\} \cup A \cup H} \frac{q_{ij}}{q_i} {}_H m_{jA}^{(p-l)} + p! \left(\frac{1}{q_i}\right)^p \sum_{j \in A \setminus \{i\}} \frac{q_{ij}}{q_i}$$
$$= \frac{p}{q_i} \left(\sum_{l=1}^{p} C_{p-1}^{l-1} (l-1)! \left(\frac{1}{q_i}\right)^{l-1} \sum_{j \notin \{i\} \cup A \cup H} \frac{q_{ij}}{q_i} {}_H m_{jA}^{((p-1)(l-1))}\right.$$

$$+(p-1)!\left(\frac{1}{q_i}\right)^{p-1}\sum_{j\in A\setminus\{i\}}\frac{q_{ij}}{q_i}\Bigg)$$

$$=\frac{p}{q_i}\left(\sum_{l=0}^{p-1}C_{p-1}^{l}\,l!\left(\frac{1}{q_i}\right)^{l}\sum_{j\notin\{i\}\cup A\cup H}\frac{q_{ij}}{q_i}\,{}_Hm_{jA}^{(p-1-l)}\right.$$

$$\left.+(p-1)!\left(\frac{1}{q_i}\right)^{p-1}\sum_{j\in A\setminus\{i\}}\frac{q_{ij}}{q_i}\right)=\frac{p}{q_i}\,{}_Hm_{iA}^{(p-1)}.\ \square \qquad (9.3.53)$$

Theorem 9.3.4 *If*

$$_Hm_{iA}\leqslant c_H f_{iA}^{*}\qquad (i\in E), \qquad (9.3.54)$$

then

$$_Hm_{iA}^{(p)}\leqslant p!\,c^{p}\,{}_Hf_{iA}^{*}\qquad (p\geqslant 1,\quad i\in E). \qquad (9.3.55)$$

Proof. It is similar to the proof of Theorem 6.3.4. □

§ 9.4 Criterion for the positive recurrence

Let

$$r_{ij}=\begin{cases}0, & i=j,\\ \dfrac{q_{ij}}{q_i}, & i\neq j,\end{cases} \qquad (9.4.1)$$

$$R=(r_{ij},\ i,\ j\in E). \qquad (9.4.2)$$

Suppose s is an element of E. Let

$$D(s)=\{i: s\underset{R}{\rightsquigarrow} i\}\setminus\{s\}. \qquad (9.4.3)$$

Definition 9.4.1 *If*

$$f_{ss}^{*}=1, \qquad (9.4.4)$$

then s is called a recurrent state of the minimal Q-process $\{x(t,\ \omega),\ t<\tau(\omega)\}$. If s is a recurrent state of the minimal Q-process $\{x(t,\ \omega),\ t<\tau(\omega)\}$ and

$$m_{ss}<+\infty, \qquad (9.4.5)$$

then s is called a positively recurrent state of $\{x(t,\ \omega),\ t<\tau(\omega)\}$.

Lemma 9.4.1 *If s is a recurrent state of $\{x(t, \omega), t < \tau(\omega)\}$, then $\{m_{is}, i \in D(s)\}$ is the minimal nonnegative solution of the first-type system of l-bounded equations*

$$x_i = \sum_{j \in D(s) \setminus \{i\}} \frac{q_{ij}}{q_i} x_j + \frac{1}{q_i} \qquad (i \in D(s)), \tag{9.4.6}$$

and

$$m_{ss} = \sum_{j \in D(s)} \frac{q_{si}}{q_s} m_{js} + \frac{1}{q_s}. \tag{9.4.7}$$

Proof. If s is a recurrent state of $\{x(t, \omega), t < \tau(\omega)\}$, then

$$f_{ss}^* = 1. \tag{9.4.8}$$

We know from (9.4.8), Theorems 9.3.2 and 5.6.2 that

$$f_{is}^* = 1 \qquad (i \in D(s)). \tag{9.4.9}$$

The lemma can be deduced from (9.4.8), (9.4.9), Theorem 9.3.3 and Corollary 3.4.1. □

The criteria for recurrence and positive recurrence of the states of a minimal Q-process were given in [17]. In some special cases, by using the results of the above section, we may practically calculate f_{ss}^* and m_{ss}^* so as to determine whether s is recurrent and positively recurrent. We will not go further into this, but only give a new "criterion for positive recurrence" here.

Theorem 9.4.1 *If s is a recurrent state of a minimal Q-process $\{x(t, \omega), t < \tau(\omega)\}$, then a necessary and sufficient condition for s to be a positively recurrent state is, that the system of inequalities*

$$x_i \geqslant \sum_{j \in D(s) \setminus \{i\}} \frac{q_{ij}}{q_i} x_j + \frac{1}{q_i} \qquad (i \in D(s)) \tag{9.4.10}$$

has a nonnegative solution $x_i (i \in D(s))$ satisfying the condition

$$\sum_{j \in D(s)} \frac{q_{si}}{q_s} x_j < +\infty. \tag{9.4.11}$$

Proof. By Lemma 9.4.1 and Corollary 3.3.2, we get the theorem easily. □

§ 9.5 Distribution and moments of integral-type functionals

Suppose A is a subset of E (A may be empty. If it is, A will be omitted from the notation), and $H_N = \{N+1, N+2, \cdots\}$. Let

$$\tau_A(\omega) = \begin{cases} \inf\{t: \tau^{(1)}(\omega) < t < \tau(\omega), x(t, \omega) \in A\}, \\ \tau(\omega), \text{ if the above set is empty,} \end{cases} \tag{9.5.1}$$

and

$$\tau_A^{(n)}(\omega) = \min(\tau^{(n)}(\omega), \tau_A(\omega)). \tag{9.5.2}$$

Obviously $\tau_A^{(n)}(\omega)$ and $\tau_A(\omega)$ are optional random variables.

Suppose $v(i)$ $(i \in E)$ is a nonnegative finite-valued function on E. Let

$$\xi_A^{(n)}(\omega) = \int_0^{\tau_A^{(n)}(\omega)} v(x(t, \omega))dt \tag{9.5.3}$$

and

$$\xi_A(\omega) = \int_0^{\tau_A(\omega)} v(x(t, \omega))dt. \tag{9.5.4}$$

It is easy to show that $\xi_A^{(n)}(\omega)$ and $\xi_A(\omega)$ are random variables and

$$\xi_A^{(n)}(\omega) \uparrow \xi_A(\omega) \qquad (n \uparrow +\infty), \tag{9.5.5}$$

$$\xi_{A \cup H_N}(\omega) \uparrow \xi_A(\omega) \qquad (N \uparrow +\infty) \tag{9.5.6}$$

with probability one. Let

$$F_{iA}^{(n)}(t) = P(\xi_A^{(n)}(\omega) \leqslant t \mid x_0 = i), \tag{9.5.7}$$

$$F_{iA}(t) = P(\xi_A(\omega) \leqslant t \mid x_0 = i), \tag{9.5.8}$$

$$\varphi_{iA}^{(n)}(\lambda) = \int_0^\infty e^{-\lambda t} dF_{iA}^{(n)}(t) \qquad (\lambda > 0), \tag{9.5.9}$$

$$\varphi_{iA}(\lambda) = \int_0^\infty e^{-\lambda t} dF_{iA}(t) \qquad (\lambda > 0), \tag{9.5.10}$$

$$\psi_{iA}^{(n)}(\lambda) = 1 - \varphi_{iA}^{(n)}(\lambda), \tag{9.5.11}$$

$$\psi_{iA}(\lambda) = 1 - \varphi_{iA}(\lambda), \tag{9.5.12}$$

$$T_{iA}^{(n,p)} = M\{[\xi_A^{(n)}(\omega)]^p | x_0 = i\}, \tag{9.5.13}$$

$$T_{iA}^{(p)} = M\{[\xi_A(\omega)]^p | x_0 = i\}, \tag{9.5.14}$$

where $p = 0, 1, \cdots$. Obviously

$$T_{iA}^{(n,0)} = T_{iA}^{(0)} = 1. \tag{9.5.15}$$

Henceforth we stipulate that $p = 1, 2, \cdots$.

Sometimes we write T_{iA} for $T_{iA}^{(1)}$. As mentioned above, if $A = \varnothing$, then A may be omitted from the notations. For example, $T_{i\varnothing}^{(p)} = T_i^{(p)}$.

From (9.5.5) and (9.5.6) we obtain

$$\psi_{iA}^{(n)}(\lambda) \uparrow \psi_{iA}(\lambda) \qquad (n \uparrow +\infty), \tag{9.5.16}$$

$$\psi_{iA \cup H_N}(\lambda) \uparrow \psi_{iA}(\lambda) \qquad (N \uparrow +\infty), \tag{9.5.17}$$

$$T_{iA}^{(n,p)} \uparrow T_{iA}^{(p)} \qquad (n \uparrow +\infty), \tag{9.5.18}$$

$$T_{iA \cup H_N}^{(p)} \uparrow T_{iA}^{(p)} \qquad (N \uparrow +\infty). \tag{9.5.19}$$

Let

$$\chi_n(\omega) = x(\tau^{(n)}(\omega), \omega), \tag{9.5.20}$$

$$\hat{\tau}_A(\omega) = \begin{cases} \inf\{n:\ \chi_n(\omega) \in A,\ n \geqslant 1\}, \\ +\infty, \quad \text{if the above set is empty.} \end{cases} \tag{9.5.21}$$

$$\hat{\tau}_A^{(n)}(\omega) = \min(n,\ \hat{\tau}_A(\omega)), \tag{9.5.22}$$

$$\hat{\xi}_A(\omega) = \sum_{k=1}^{\hat{\tau}_A^{(n)}(\omega)} \frac{v(\chi_{k-1}(\omega))}{q_{\chi_{k-1}(\omega)}}, \tag{9.5.23}$$

$$\hat{F}_{iA}^{(n)}(t) = p(\hat{\xi}_A(\omega) \leqslant t | x_0 = i), \tag{9.5.24}$$

$$\hat{F}_{iA}(t) = p(\hat{\xi}_A(\omega) \leqslant t | x_0 = i), \tag{9.5.25}$$

$$\hat{\varphi}_{iA}^{(n)}(\lambda) = \int_0^\infty e^{-\lambda t} d\hat{F}_{iA}^{(n)}(t), \tag{9.5.26}$$

$$\hat{\varphi}_{iA}(\lambda) = \int_0^\infty e^{-\lambda t} d\hat{F}_{iA}(t), \tag{9.5.27}$$

$$\hat{T}_{iA}^{(p)} = M\{[\hat{\xi}_A(\omega)]^p | x_0 = i\} \qquad (p = 0, 1, \cdots). \tag{9.5.28}$$

The investigation of distribution and moments of an integral-type functional $\xi_A(\omega)$ (the aim of introducing $\xi_A^{(n)}(\omega)$ is for consideration of $\xi_A(\omega)$, particularly $\xi(\omega)$) is due to Z. K. Wang who successfully solved the calculation of the distribution and moments of $\xi_{H_N}(\omega)$ and $\xi(\omega)$ for birth and death processes in [18]. Thereafter L. D. Wu [19] and C. Q. Yang [11] further studied the problem for more general and the most general homogeneous denumerable Markov processes respectively, and obtained some intermediate results. Starting from their works we have, in this chapter, solved thoroughly the calculation of the distribution and moments of $\xi_A(\omega)$, and their relations with distribution and moments of $\hat{\xi}_A(\omega)$ by using "the theory of the minimal nonnegative solution of system of nonnegative linear equations".

Lemma 9.5.1 *$\psi_{iA}^{(n)}(\lambda)$ is determined uniquely by the following recursion formulae:*

$$\left.\begin{aligned} &\psi_{iA}^{(0)}(\lambda) \equiv 0 \qquad (i \in E), \\ &\psi_{iA}^{(n+1)}(\lambda) = \sum_{j \notin \{i\} \cup A} \frac{q_{ij}}{\lambda v(i) + q_i} \psi_{jA}^{(n)}(\lambda) + \frac{\lambda v(i)}{\lambda v(i) + q_i} \qquad (n \geq 0, \quad i \in E). \end{aligned}\right\} \tag{9.5.29}$$

Proof. By (9.5.11), it suffices to prove that $\varphi_{iA}^{(n)}(\lambda)$ satisfy the following recursion formulae:

$$\left.\begin{aligned} &\varphi_{iA}^{(0)}(\lambda) \equiv 1 \qquad (i \in E), \\ &\varphi_{iA}^{(n+1)}(\lambda) = \sum_{j \in \{i\} \cup A} \frac{q_{ij}}{\lambda v(i) + q_i} \varphi_{jA}^{(n)}(\lambda) \\ &\qquad + \sum_{j \in A \setminus \{i\}} \frac{q_{ij}}{\lambda v(i) + q_i} \qquad (n \geq 0, \ i \in E). \end{aligned}\right\} \tag{9.5.30}$$

And this is similar to the relevant part of [11, Theorem 2.1], so its proof is omitted here. □

Lemma 9.5.2 *$T_{iA}^{(n,p)}$ is determined uniquely by the following recursion formulae:*

$$\left.\begin{aligned} &T_{iA}^{(1,p)} = p! \left[\frac{v(i)}{q_i}\right]^p \qquad (i \in E), \\ &T_{iA}^{(n+1,p)} = \sum_{l=0}^{p} C_p^l \sum_{j \notin \{i\} \cup A} \frac{q_{ij}}{q_i} l! \left[\frac{v(i)}{q_i}\right]^l T_{jA}^{(n,p-l)} \\ &\qquad + \sum_{j \in A \setminus \{i\}} \frac{q_{ij}}{q_i} p! \left[\frac{v(i)}{q_i}\right]^p \qquad (n \geq 1, \ i \in E). \end{aligned}\right\} \tag{9.5.31}$$

Proof. The proof is similar to that of [11, Theorem 3.1]. □

Theorem 9.5.1 *$\{\psi_{iA}(\lambda),\ i\in E\}$ is the minimal nonnegative solution of the pseudo-normal system of equations*

$$x_i = \sum_{j\notin\{i\}\cup A} \frac{q_{ij}}{\lambda v(i) + q_i} x_j + \frac{\lambda v(i)}{\lambda v(i) + q_i} \qquad (i\in E). \tag{9.5.32}$$

Proof. The assertion follows from Lemma 9.5.1 and Theorem 3.2.1. □

Theorem 9.5.2 *For $p \geqslant 1$, $\{T_{iA}^{(p)},\ i\in E\}$ is the minimal nonnegative solution of the first-type system of 1-bounded equations*

$$x_i = \sum_{j\notin\{i\}\cup A} \frac{q_{ij}}{q_i} x_j + \frac{pv(i)}{q_i} T_{iA}^{(p-1)} \qquad (i\in E). \tag{9.5.33}$$

Proof. It is easy to conclude this theorem by (9.5.18), Lemma 9.5.2, Theorem 3.2.2 and referring to the proof of Theorem 9.3.3. □

Theorem 9.5.3 *If*

$$T_{iA} \leqslant c < +\infty \qquad (i\in E), \tag{9.5.34}$$

then

$$T_{iA}^{(p)} \leqslant p!\,c^p \qquad (i\in E). \tag{9.5.35}$$

Proof. The proof is similar to that of Theorem 6.7.6. □

Remark 9.5.1 The condition (9.5.34) in Theorem 9.5.3 may be satisfied in some cases, for instance, in the case of birth and death processes [18]. In fact, it is not difficult to prove from [18, Theorem 4] and [20, Theorem 5] that the condition (9.5.34) holds when E can be partitioned into a finite number of atomic kernels.

Theorem 9.5.4 *We let*

$$F_{iA}(+\infty) = \lim_{t\to+\infty} F_{iA}(t) \tag{9.5.36}$$

and

$$\hat{F}_{iA}(+\infty) = \lim_{t\to+\infty} \hat{F}_{iA}(t), \tag{9.5.37}$$

then

$$F_{iA}(+\infty) = \hat{F}_{iA}(+\infty). \tag{9.5.38}$$

Proof. By virtue of (9.5.30) we have

$$\varphi_{iA}^{(0)}(\lambda) = 1 \qquad (i \in E),$$

$$\varphi_{iA}^{(n+1)}(\lambda) = \sum_{j \notin \{i\} \cup A} \frac{q_{ij}}{q_i}\left(1 + \frac{\lambda v(i)}{q_i}\right)^{-1} \varphi_{jA}^{(n)}(\lambda) \tag{9.5.39}$$

$$+ \sum_{j \in A \setminus \{i\}} \frac{q_{ij}}{q_i}\left(1 + \frac{\lambda v(i)}{q_i}\right)^{-1} \qquad (n \geqslant 0,\ i \in E).$$

And from Theorem 6.7.1 we deduce

$$\hat{\varphi}_{iA}^{(0)}(\lambda) = 1 \qquad (i \in E),$$

$$\hat{\varphi}_{iA}^{(n+1)}(\lambda) = \sum_{j \notin \{i\} \cup A} \frac{q_{ij}}{q_i} e^{-\lambda v(i)/q} \varphi_{jA}^{(n)}(\lambda) \tag{9.5.40}$$

$$+ \sum_{j \in A \setminus \{i\}} \frac{q_{ij}}{q_i} e^{-\lambda v(i)/q_i} \qquad (n \geqslant 0,\ i \in E).$$

It follows from (9.5.39), (9.5.40) and

$$\left(1 + \frac{\lambda v(i)}{q_i}\right)^{-1} \geqslant e^{-\lambda v(i)/q_i} \qquad (\lambda > 0,\ i \in E) \tag{9.5.41}$$

that

$$\varphi_{iA}(\lambda) \geqslant \hat{\varphi}_{iA}^{(n)}(\lambda) \qquad (i \in E). \tag{9.5.42}$$

Hence

$$F_{iA}(+\infty) \geqslant \hat{F}_{iA}(+\infty) \qquad (i \in E). \tag{9.5.43}$$

Consequently we need only prove

$$\hat{F}_{iA}(+\infty) \geqslant F_{iA}(+\infty) \qquad (i \in E). \tag{9.5.44}$$

To do so, let

$$w(i) = \min\left(1, \frac{v(i)}{q_i}\right) \qquad (i \in E), \tag{9.5.45}$$

$$\zeta_A^{(n)}(\omega) = \sum_{k=1}^{\hat{\tau}_A^{(n)}(\omega)} w(x_{k-1}(\omega)), \tag{9.5.46}$$

$$\xi_A(\omega) = \sum_{k=1}^{\hat{\tau}_A(\omega)} w(x_{k-1}(\omega)), \tag{9.5.47}$$

$$\check{F}_{iA}^{(n)}(t) = p(\xi_A^{(n)}(\omega) \leqslant t | x_0 = i), \tag{9.5.48}$$

$$\check{F}_{iA}(t) = p(\xi_A(\omega) \leqslant t | x_0 = i), \tag{9.5.49}$$

$$\check{\varphi}_{iA}^{(n)}(\lambda) = \int_0^\infty e^{-\lambda t} d\check{F}_{iA}^{(n)}(t), \tag{9.5.50}$$

$$\check{\varphi}_{iA}(\lambda) = \int_0^\infty e^{-\lambda t} d\check{F}_{iA}(t), \tag{9.5.51}$$

$$\check{F}_{iA}(+\infty) = \lim_{t \to \infty} \check{F}_{iA}(t). \tag{9.5.52}$$

Obviously

$$\check{F}_{iA}(+\infty) = \check{F}_{iA}(+\infty) \qquad (i \in E). \tag{9.5.53}$$

Hence it is enough to prove

$$\check{F}_{iA}(+\infty) \geqslant F_{iA}(+\infty) \qquad (i \in E). \tag{9.5.54}$$

Since

$$\begin{aligned}
e^{-\lambda w(i)} &= \frac{1}{1 + \lambda w(i) + \sum_{n=1}^{\infty} \frac{1}{(n+1)!} \lambda^{n+1} \omega(i)^{n+1}} \\
&= \frac{1}{1 + \lambda w(i) + w(i) \sum_{n=1}^{\infty} \frac{1}{(n+1)!} \lambda^{n+1} w(i)^n} \\
&\geqslant \frac{1}{1 + \lambda w(i) + w(i) \sum_{n=1}^{\infty} \frac{\lambda^{n+1}}{(n+1)!}} \\
&= \frac{1}{1 + \lambda w(i) + [e^\lambda - (\lambda + 1)] w(i)} \\
&= \frac{1}{1 + (e^\lambda - 1) w(i)} \geqslant \frac{1}{1 + (e^\lambda - 1) \frac{v(i)}{q_i}} \qquad (i \in E),
\end{aligned} \tag{9.5.55}$$

we see

$$\check{\varphi}_{iA}^{(n)}(\lambda) \geqslant \varphi_{iA}^{(n)}(e^\lambda - 1) \qquad (i \in E). \tag{9.5.56}$$

Hence

$$\check{\varphi}_{iA}(\lambda) \geqslant \varphi_{iA}(e^{\lambda} - 1) \qquad (i \in E). \tag{9.5.57}$$

Consequently (9.5.54) holds. □

Corollary 9.5.1 *For an arbitrary $i \in E$, necessary and sufficient conditions for*

$$F_{iA}(+\infty) = 0, \tag{9.5.58}$$

$$F_{iA}(+\infty) = 1, \tag{9.5.59}$$

and

$$0 < F_{iA}(+\infty) < 1 \tag{9.5.60}$$

are that

$$\hat{F}_{iA}(+\infty) = 0, \tag{9.5.61}$$

$$\hat{F}_{iA}(+\infty) = 1, \tag{9.5.62}$$

and

$$0 < F_{iA}(+\infty) < 1, \tag{9.5.63}$$

respectively. □

Theorem 9.5.5

$$\hat{T}_{iA}^{(p)} \leqslant T_{iA}^{(p)} \leqslant p!\ \hat{T}_{iA}^{(p)} \qquad (i \in E). \tag{9.5.64}$$

Proof. We know from Theorems 6.7.4 and 9.5.2 that

$$T_{iA} = \hat{T}_{iA} \qquad (i \in E). \tag{9.5.65}$$

Thus this theorem is valid for $p = 1$. Now we assume that this theorem is valid for $p - 1$, i.e.,

$$\hat{T}_{iA}^{(p-1)} \leqslant T_{iA}^{(p-1)} \leqslant (p-1)!\ \ \hat{T}_{iA}^{(p-1)} \qquad (i \in E). \tag{9.5.66}$$

We will prove that it is also valid for p.

It is easy to see that $\{\hat{T}_{iA}^{(p)}, i \in E\}$ is the minimal nonnegative solution of the first-type system of 1-bounded equations

$$x_i = \sum_{j \in \{i\} \cup A} \frac{q_{ij}}{q_i} x_j + \sum_{l=1}^{p} C_p^l \left[\frac{v(i)}{q_i}\right]^l \sum_{j \notin \{i\} \cup A} \frac{q_{ij}}{q_i} \hat{T}_{jA}^{(p-l)}$$

$$+\left[\frac{v(i)}{q_i}\right]^p \sum_{j\in A\setminus\{i\}} \frac{q_{ij}}{q_i} \qquad (i\in E). \tag{9.5.67}$$

Thus from

$$C_{p-1}^{l-1} \leqslant \frac{p}{l} C_{p-1}^{l-1} = C_p^l \leqslant p C_{p-1}^{l-1} \qquad (p \geqslant l \geqslant 1), \tag{9.5.68}$$

we obtain

$$\frac{v(i)}{q_i} \hat{T}_{iA}^{(p-1)} \leqslant \sum_{l=1}^{p} C_p^l \left[\frac{v(i)}{q_i}\right]^l \sum_{j\notin\{i\}\cup A} \frac{q_{ij}}{q_i} \hat{T}_{jA}^{(p-l)}$$

$$+\left[\frac{v(i)}{q_i}\right]^p \sum_{j\in A\setminus\{i\}} \frac{q_{ij}}{q_i} \leqslant \frac{pv(i)}{q_i} \hat{T}_{iA}^{(p-1)} \qquad (i\in E). \tag{9.5.69}$$

Hence if let $\{\check{T}_{iA}, i\in E\}$ stand for the minimal nonnegative solution of the first-type system of 1-bounded equations

$$x_i = \sum_{j\notin\{i\}\cup A} \frac{q_{ij}}{q_i} x_j + \frac{pv(i)}{q_i} \hat{T}_{iA}^{(p-1)} \qquad (i\in E), \tag{9.5.70}$$

then

$$\hat{T}_{iA}^{(p)} \leqslant \check{T}_{iA}^{(p)} \qquad (i\in E). \tag{9.5.71}$$

From the former half of (9.5.66) and Theorem 9.5.2, we have

$$\check{T}_{iA}^{(p)} \leqslant T_{iA}^{(p)} \qquad (i\in E). \tag{9.5.72}$$

We know from (9.5.71) and (9.5.72) that the former half of (9.5.64) holds for p. If $\{\dot{T}_{iA}^{(p)}, i\in E\}$ stands for the minimal nonnegative solution of the first-type system of 1-bounded equations

$$x_i = \sum_{j\notin\{i\}\cup A} \frac{q_{ij}}{q_i} x_j + \frac{p!\,v(i)}{q_i} T_{iA}^{(p-1)} \qquad (i\in E), \tag{9.5.73}$$

then by (9.5.69) we have

$$p!\ \hat{T}_{iA}^{(p)} \geqslant \dot{T}_{iA}^{(p)} \qquad (i\in E). \tag{9.5.74}$$

From the latter half of (9.5.66), it can be seen that

$$p!\frac{v(i)}{q_i} \hat{T}_{iA}^{(p-1)} \geqslant p\frac{v(i)}{q_i} T_{iA}^{(p-1)} \qquad (i\in E). \tag{9.5.75}$$

Therefore, by Theorem 9.5.2,

$$\dot{T}^{(p)}_{iA} \geqslant T^{(p)}_{iA} \qquad (i \in E). \tag{9.5.76}$$

We obtain from (9.5.74) and (9.5.76) that the latter half of (9.5.64) also holds for p. Therefore this theorem is valid by induction. □

Corollary 9.5.2 *For each* $i \in E$,

$$T^{(p)}_{iA} < +\infty \tag{9.5.77}$$

iff

$$\hat{T}^{(p)}_{iA} < +\infty. \ \square \tag{9.5.78}$$

Equations (9.5.17) and (9.5.19) reduce the calculation of the distribution and moments of $\xi_A(\omega)$ and $\xi(\omega)$ to that of the distribution and moments of $\xi_{H_N}(\omega)$. So we will consider the calculation of the distribution and moments of $\xi_{H_N}(\omega)$. The following results in this section are immediate consequences of Chapter V, so the proofs are omitted.

Let

$$\lim_{\lambda \to +0} \psi_{iH_N}(\lambda) = \psi_{iH_N}(0) \qquad (i = 1, 2, \cdots, N), \tag{9.5.79}$$

$$D^{(N)} = \{1, 2, \cdots, N\} \cap \{i: i \underset{R}{\not\rightsquigarrow} H_N\}, \tag{9.5.80}$$

$$\bar{D}^{(N)} = \{1, 2, \cdots, N\} \cap \{i: i \underset{R}{\rightsquigarrow} H_N\}, \tag{9.5.81}$$

$$\hat{D}^{(N)} = D^{(N)} \cap \{\text{the essential set of subscripts of } R\}, \tag{9.5.82}$$

$$\check{D}^{(N)} = D^{(N)} \cap \{\text{the nonessential set of subscripts of } R\}, \tag{9.5.83}$$

$$t_{ij} = \begin{cases} q_{ij}/q_i, & i \neq j, \quad i, \quad j \in \hat{D}^{(N)}, \\ 0, & i = j, \quad i, \quad j \in \hat{D}^{(N)} \quad \text{or} \quad i = 0, \quad j \in \hat{D}^{(N)}, \\ v(i), & i \in \hat{D}^{(N)}, \qquad j = 0, \\ 1, & i = j = 0. \end{cases} \tag{9.5.84}$$

We define a matrix

$$T = (t_{ij}, \ i, \ j \in \{0\} \cup \hat{D}^{(N)}), \tag{9.5.85}$$

and put

$$\hat{D}^{(N)}_1 = \{i: \quad i \in \hat{D}^{(N)} \quad \text{and} \quad i \underset{T}{\not\rightsquigarrow} 0\}, \tag{9.5.86}$$

$$\hat{D}^{(N)}_2 = \{i: \quad i \in \hat{D}^{(N)} \quad \text{and} \quad i \underset{T}{\rightsquigarrow} 0\}, \tag{9.5.87}$$

$$\breve{D}_1^{(N)} = \{i: \quad i \in \breve{D}^{(N)} \quad \text{and} \quad i \underset{R}{\not\rightsquigarrow} \hat{D}_2^{(N)}\}, \tag{9.5.88}$$

$$\breve{D}_2^{(N)} = \{i: \quad i \in \breve{D}^{(N)} \quad \text{and} \quad i \underset{R}{\rightsquigarrow} \hat{D}_2^{(N)}\}, \tag{9.5.89}$$

$$\bar{D}_1^{(N)} = \{i: \quad i \in \bar{D}^{(N)} \quad \text{and} \quad i \underset{R}{\overset{H_N}{\not\rightsquigarrow}} \hat{D}_2^{(N)}\}, \tag{9.5.90}$$

$$\bar{D}_2^{(N)} = \{i: \quad i \in \bar{D}^{(N)} \quad \text{and} \quad i \underset{R}{\overset{H_N}{\rightsquigarrow}} \hat{D}_2^{(N)}\}. \tag{9.5.91}$$

Theorem 9.5.6 $\psi_{iH_N}(\lambda)$ $(\lambda > 0, \quad i = 1, 2, \cdots, N)$ *is the minimal nonnegative solution of the pseudo-normal system of equations*

$$x_i = \sum_{\substack{j=1 \\ j \neq i}}^{N} \frac{q_{ij}}{\lambda v(i) + q_i} x_i + \frac{\lambda v(i)}{\lambda v(i) + q_i} \qquad (i = 1, 2, \cdots, N). \tag{9.5.92}$$

In detail, $\psi_{iH_N}(\lambda)$ $(\lambda \geqslant 0, \quad i = 1, 2, \cdots, N)$ *is determined uniquely as follows:*

(i) $\psi_{iH_N}(\lambda)$ $(\lambda \geqslant 0, \quad i \in \hat{D}_1^{(N)})$ *is the minimal nonnegative solution (i. e., zero solution) of the first-type blockable random system of homogeneous equations*

$$x_i = \sum_{j \in \hat{D}_1^{(N)} \setminus \{i\}} \frac{q_{ij}}{q_i} x_j \qquad (i \in \hat{D}_1^{(N)}). \tag{9.5.93}$$

Hence

$$\psi_{iH_N}(\lambda) \equiv 0 \qquad (\lambda \geqslant 0, \quad i \in \hat{D}_1^{(N)}). \tag{9.5.94}$$

(ii) $\psi_{iH_N}(\lambda)$ $(\lambda \geqslant 0, \quad i \in \breve{D}_1^{(N)})$ *is the minimal nonnegative solution (i. e., the unique (ordinary) solution) of the first-type leading-outside system of equations*

$$x_i = \sum_{j \in \breve{D}_1^{(N)} \setminus \{i\}} \frac{q_{ij}}{\lambda v(i) + q_i} x_j + \frac{\lambda v(i)}{\lambda v(i) + q_i} \qquad (i \in \breve{D}_1^{(N)}). \tag{9.5.95}$$

(iii) $\psi_{iH_N}(\lambda)$ $(\lambda \geqslant 0, i \in \bar{D}_1^{(N)})$ *is the minimal nonnegative solution (i. e., the unique (ordinary) solution) of the first-type leading-outside system of equations*

$$x_i = \sum_{j \in \bar{D}_1^{(N)} \setminus \{i\}} \frac{q_{ij}}{\lambda v(i) + q_i} x_j + \sum_{j \in \breve{D}_1^{(N)} \setminus \{i\}} \frac{q_{ij}}{\lambda v(i) + q_i} \psi_{jH_N}(\lambda)$$
$$+ \frac{\lambda v(i)}{\lambda v(i) + q_i} \qquad (i \in \bar{D}_1^{(N)}). \tag{9.5.96}$$

(iv) $\psi_{iH_N}(\lambda)$ $(\lambda > 0,\ i \in \hat{D}_2^{(N)})$ *is the minimal nonnegative solution (i. e., the unique (ordinary) solution) of the normal leading-outsidesystem of equations*

$$x_i = \sum_{j \in \hat{D}_2^{(N)} \setminus \{i\}} \frac{q_{ij}}{\lambda v(i) + q_i} x_j + \frac{\lambda v(i)}{\lambda v(i) + q_i} \qquad (i \in \hat{D}_2^{(N)}). \tag{9.5.97}$$

Hence

$$\psi_{iH_N}(\lambda) \equiv 1 \qquad (\lambda > 0, \quad i \in \hat{D}_2^{(N)}). \tag{9.5.98}$$

When $\lambda = 0$, *the corresponding conclusion does not hold. In detail, when* $\lambda = 0$, (9.5.97) *is transformed to the following first-type blockable random system of homogeneous equations*

$$x_i = \sum_{j \in \hat{D}_2^{(N)} \setminus \{i\}} \frac{q_{ij}}{q_i} x_j \qquad (i \in \hat{D}_2^{(N)}). \tag{9.5.99}$$

Hence the minimal nonnegative solution x_i^* $(i \in \hat{D}_2^{(N)})$ *of* (9.5.99) *is the zero solution, namely*

$$x_i^* \equiv 0 \qquad (i \in \hat{D}_2^{(N)}). \tag{9.5.100}$$

Though

$$\psi_{iH_N}(0) \equiv 1 \qquad (i \in \hat{D}_2^{(N)}) \tag{9.5.101}$$

is a nonnegative solution of (9.5.99), *it is not the minimal nonnegative solution, i. e., not the zero solution.*

(v) $\psi_{iH_N}(\lambda)$ $(\lambda \geqslant 0,\ i \in \check{D}_2^{(N)})$ *is the minimal nonnegative solution (i. e., the unique (ordinary) solution) of the first-type leading-outside system of strictly nonhomogeneous equations*

$$\begin{aligned} x_i = & \sum_{j \in \hat{D}_2^{(N)} \setminus \{i\}} \frac{q_{ij}}{\lambda v(i) + q_i} x_j + \sum_{j \in \check{D}_2^{(N)} \setminus \{i\}} \frac{q_{ij}}{\lambda v(i) + q_i} \\ & + \sum_{j \in \check{D}_1^{(N)} \setminus \{i\}} \frac{q_{ij}}{\lambda v(i) + q_i} \psi_{jH_N}(\lambda) + \frac{\lambda v(i)}{\lambda v(i) + q_i} \qquad (i \in \check{D}_2^{(N)}). \end{aligned} \tag{9.5.102}$$

Hence

$$\psi_{iHN}(\lambda) > 0 \qquad (\lambda \geqslant 0, \quad i \in \check{D}_2^{(N)}). \tag{9.5.103}$$

(vi) $\psi_{iH_N}(\lambda)$ $(\lambda \geqslant 0, \quad i \in \bar{D}_2^{(N)})$ *is the minimal nonnegative solution (i. e., the unique (ordinary) solution) of the first-type leading-outside system of strictly nonhomogeneous equations*

$$x_i = \sum_{j \in \bar{D}_2^{(N)} \setminus \{i\}} \frac{q_{ij}}{\lambda v(i) + q_i} x_j + \sum_{j \in \hat{D}_2^{(N)} \setminus \{i\}} \frac{q_{ij}}{\lambda v(i) + q_i}$$

$$+ \sum_{j \in \check{D}^{(N)} \cup \bar{D}_1^{(N)} \setminus \{i\}} \frac{q_{ij}}{\lambda v(i) + q_i} \psi_{jH_N}(\lambda)$$

$$+ \frac{\lambda v(i)}{\lambda v(i) + q_i} \qquad (i \in \bar{D}_2^{(N)}). \tag{9.5.104}$$

Hence

$$\psi_{iH_N}(\lambda) > 0 \qquad (\lambda \geqslant 0, \quad i \in \bar{D}_2^{(N)}). \ \square \tag{9.5.105}$$

Remark 9.5.2 From Theorem 9.5.6, we have

$$\psi_{iH_N}(\lambda) \equiv 0 \qquad (\lambda \geqslant 0, \quad i \in \hat{D}_1^{(N)} \cup \check{D}_1^{(N)} \cup \bar{D}_1^{(N)}) \tag{9.5.106}$$

and

$$\psi_{iH_N}(\lambda) > 0 \qquad (\lambda \geqslant 0, \quad i \in \hat{D}_2^{(N)} \cup \check{D}_2^{(N)} \cup \bar{D}_2^{(N)}). \tag{9.5.107}$$

Theorem 9.5.7 T_{iH_N} $(i = 1, 2, \cdots, N)$ *is the minimal nonnegative solution of the first-type regular system of equations*

$$x_i = \sum_{\substack{j=1 \\ j \neq i}}^{N} \frac{q_{ij}}{q_i} x_j + \frac{v(i)}{q_i} \qquad (i = 1, 2, \cdots, N). \tag{9.5.108}$$

To be specific,

(i) $T_{iH_N} (i \in \hat{D}_1^{(N)})$ *is the zero solution of the blockable random system of homogeneous equations*

$$x_i = \sum_{j \in \hat{D}_1^{(N)} \setminus \{i\}} \frac{q_{ij}}{q_i} x_j \qquad (i \in \hat{D}_1^{(N)}). \tag{9.5.109}$$

Hence

$$T_{iH_N} \equiv 0 \qquad (i \in \hat{D}_1^{(N)}). \tag{9.5.110}$$

(ii) T_{iH_N} $(i \in \check{D}_1^{(N)} \cup \bar{D}_1^{(N)})$ *is the unique (ordinary) solution of the first-type leading-outside system of equations*

$$x_i = \sum_{j \in \check{D}_1^{(N)} \cup \bar{D}_1^{(N)} \setminus \{i\}} \frac{q_{ij}}{q_i} x_j + \frac{v(i)}{q_i} \qquad (i \in \check{D}_1^{(N)} \cup \bar{D}_1^{(N)}). \tag{9.5.111}$$

(iii) $T_{iH_N} (i \in \hat{D}_2^{(N)} \cup \check{D}_2^{(N)} \cup \bar{D}_2^{(N)})$ *is the minimal nonnegative solution of the tailed random system of strictly nonhomogeneous equations*

$$x_i = \sum_{j \in \hat{D}_2^{(N)} \cup \check{D}_2^{(N)} \cup \bar{D}_2^{(N)} \setminus \{i\}} \frac{q_{ij}}{q_i} x_j + \sum_{j \in \check{D}_1^{(N)} \cup \bar{D}_1^{(N)} \setminus \{i\}} \frac{q_{ij}}{q_i} T_{jH_N} + \frac{v(i)}{q_i}$$

$$\left(i \in \hat{D}_2^{(N)} \bigcup \check{D}_2^{(N)} \bigcup \bar{D}_2^{(N)}\right). \tag{9.5.112}$$

Hence

$$T_{iH_N} = +\infty \qquad \left(i \in \hat{D}_2^{(N)} \bigcup \check{D}_2^{(N)} \bigcup \bar{D}_2^{(N)}\right). \ \square \tag{9.5.113}$$

Theorem 9.5.8 $T_{iH_N}^{(p)}$ $(i = 1, 2, \cdots, N)$ *is the minimal nonnegative solution of the first-type regular system of equations*

$$x_i = \sum_{\substack{j=1 \\ j \neq i}}^{N} \frac{q_{ij}}{q_i} x_j + \frac{pv(i)}{q_i} T_{iH_N}^{(p-1)} \qquad (i = 1, 2, \cdots, N). \tag{9.5.114}$$

To be specific,

(i) $T_{iH_N}^{(p)}$ $\left(i \in \hat{D}_1^{(N)}\right)$ *is the zero solution of the first-type blockable random system of homogeneous equations*

$$x_i = \sum_{j \in \hat{D}_1^{(N)} \setminus \{i\}} \frac{q_{ij}}{q_i} x_j \qquad \left(i \in \hat{D}_1^{(N)}\right). \tag{9.5.115}$$

Hence

$$T_{iH_N}^{(p)} \equiv 0 \qquad \left(i \in \hat{D}_1^{(N)}\right). \tag{9.5.116}$$

(ii) $T_{iH_N}^{(p)}$ $\left(i \in \check{D}_1^{(N)} \bigcup \bar{D}_1^{(N)}\right)$ *is the unique (ordinary) solution of the first-type leading-outside system of equations*

$$x_i = \sum_{j \in \check{D}_1^{(N)} \cup \bar{D}_1^{(N)} \setminus \{i\}} \frac{q_{ij}}{q_i} x_j + \frac{pv(i)}{q_i} T_{iH_N}^{(p-1)} \qquad \left(i \in \check{D}_1^{(N)} \bigcup \bar{D}_1^{(N)}\right), \tag{9.5.117}$$

and

$$T_{iH_N}^{(p)} \leqslant p! \quad c^p \qquad \left(i \in \check{D}_1^{(N)} \bigcup \bar{D}_1^{(N)}\right). \tag{9.5.118}$$

(iii) *When* $p \geqslant 2$, $T_{iH_N}^{(p)}$ $\left(i \in \hat{D}_2^{(N)} \bigcup \check{D}_2^{(N)} \bigcup \bar{D}_2^{(N)}\right)$ *is the minimal nonnegative solution of the system of equations with essentially infinite constant terms*

$$x_i = \sum_{j \in \hat{D}_2^{(N)} \bigcup \check{D}_2^{(N)} \bigcup \bar{D}_2^{(N)} \setminus \{i\}} \frac{q_{ij}}{q_i} x_j + \sum_{j \in \check{D}_1^{(N)} \bigcup \bar{D}_1^{(N)} \setminus \{i\}} \frac{q_{ij}}{q_i} T_{jH_N}^{(p)} + \frac{pv(i)}{q_i} T_{iH_N}^{(p-1)}$$

$$\left(i \in \hat{D}_2^{(N)} \bigcup \check{D}_2^{(N)} \bigcup \bar{D}_2^{(N)}\right). \tag{9.5.119}$$

Hence

$$T_{iH_N}^{(p)} = +\infty \qquad (i \in \hat{D}_2^{(N)} \bigcup \check{D}_2^{(N)} \bigcup \bar{D}_2^{(N)}). \ \square \tag{9.5.120}$$

Remark 9.5.3 We have studied the calculation of the distribution and moments of $\xi_A(\omega)$. In the next section we shall present a method for calculating moments

$$\int_{-0}^{\infty} t^p dP(\xi_A(\omega) \leqslant t | x_0 = i) \tag{9.5.121}$$

of the distribution of $\xi_A(\omega)$. It is necessary when

$$P(\xi_A(\omega) = +\infty) > 0, \tag{9.5.122}$$

since in this case

$$T_{iA}^{(p)} = +\infty \qquad (i \in E, \quad p \geqslant 1) \tag{9.5.123}$$

and it is not enough only to study the distribution and moments of $\xi_A(\omega)$.

§ 9.6 Distribution and moments of integral-type functionals on pseudo-translatable sets

Definition 9.6.1 *A set $\Lambda \in \mathscr{F}\{x_t, t < \sigma\}$ is called a translatable set of the process $X(\omega) = \{x_t, t < \sigma\}$ if*

$$\theta_{\tau^{(1)}} \Lambda \doteq \Lambda, \tag{9.6.1}$$ [1]

where $\theta_{\tau^{(1)}}$ is the translation operator defined in [6].

Definition 9.6.2 *A set $\Lambda \in \mathscr{F}\{x_t, t < \sigma\}$ is called a pseudo-translatable set of the process $X(\omega) = \{x_t, t < \sigma\}$, if there exists a decomposition of E, $E = E^{(1)} \bigcup E^{(2)}$, $E^{(1)} \bigcap E^{(2)} = \varnothing$, such that:*

(i) *for any $j \in E^{(1)}$,*

$$(x_{\tau^{(1)}} = j, \ \Lambda) \doteq (x_{\tau^{(1)}} = j, \ \theta_{\tau^{(1)}} \Lambda); \tag{9.6.2}$$

(ii) *for any $j \in E^{(2)}$,*

$$(x_{\tau^{(1)}} = j, \ \Lambda) \doteq (x_{\tau^{(1)}} = j, \ \Lambda_j), \tag{9.6.3}$$

where Λ_j $(j \in E^{(2)})$ is a translatable set of the process.

---

1 In general, the right side of this expression should be $\Lambda \bigcap \{\tau^{(1)} < \sigma\}$. But under the assumption of (9.1.1), we have $P(\tau^{(1)} < \sigma) = 1$, there fore $\Lambda \bigcap \{\tau^{(1)} < \sigma\}$ can be replaced by Λ.

Clearly, a pseudo-translatable set becomes a translatable set when $E^{(2)} = \varnothing$

Suppose Λ is a pseudo-translatable set. The definitions of $\xi_A^{(n)}(\omega)$ and $\xi_A(\omega)$ are as in §9.5. Let

$$\overset{(\Lambda)}{F}{}_{iA}^{(n)}(t) = P(\xi_A^{(n)}(\omega) \leqslant t,\ \Lambda | x_0 = i), \tag{9.6.4}$$

$$\overset{(\Lambda)}{F}{}_{iA}(t) = P(\xi_A(\omega) \leqslant t,\ \Lambda | x_0 = i), \tag{9.6.5}$$

$$\overset{(\Lambda)}{\varphi}{}_{iA}^{(n)}(\lambda) = \int_{-0}^{\infty} e^{-\lambda t} d\overset{(\Lambda)}{F}{}_{iA}^{(n)}(t) \qquad (\lambda > 0), \tag{9.6.6}$$

$$\overset{(\Lambda)}{\varphi}{}_{iA}(\lambda) = \int_{-0}^{\infty} e^{-\lambda t} d\overset{(\Lambda)}{F}{}_{iA}(t) \qquad (\lambda > 0), \tag{9.6.7}$$

$$\overset{(\Lambda)}{\psi}{}_{iA}^{(n)}(\lambda) = P_i(\Lambda) - \overset{(\Lambda)}{\varphi}{}_{iA}^{(n)}(\lambda), \tag{9.6.8}$$

$$\overset{(\Lambda)}{\psi}{}_{iA}(\lambda) = P_i(\Lambda) - \overset{(\Lambda)}{\varphi}{}_{iA}(\lambda), \tag{9.6.9}$$

$$\overset{(\Lambda)}{T}{}_{iA}^{(n,p)} = M\{[\xi_A^{(n)}(\omega)]^p I_\Lambda(\omega) | x_0 = i\}, \tag{9.6.10}$$

$$\overset{(\Lambda)}{T}{}_{iA}^{(p)} = M\{[\xi_A(\omega)]^p I_\Lambda(\omega) | x_0 = i\}, \tag{9.6.11}$$

where

$$I_\Lambda(\omega) = \begin{cases} 1, & \omega \in \Lambda, \\ 0, & \omega \notin \Lambda \end{cases} \tag{9.6.12}$$

is the characteristic function of set Λ.

In this section we will present methods for calculating $\overset{(\Lambda)}{\psi}{}_{iA}(\lambda)$ and $\overset{(\Lambda)}{T}{}_{iA}^{(p)}$. It is apparent that

$$\text{V} - \lim_{n \to \infty} \overset{(\Lambda)}{\psi}{}_{iA}^{(n)}(\lambda) = \overset{(\Lambda)}{\psi}{}_{iA}(\lambda), \tag{9.6.13}$$

$$\text{V} - \lim_{n \to \infty} \overset{(\Lambda)}{T}{}_{iA}^{(n,p)} = \overset{(\Lambda)}{T}{}_{iA}^{(p)}, \tag{9.6.14}$$

$$\overset{(\Lambda)}{T}{}_{iA}^{(n,0)} = \overset{(\Lambda)}{T}{}_{iA}^{(0)} = P_i(\Lambda). \tag{9.6.15}$$

Lemma 9.6.1 *Suppose Λ is a pseudo-translatable set. Then*

$$P_i(\Lambda) = \sum_{j \in E^{(1)} \setminus \{i\}} \frac{q_{ij}}{q_i} P_j(\Lambda) + \sum_{j \in E^{(2)} \setminus \{i\}} \frac{q_{ij}}{q_i} P_j(\Lambda_j) \qquad (i \in E). \tag{9.6.16}$$

Proof. First, we show the following two equalities:

$$P(x_{\tau(1)}=j,\ \Lambda|x_0=i)=\frac{q_{ij}}{q_i}P_j(\Lambda)\qquad (j\in E^{(1)}\backslash\{i\}),\tag{9.6.17}$$

$$P(x_{\tau(1)}=j,\ \Lambda|x_0=i)=\frac{q_{ij}}{q_i}P_j(\Lambda_j)\qquad (j\in E^{(2)}\backslash\{i\}).\tag{9.6.18}$$

Equations (9.6.17) and (9.6.18) obviously hold when $j\in(k:\ q_{ik}=0)$, because in this case both sides of (9.6.17) and (9.6.18) are zero. Therefore in what follows we let $i\in(k:\ q_{ik}\neq 0)$. If $j\in E^{(1)}\bigcap\{k:\ q_{ik}\neq 0\}\backslash\{i\}$, then by (9.6.2)

$$\begin{aligned}P(x_{\tau(1)}=j,\ \Lambda|x_0=i)&=P(x_{\tau(1)}=j,\ \theta_{\tau(1)}\Lambda|x_0=i)\\&=P(x_{\tau(1)}=j|x_0=i)=P(\theta_{\tau(1)}\Lambda|x_0=i,\ x_{\tau(1)}=j)\\&=\frac{q_{ij}}{q_i}P(\Lambda|x_0=j)=\frac{q_{ij}}{q_i}P_j(\Lambda).\end{aligned}\tag{9.6.19}$$

If $j\in E^2\bigcap\{k:\ q_{ik}\neq 0\}\backslash\{i\}$, then by (9.6.1) and (9.6.3)

$$\begin{aligned}P(x_{\tau(1)}=j,\ \Lambda|x_0=i)&=P(x_{\tau(1)}=j,\ \Lambda_j|x_0=i)\\&=P(x_{\tau(1)}=j,\ \theta_{\tau(1)}\Lambda_j|x_0=i)=\frac{q_{ij}}{q_i}P_j(\Lambda_j).\end{aligned}\tag{9.6.20}$$

Thus (9.6.17) and (9.6.18) hold.
Equations (9.6.17), (9.6.18) and

$$P_i(\Lambda)=\sum_{j\neq i}P(x_{\tau(1)}=j,\ \Lambda|x_0=i)\tag{9.6.21}$$

imply the lemma. □

Lemma 9.6.2 $\overset{(\Lambda)}{F}{}^{(n)}_{iA}(t)$ *satisfies the following relations*

$$\left.\begin{aligned}\overset{(\Lambda)}{F}{}^{(1)}_{iA}(t)&=P_i(\Lambda)P(v(i)\tau^{(1)}\leqslant t|x_0=i)\qquad(i\in E),\\ \overset{(\Lambda)}{F}{}^{(n+1)}_{iA}(t)&=\sum_{j\in\bar{A}\bigcap E^{(1)}\backslash\{i\}}\frac{q_{ij}}{q_i}\int_{-0}^{t}\overset{(\Lambda)}{F}{}^{(n)}_{jA}(t-u)dP(v(i)\tau^{(1)}\leqslant u|x_0=i)\\&\quad+\sum_{j\in\bar{A}\bigcap E^{(2)}\backslash\{i\}}\frac{q_{ij}}{q_i}\int_{-0}^{t}\overset{(\Lambda_j)}{F}{}^{(n)}_{jA}(t-u)dP(v(i)\tau^{(1)}\leqslant u|x_0=i)\end{aligned}\right|$$

$$\left.\begin{aligned} &+ \sum_{j\in A\cap E^{(1)}\setminus\{i\}} \frac{q_{ij}}{q_i} P(v(i)\tau^{(1)} \leqslant t | x_0 = i) P_j(\Lambda) \\ &+ \sum_{j\in A\cap E^{(2)}\setminus\{i\}} \frac{q_{ij}}{q_i} P(v(i)\tau^{(1)} \leqslant t | x_0 = i) + P_j(\Lambda_j) \\ &\qquad (n \geqslant 1, \quad i\in E). \end{aligned}\right\} \tag{9.6.22}$$

Proof. It is easy to complete the proof by referring to the proofs of Lemmas 9.3.7 and 9.6.1. □

Lemma 9.6.3 $\overset{(\Lambda)}{\varphi}{}^{(n)}_{iA}(\lambda)$ *satisfies the following relations:*

$$\left.\begin{aligned} \overset{(\Lambda)}{\varphi}{}^{(1)}_{iA}(\lambda) &= \frac{q_i P_i(\Lambda)}{\lambda v(i) + q_i} \qquad (i\in E), \\ \overset{(\Lambda)}{\varphi}{}^{(n+1)}_{iA}(\lambda) &= \sum_{j\in \bar{A}\cap E^{(1)}\setminus\{i\}} \frac{q_{ij}}{\lambda v(i) + q_i} \overset{(\Lambda)}{\varphi}{}^{(n)}_{jA}(\lambda) \\ &+ \sum_{j\in \bar{A}\cap E^{(2)}\setminus\{i\}} \frac{q_{ij}}{\lambda v(i) + q_i} \overset{(\Lambda_j)}{\varphi}{}^{(n)}_{jA}(\lambda) \\ &+ \sum_{j\in A\cap E^{(1)}\setminus\{i\}} \frac{q_{ij}}{\lambda v(i) + q_i} P_j(\Lambda) \\ &+ \sum_{j\in A\cap E^{(2)}\setminus\{i\}} \frac{q_{ij}}{\lambda v(i) + q_i} P_j(\Lambda_j) \qquad (n \geqslant 1, \quad i\in E). \end{aligned}\right\} \tag{9.6.23}$$

Proof. Noticing Lemmas 9.3.1 and 9.3.3, we take Laplace-Stieltjes transforms on both sides of (9.6.22), to obtain this lemma. □

Lemma 9.6.4 $\overset{(\Lambda)}{\psi}{}^{(n)}_{iA}(\lambda)$ *satisfies the following:*

$$\left.\begin{aligned} \overset{(\Lambda)}{\psi}{}^{(1)}_{iA}(\lambda) &= \frac{\lambda v(i) P_i(\Lambda)}{\lambda v(i) + q_i} \qquad (i\in E), \\ \overset{(\Lambda)}{\psi}{}^{(n+1)}_{iA}(\lambda) &= \sum_{j\in \bar{A}\cap E^{(1)}\setminus\{i\}} \frac{q_{ij}}{\lambda v(i) + q_i} \overset{(\Lambda)}{\psi}{}^{(n)}_{jA}(\lambda) \\ &+ \sum_{j\in \bar{A}\cap E^{(2)}\setminus\{i\}} \frac{q_{ij}}{\lambda v(i) + q_i} \overset{(\Lambda_j)}{\psi}{}^{(n)}_{jA}(\lambda) \\ &+ \frac{\lambda v(i) P_i(\Lambda)}{\lambda v(i) + q_i} \qquad (n \geqslant 1, \quad i\in E). \end{aligned}\right\} \tag{9.6.24}$$

Proof. This lemma follows from (9.6.8) and (9.6.23). □

Lemma 9.6.5 $\overset{(\Lambda)}{T}{}_{iA}^{(n,p)}$ *satisfies the following:*

$$\left.\begin{aligned}
&\overset{(\Lambda)}{T}{}_{iA}^{(1,p)} = P_i(\Lambda)p!\left(\frac{v(i)}{q_i}\right)^p \qquad (i\in E),\\
&\overset{(\Lambda)}{T}{}_{iA}^{(n+1,p)} = \sum_{j\in\bar{A}\cap E^{(1)}\setminus\{i\}} \frac{q_{ij}}{q_i}\sum_{l=0}^{p} C_p^l l!\left(\frac{v(i)}{q_i}\right)^l \overset{(\Lambda)}{T}{}_{jA}^{(n,p-l)}\\
&\quad + \sum_{j\in\bar{A}\cap E^{(2)}\setminus\{i\}} \frac{q_{ij}}{q_i}\sum_{l=0}^{p} C_p^l l!\left(\frac{v(i)}{q_i}\right)^l \overset{(\Lambda_j)}{T}{}_{jA}^{(n,p-l)}\\
&\quad + \Bigg(\sum_{j\in A\cap E^{(1)}\setminus\{i\}} \frac{q_{ij}}{q_i}P_j(\Lambda)\\
&\quad + \sum_{j\in A\cap E^{(2)}\setminus\{i\}} \frac{q_{ij}}{q_i}P_j(\Lambda_j)\Bigg)p!\left(\frac{v(i)}{q_i}\right) \qquad (n\geqslant 1,\ i\in E).
\end{aligned}\right\} \tag{9.6.25}$$

Proof. From

$$P(\xi_A^{(n)} < +\infty) = 1, \tag{9.6.26}$$

we see that

$$\overset{(\Lambda)}{T}{}_{iA}^{(n,p)} = \int_{-0}^{\infty} t^p d\overset{(\Lambda)}{F}{}_{iA}^{(n)}(t). \tag{9.6.27}$$

This lemma follows from (9.6.27), Lemmas 9.6.2 and 9.3.4. □

Theorem 9.6.1 *Suppose* Λ *is a translatable set, then* $\{\overset{(\Lambda)}{\psi}{}_{iA}(\lambda),\ i\in E\}$ *is the minimal nonnegative solution of the pseudo-normal system of equations*

$$x_i = \sum_{j\in\bar{A}\setminus\{i\}} \frac{q_{ij}}{\lambda v(i)+q_i}x_j + \frac{\lambda v(i)P_i(\Lambda)}{\lambda v(i)+q_i} \qquad (i\in E). \tag{9.6.28}$$

Proof. It follows from (9.6.13), Lemma 9.6.4 and Theorem 3.2.1. □

Theorem 9.6.2 *Suppose* Λ *is a translatable set. Then for* $p\geqslant 1$, $\{\overset{(\Lambda)}{T}{}_{iA}^{(p)},\ i\in E\}$ *is the minimal nonnegative solution of the first-type system of 1-bounded equations*

$$x_i = \sum_{j\in\bar{A}\setminus\{i\}} \frac{q_{ij}}{q_i}x_j + \frac{pv(i)}{q_i}\overset{(\Lambda)}{T}{}_{iA}^{(p-1)} \qquad (i\in E). \tag{9.6.29}$$

Proof. It is easy to finish the proof of this theorem by referring to the proof of Theorem 9.3.3. □

Theorem 9.6.3 *Suppose Λ is a pseudo-translatable set, then $\{\overset{(\Lambda)}{\psi}_{iA}(\lambda), i \in E\}$ is the minimal nonnegative solution of the pseudo-normal system of equations*

$$x_i = \sum_{j \in \bar{A} \cap E^{(1)} \setminus \{i\}} \frac{q_{ij}}{\lambda v(i) + q_i} x_j + \sum_{j \in \bar{A} \cap E^{(2)} \setminus \{i\}} \frac{q_{ij}}{\lambda v(i) + q_i} \overset{(\Lambda_j)}{\psi}_{jA}(\lambda) + \frac{\lambda v(i) P_i(\Lambda)}{\lambda v(i) + q_i} \quad (i \in E). \tag{9.6.30}$$

Proof. It follows readily from (9.6.13), (9.6.24) and Theorem 3.2.2. □

Theorem 9.6.4 *Suppose Λ is a pseudo-translatable set. Then for $p \geqslant 1$, $\{\overset{(\Lambda)}{T}{}^{(p)}_{iA}, i \in E\}$ is the minimal nonnegative solution of the first-type system of 1-bounded equations*

$$x_i = \sum_{j \in \bar{A} \cap E^{(1)} \setminus \{i\}} \frac{q_{ij}}{q_i} x_j + \sum_{i \in \bar{A} \cap E^{(2)} \setminus \{i\}} \frac{q_{ij}}{q_i} \overset{(\Lambda_j)}{T}{}^{(p)}_{jA} + \frac{pv(i)}{q_i} \overset{(\Lambda)}{T}{}^{(p-1)}_{iA} \quad (i \in E). \tag{9.6.31}$$

Proof. By referring to the proof of Theorem 9.3.3, it is easy to complete the proof of the theorem. □

Theorem 9.6.5 *Suppose Δ is a translatable set, and let*

$$\Lambda = \{\omega: \ {}_H\sigma_A(\omega) < +\infty\} \cap \Delta. \tag{9.6.32}$$

Then Λ is a pseudo-translatable set, and $\{\overset{(\Lambda)}{\psi}_{iA}(\lambda), \ i \in E\}$ is the minimal nonnegative solution of the pseudo-normal system of equations

$$x_i = \sum_{j \in \overline{A \cup H} \setminus \{i\}} \frac{q_{ij}}{\lambda v(i) + q_i} x_j + \frac{\lambda v(i) P_i(\Lambda)}{\lambda v(i) + q_i} \quad (i \in E). \tag{9.6.33}$$

For $p \geqslant 1$, $\{\overset{(\Lambda)}{T}{}^{(p)}_{iA}; i \in E\}$ is the minimal nonnegative solution of the first-type system of 1-bounded equations

$$x_i = \sum_{j \in \overline{A \cup H} \setminus \{i\}} \frac{q_{ij}}{q_i} x_j + \frac{pv(i)}{q_i} \overset{(\Lambda)}{T}{}^{(p-1)}_{iA} \quad (i \in E). \tag{9.6.34}$$

Proof. Clearly Λ is a pseudo-translatable set, and

$$E^{(1)} = \overline{A \cup H}, \tag{9.6.35}$$

$$E^{(2)} = A \cup H, \tag{9.6.36}$$

$$(x_{\tau(1)} = j, \ \Lambda) = (x_{\tau(1)} = j, \ \Delta) \quad (j \in A), \tag{9.6.37}$$

$$\left(x_{\tau(1)} = j,\ \Lambda\right) = \left(x_{\tau(1)} = j,\ \varnothing\right) = \varnothing \qquad (j \in H). \tag{9.6.38}$$

Hence

$$\bar{A} \bigcap E^{(1)} = \overline{A \bigcup H}, \tag{9.6.39}$$

$$\bar{A} \bigcap E^{(2)} = \bar{A} \bigcap H \subset H, \tag{9.6.40}$$

$$A \bigcap E^{(1)} = \varnothing, \tag{9.6.41}$$

$$A \bigcap E^{(2)} = A, \tag{9.6.42}$$

$$\overset{(\Lambda_j)}{\psi}_{jA}(\lambda) = \overset{(\varnothing)}{\psi}_{jA}(\lambda) = 0 \qquad (j \in H), \tag{9.6.43}$$

$$\overset{(\Lambda_j)}{T}{}^{(p)}_{jA} = \overset{(\varnothing)}{T}{}^{(p)}_{jA} = 0 \qquad (p \geqslant 1,\ \ j \in H), \tag{9.6.44}$$

Consequently, our theorem follows immediately from Theorems 9.6.3 and 9.6.4. □

Theorem 9.6.6 *Suppose Δ is a translatable set, and let*

$$\Lambda = \{\omega:\ \ \sigma_A(\omega) = +\infty\} \bigcap \Delta. \tag{9.6.45}$$

Then Λ is a pseudo-translatable set. Furthermore,

$$\overset{(\Lambda)}{\psi}_{iA}(\lambda) = \overset{(\Lambda)}{\psi}_{i}(\lambda) \qquad (i \in E), \tag{9.6.46}$$

$$\overset{(\Lambda)}{T}{}^{(p)}_{iA} = \overset{(\Lambda)}{T}{}^{(p)}_{i} \qquad (i \in E) \tag{9.6.47}$$

and $\{\overset{(\Lambda)}{\psi}_{i}(\lambda),\ i \in E\}$ is the minimal nonnegative solution of the pseudo-normal system of equations

$$x_i = \sum_{j \in A \setminus \{i\}} \frac{q_{ij}}{\lambda v(i) + q_i} x_j + \frac{\lambda v(i) P_i(\Lambda)}{\lambda v(i) + q_i} \qquad (i \in E). \tag{9.6.48}$$

For $p \geqslant 1$, $\{\overset{(\Lambda)}{T}{}^{(p)}_{i},\ i \in E\}$ is the minimal nonnegative solution of the first-type system of 1-bounded equations

$$x_i = \sum_{j \in A \setminus \{i\}} \frac{q_{ij}}{q_i} x_j + \frac{p v(i)}{q_i} \overset{(\Lambda)}{T}{}^{(p-1)}_{i} \qquad (i \in E). \tag{9.6.49}$$

Proof. It is obvious that Λ is a pseudo-translatable set, and

$$E^{(1)} = \bar{A}, \tag{9.6.50}$$

$$E^{(2)} = A, \tag{9.6.51}$$

$$(x_{\tau(1)}=j,\ \Lambda)=(x_{\tau(1)}=j,\ \varnothing)=\varnothing \qquad (j\in E^{(2)}). \tag{9.6.52}$$

Moreover (9.6.46) and (9.6.47) hold obviously. Therefore this theorem follows readily from Theorems 9.6.3 and 9.6.4. □

Theorem 9.6.7 *If we let*

$$W_i^{(p)}=\int_{-0}^{\infty} t^p dP_i(\xi(\omega)\leqslant t) \qquad (p=0,\ 1,\ \cdots), \tag{9.6.53}$$

then

$$W_i^{(0)}=\lim_{\lambda\to 0}\varphi_i(\lambda) \qquad (i\in E), \tag{9.6.54}$$

and for $p\geqslant 1$, $\{W_i^{(p)},\ i\in E\}$ *is the minimal nonnegative solution of the first-type system of* 1*-bounded equations*

$$x_i=\sum_{j\neq i}\frac{q_{ij}}{q_i}x_j+\frac{pv(i)}{q_i}W_i^{(p-1)} \qquad (i\in E). \tag{9.6.55}$$

Proof. Let

$$\Delta=\{\omega:\ \ \xi(\omega)<+\infty\}. \tag{9.6.56}$$

Hence

$$W_i^{(p)}=M\{[\xi(\omega)]^p I_\Delta(\omega)|x_0=i\}. \tag{9.6.57}$$

It is obvious that Δ is a translatable set, consequently, our theorem follows from Theorem 9.6.2. □

Similarly we can also calculate the moments of the distribution of $\xi_A(\omega)$.

§ 9.7 Extensions of the results in §9.3

Suppose Δ is a translatable set, the definition of $\xi_A(\omega)$ is as in §9.5. Let

$${}_H\overset{(v)}{\sigma}_A(\omega)=\begin{cases}\xi_A(\omega), & \text{if } {}_H\sigma_A(\omega)<+\infty,\\ +\infty; & \text{if } {}_H\sigma_A(\omega)=+\infty,\end{cases} \tag{9.7.1}$$

$${}_H\overset{(\Delta,v)}{\Phi}_{iA}(\lambda)=\int_{-0}^{\infty} e^{-\lambda t}dP({}_H\overset{(v)}{\sigma}_A(\omega)\leqslant t,\ \ \Delta|x_0=i), \tag{9.7.2}$$

$${}_H\overset{(\Delta,v)}{m}{}_{iA}^{(p)}=\int_{-0}^{\infty} t^p dP({}_H\overset{(v)}{\sigma}_A(\omega)\leqslant t,\ \Delta|x_0=i), \tag{9.7.3}$$

$$ {}_{H}\overset{(\Delta,v)}{f}{}^{*}_{iA} = {}_{H}\overset{(\Delta,v)}{m}{}^{(0)}_{iA} = P({}_{H}\overset{(v)}{\sigma}{}_{A}(\omega) < +\infty,\ \Delta | x_0 = i). \tag{9.7.4} $$

The proofs of the results in this section are omitted because they can be deduced by following the example of the proofs in §9.3 and §9.6. □

Theorem 9.7.1 *$\{{}_{H}\overset{(\Delta,v)}{\Phi}{}_{iA}(\lambda),\ i\in E\}$ is the minimal nonnegative solution of the pseudo-normal system of equations*

$$ x_i = \sum_{j\notin\{i\}\cup A\cup H} \frac{q_{ij}}{\lambda v(i) + q_i} x_j + \sum_{j\in A\setminus\{i\}} \frac{q_{ij}P_j(\Lambda)}{\lambda v(i) + q_i} \qquad (i\in E).\ \square \tag{9.7.5} $$

Theorem 9.7.2 *$\{{}_{H}\overset{(\Delta,v)}{f}{}^{*}_{iA},\ i\in E\}$ is the minimal nonnegative solution of the pseudo-normal system of equations*

$$ x_i = \sum_{j\notin\{i\}\cup A\cup H} \frac{q_{ij}}{q_i} x_j + \sum_{j\in A\setminus\{i\}} \frac{q_{ij}P_j(\Lambda)}{q_i} \qquad (i\in E).\ \square \tag{9.7.6} $$

Corollary 9.7.1 *If s is an element of E, then*

$$ {}_{H}\overset{(\Delta,v)}{f}{}^{*}_{is} = {}_{H}\overset{(\Omega,v)}{f}{}^{*}_{is} P_s(\Delta) \qquad (i\in E), \tag{9.7.7} $$

$$ {}_{H}\overset{(\Delta,v)}{f}{}^{*}_{iA} = \sum_{j\in A} {}_{H\cup A}\overset{(\Omega,v)}{f}{}^{*}_{ij} P_j(\Delta) \qquad (i\in E).\ \square \tag{9.7.8} $$

Theorem 9.7.3 *For $p \geqslant 1$, $\{{}_{H}\overset{(\Delta,v)}{m}{}^{(p)}_{iA},\ i\in E\}$ is the minimal nonnegative solution of the first-type system of 1-bounded equations*

$$ x_i = \sum_{j\notin\{i\}\cup A\cup H} \frac{q_{ij}}{q_i} x_j + \frac{pv(i)}{q_i}\, {}_{H}\overset{(\Delta,v)}{m}{}^{(p-1)}_{iA} \qquad (i\in E).\ \square \tag{9.7.9} $$

Corollary 9.7.2

$$ {}_{H}\overset{(\Delta,v)}{m}{}^{(p)}_{is} = {}_{H}\overset{(\Omega,v)}{m}{}^{(p)}_{is} P_s(\Delta) \qquad (i\in E), \tag{9.7.10} $$

$$ {}_{H}\overset{(\Delta,v)}{m}{}^{(p)}_{iA} = \sum_{j\in A} {}_{H\cup A}\overset{(\Omega,v)}{m}{}^{(p)}_{ij} P_j(\Delta) \qquad (i\in E).\ \square \tag{9.7.11} $$

Remark 9.7.1 Theorem 9.7.3 can also be deduced from Theorem 9.6.5.

CHAPTER X
Q-Processes of Order One

§ 10.1 Introduction

Part I of this book, indicates that the minimal Q-process and Q-processes of order one are the bases for the study of general Q-processes. In the preceding chapter we studied the minimal Q-process, and in this chapter, we shall proceed to investigate Q-processes of order one.

Henceforth we are often concerned with the Martin exit boundary theory of Q-processes, which is not difficult for general Q-processes on the basis of the Martin boundary theory for Markov chains studied in Chapter VII. For brevity, we shall omit the details and quote the results on Martin boundary theory and some related conclusions directly from [14]. There is no loss of generality because although in [14], the Martin boundary theory is set up with a restriction (denoted there by (P. 5)) on Q-processes, this is not essential, as the author pointed out. In fact, we have shown in Chapter VII that the restriction is not needed: In what follows we will only quote those conclusions which are actually independent of Condition (P.5) in [14] and therefore the restriction (P.5) on Q-processes will not be mentioned.

We must point out in advance that if it occurs that our statements only hold with the exception of a set whose measure is zero, or a common zero measure set with respect to a series of measures, this is not a result of the authors' negligence but is intended to simplify the presentation. By some usual methods, it is easy to make our statement strict without changing the transition probability of the process (for example, by purifying the phase space or selecting a representative from an "equivalent class" of functions).

Suppose $X(\omega)=\{x(t, \omega), t<\sigma(\omega)\}$ is a Q-process of order one, $\tau_1(\omega)$ denotes the first flying point of $x(\cdot, \omega)$, and $(\partial X)_e$ denotes the exit boundary (of class one) of $X(\omega)$ defined in [14]. Let

$$\Pi(b, j) = P(x(\tau_1) = j | x(\tau_1 - 0) = b) \qquad (b \in (\partial X)_e,\ j \in E). \tag{10.1.1}$$

Definition 10.1.1 $\Pi_{(\partial X)_e \times E} = (\Pi(b, j),\ b \in (\partial X)_e, j \in E)$ *is called the flying probability matrix*[1] *of a Q-process $X(\omega)$ of order one.*

---

1 When $(\partial X)_e$ is nondenumerable, it seems improper that $\Pi_{(\partial X)_e \times E}$ is called a matrix, but it is proper if you get used to.

It is seen from [14], the flying probability matrix $\Pi_{(\partial X)_e\times E}=(\Pi(b,j), b\in(\partial X)_e, j\in E)$ has the following properties:

(1) For each fixed $j\in E$, $\Pi(b,j)$ $(b\in(\partial X)_e)$ is a Borel measurable function on $(\partial X)_e$;

(2)
$$\Pi(b,j)\geqslant 0 \qquad (b\in(\partial X)_e,\ j\in E); \tag{10.1.2}$$

$$\sum_{j\in E}\Pi(b,j)\leqslant 1, \qquad (b\in(\partial X)_e). \tag{10.1.3}$$

Conversely, by [14, §7], we have

Lemma 10.1.1 *Suppose the matrix* $\Pi_{(\partial X)_e\times E}=(\Pi(b,j), b\in(\partial X)_e, j\in E)$ *satisfies Conditions* (1) *and* (2). *Then there exists the complete probability space, on which we may define a Q-process of order one such that its flying probability matrix is* $\Pi_{(\partial X)_e\times E}$. □

The above lemma will be applied in Chapter XIII.

Q-processes of order one are called *instantaneous return processes of class one* in [14].

We know from Theorem 10.2.1 below that the transition probability matrix of a Q-process of order one is determined uniquely by its Q-matrix and flying probability matrix, so we introduce

Definition 10.1.2 *The Q-processes of order one with flying probability matrix* $\Pi_{(\partial X)_e\times E}$ *are also called* $(Q, \Pi_{(\partial X)_e\times E})$*-processes.*

Remark 10.1.1 Suppose $D\subset E$. If $\Pi(a,j)=0$ $(a\in(\partial X)_e, j\notin D)$, then $\Pi_{(\partial X)_e\times E}$ is often noted instead of $\Pi_{(\partial X)_e\times D}$.

In what follows we shall investigate Q-processes of order one, i.e., the transition probability, and the distribution and moments of the first passage time of $(Q, \Pi_{(\partial X)_e\times E})$-processes. The details of both integraltype functionals and the classification of states will be omitted since the study of integral-type functionals is similar to that of the first passage time, and the classification of states is an immediate consequence of the distribution and moments of the first passage time.

§ 10.2 Transition probabilities

Suppose $X(\omega)=\{x(t,\omega), t<\sigma(\omega)\}$ is a $(Q, \Pi_{(\partial X)_e\times E})$-process, and let $\mathscr{B}_e$ represent the σ-algebra on $(\partial X)_e$ generated by the family of open sets in $(\partial X)_e$,

$$h(i, B) = P_i(x(\tau_1 - 0) \in B), \qquad B \in \mathscr{B}_e, \tag{10.2.1}$$

$$h(i, B, t) = P_i(x(\tau_1 - 0) \in B, \tau_1 \leqslant t), \qquad B \in \mathscr{B}_e, \tag{10.2.2}$$

$$h_\lambda(i, B) = \int_0^\infty e^{-\lambda t} dh(i, B, t) \qquad (\lambda \geqslant 0), \tag{10.2.3}$$

$$T_{ij}(t) = \int_{(\partial X)_e} h(i, da, t) \Pi(a, j), \tag{10.2.4}$$

$$T_{ij}^{(1)}(t) = T_{ij}(t), \tag{10.2.5}$$

$$T_{ij}^{(n+1)}(t) = \sum_{k \in E} [T_{ik}^{(n)}(\cdot) * T_{kj}^{(1)}(\cdot)](t) \ (* \text{ denotes convolution}), \tag{10.2.6}$$

$$K_{ij}(t) = \sum_{n=1}^{\infty} T_{ij}^{(n)}(t), \tag{10.2.7}$$

$$W_{ij}(\lambda) = \int_0^\infty e^{-\lambda t} dT_{ij}(t) \qquad (\lambda \geqslant 0), \tag{10.2.8}$$

$$W_{ij}^{(1)}(\lambda) = W_{ij}(\lambda), \tag{10.2.9}$$

$$W_{ij}^{(n+1)}(\lambda) = \sum_{k \in E} W_{ik}^{(n)}(\lambda) W_{kj}^{(1)}(\lambda), \tag{10.2.10}$$

$$p_{ij}^{\min}(t) = P_i(x(t) = j, t < \tau_1), \tag{10.2.11}$$

$$p_{ij}(t) = P_i(x(t) = j), \tag{10.2.12}$$

$$p_{ij}^{\min}(\lambda) = \int_0^\infty e^{-\lambda t} p_{ij}^{\min}(t) dt, \tag{10.2.13}$$

$$p_{ij}(\lambda) = \int_0^\infty e^{-\lambda t} p_{ij}(t) dt. \tag{10.2.14}$$

Lemma 10.2.1 $\{h(i, B) - h_\lambda(i, B), i \in E\}$ *is the minimal nonnegative solution of the pseudo-normal system of equations*

$$x_i = \sum_{j \in E \setminus \{i\}} \frac{q_{ij}}{\lambda + q_i} x_j + \frac{\lambda h(i, B)}{\lambda + q_i} \qquad (i \in E). \tag{10.2.15}$$

Proof. In Theorem 9.6.1, let $A = \varnothing$, $V(i) \equiv 1 \, (i \in E)$, $\Lambda = (x(\tau_1 - 0) \in B)$. Then we obtain this lemma immediately. □

Theorem 10.2.1 $\{p_{ij}(\lambda), i\in E\}$ *is the minimal nonnegative solution of the first-type system of* 1*-bounded equations*

$$x_i = \sum_{k\in E}\left(\int_{(\partial X)_e} h_\lambda(i, da)\Pi(a, k)\right)x_k + p_{ij}^{\min}(\lambda) \qquad (i\in E). \tag{10.2.16}$$

Proof. It follows by (10.2.4) and (10.2.8) that

$$W_{ij}(\lambda) = \int_{(\partial X)_e} h_\lambda(i, db)\Pi(b, j). \tag{10.2.17}$$

From Formulae 10 in the Appendix of [14] we obtain

$$p_{ij}(\lambda) = p_{ij}^{\min}(\lambda) + \int_{(\partial X)_e} h_\lambda(i, da)\left(\sum_{n=0}^{\infty} V_\lambda^{(n)}(a, db)\right)\sum_{k\in E}\Pi(b, k)p_{kj}^{\min}(\lambda)$$

$$= p_{ij}^{\min}(\lambda) + \sum_{k\in E}\sum_{n=0}^{\infty}\int_{(\partial X)_e}\int_{(\partial X)_e} h_\lambda(i, da)V_\lambda^{(n)}(a, db)\Pi(b, k)p_{kj}^{\min}(\lambda), \tag{10.2.18}$$

where

$$V_\lambda^{(0)}(a, B) = \begin{cases} 0, & a\notin B, \\ 1, & a\in B, \end{cases} \tag{10.2.19}$$

$$V_\lambda^{(1)}(a, B) = \sum_{l\in E}\Pi(a, l)h_\lambda(l, B), \tag{10.2.20}$$

$$V_\lambda^{(n+1)}(a, B) = \int_{(\partial X)_e} V_\lambda^{(n)}(a, db)V_\lambda^{(1)}(b, B). \tag{10.2.21}$$

We will now prove

$$W_{ik}^{(n+1)}(\lambda) = \int_{(\partial X)_e}\int_{(\partial X)_e} h_\lambda(i, da)V_\lambda^{(n)}(a, db)\Pi(b, k) \qquad (n\geqslant 0). \tag{10.2.22}$$

In fact

$$\int_{(\partial X)_e}\int_{(\partial X)_e} h_\lambda(i, da)V_\lambda^{(0)}(a, db)\Pi(b, k)$$

$$= \int_{(\partial X)_e} h_\lambda(i, da)\int_{(\partial X)_e} V_\lambda^{(0)}(a, db)\Pi(b, k)$$

$$= \int_{(\partial X)_e} h_\lambda(i, da)\Pi(a, k)$$

$$= \int_{(\partial X)_e} \left[\int_0^\infty e^{-\lambda t} dh(i, da, t)\right]\Pi(a, k)$$

$$= \int_0^\infty e^{-\lambda t} \int_{(\partial X)_e} dh(i, da, t)\Pi(a, k)$$

$$= \int_0^\infty e^{-\lambda t} d \int_{(\partial X)_e} h(i, da, t)\Pi(a, k)$$

$$= \int_0^\infty e^{-\lambda t} dT_{ik}(t) = W_{ik}^{(1)}(\lambda). \tag{10.2.23}$$

So (10.2.22) holds when $n = 0$. We now assume that (10.2.22) holds for n, thus

$$\int_{(\partial X)_e} \int_{(\partial X)_e} h_\lambda(i, da) V_\lambda^{(n)}(a, db)\Pi(b, k)$$

$$= \int_{(\partial X)_e} \int_{(\partial X)_e} h_\lambda(i, da)\left[\int_{(\partial X)_e} V_\lambda^{(n-1)}(a, dc) V_\lambda^{(1)}(c, db)\right]\Pi(b, k)$$

$$= \int_{(\partial X)_e} \int_{(\partial X)_e} h_\lambda(i, da) \int_{(\partial X)_e} V_\lambda^{(n-1)}(a, dc) \sum_{l\in E} \Pi(c, l) h_\lambda(l, db)\Pi(b, k)$$

$$= \sum_{l\in E} \int_{(\partial X)_e} \int_{(\partial X)_e} h_\lambda(i, da) V_\lambda^{(n-1)}(a, dc)\Pi(c, l) \int_{(\partial X)_e} h_\lambda(l, db)\Pi(b, k)$$

$$= \sum_{l\in E} W_{il}^{(n)}(\lambda) W_{lk}^{(1)} = W_{ik}^{(n+1)}(\lambda). \tag{10.2.24}$$

Therefore (10.2.22) also holds for $n+1$, (10.2.22) holds universally by induction.

It follows by (10.2.18) and (10.2.22) that

$$p_{ij}(\lambda) = p_{ij}^{\min}(\lambda) + \sum_{k\in E}\left(\sum_{n=1}^{\infty} W_{ik}^{(n)}(\lambda)\right) p_{kj}^{\min}(\lambda). \tag{10.2.25}$$

Our theorem follows readily from (10.2.17) and Theorem 3.7.1.□

Definition 10.2.1 *Suppose $k(\cdot, \Delta) \geqslant 0$ and $g(\cdot) \geqslant 0$ are measurable functions on $(\partial X)_e$, $k(a, \cdot)$ is a measure on $(\partial X)_e$. The nonnegative solution $0 \leqslant \xi^{(a)} \leqslant +\infty$ $(a \in (\partial X)_e)$ of the integral equation*

$$\xi^{(a)} = \int_{(\partial X_e)} k(a, db)\xi^{(b)} + g(a) \qquad (a \in (\partial X)_e) \tag{10.2.26}$$

is called the minimal nonnegative solution, if for any nonnegative solution $0 \leqslant \hat{\xi}^{(a)} \leqslant +\infty$ $(a \in (\partial X)_e)$ *of* (10.2.26) *we have*

$$\xi^{(a)} \leqslant \hat{\xi}^{(a)} \qquad (a \in (\partial X)_e). \tag{10.2.27}$$

Following the proof of Theorem 3.2.1, it is easy to obtain:

Lemma 10.2.2 *The minimal nonnegative solution of the integral equation* (10.2.26) *exists uniquely.* □

Equation (10.2.18) yields the following:

Theorem 10.2.2

$$p_{ij}(\lambda) = p_{ij}^{\min}(\lambda) + \int_{(\partial X)_e} h_\lambda(i, da) \sum_{n=0}^{\infty} \int_{(\partial X)_e} V_\lambda^{(n)}(a, db) \sum_{k \in E} \Pi(b, k) p_{kj}^{\min}(\lambda)$$

$$(i, j \in E). \ \square \tag{10.2.28}$$

Theorem 10.2.3

$$p_{ij}(\lambda) = p_{ij}^{\min}(\lambda) + \int_{(\partial X)_e} h_\lambda(i, da) \xi_j^{(a)}(\lambda) \qquad (i, j \in E), \tag{10.2.29}$$

where

$$\xi_j^{(a)}(\lambda) = \sum_{k \in E} \Pi(a, k) p_{kj}(\lambda)$$

$$= \int_0^\infty e^{-\lambda t} P(x(t + \tau_1) = j | x(\tau_1 - 0) = a) dt \qquad (a \in (\partial X)_e) \tag{10.2.30}$$

is the minimal nonnegative solution of the integral equation

$$\xi^{(a)} = \int_{(\partial X)_e} \left(\sum_{i \in E} \Pi(a, i) h_\lambda(i, db) \right) \xi^{(b)}$$

$$+ \sum_{k \in E} \Pi(b, k) p_{kj}^{\min}(\lambda) \qquad (a \in (\partial X)_e). \tag{10.2.31}$$

Proof. It can be seen by (10.2.28) that

$$\xi_j^{(a)} = \sum_{n=0}^{\infty} \int_{(\partial X)_e} V_\lambda^{(n)}(a, db) \sum_{k\in E} \Pi(b, k) p_{kj}^{\min}(\lambda) \qquad (a\in(\partial X)_e). \tag{10.2.32}$$

It is easy to show that $\xi_j^{(a)}(\lambda)$ $(a\in(\partial X)_e)$ is the minimal nonnegative solution of the integral equation (10.2.31) by (10.2.20), (10.2.32) in a manner similar to the proofs of Theorems 3.2.1 and 3.7.1. Therefore it follows from (10.2.28) and (10.2.32) that

$$\begin{aligned}
\xi_j^{(a)}(\lambda) &= \int_{(\partial X)_e} \left(\sum_{i\in E} \Pi(a, i) h_\lambda(i, db) \right) \sum_{n=0}^{\infty} \int_{(\partial X)_e} V_\lambda^{(n)}(b, dc) \\
&\quad \times \sum_{k\in E} \Pi(c, k) p_{kj}^{\min}(\lambda) + \sum_{i\in E} \Pi(a, i) p_{ij}^{\min}(\lambda) \\
&= \sum_{i\in E} \Pi(a, i) \Bigg(p_{ij}^{\min}(\lambda) + \int_{(\partial X)_e} h_\lambda(i, db) \\
&\quad \times \sum_{n=0}^{\infty} \int_{(\partial X)_e} V_\lambda^{(n)}(b, dc) \sum_{k\in E} \Pi(c, k) p_{kj}^{\min}(\lambda) \Bigg) \\
&= \sum_{i\in E} \Pi(a, i) p_{ij}(\lambda) \qquad (a\in(\partial X)_e).
\end{aligned} \tag{10.2.33}$$

Taking $\Delta\in\mathscr{B}_e$ and denoting by $P_{x(\tau_1-0)}$ the distribution of $x(\tau_1-0)$, we have

$$\begin{aligned}
&\int_\Delta P(x(t+\tau_1)=j,\, x(\tau_1)=k \mid x(\tau_1-0)=a) P_{x(\tau_1-0)}(da) \\
&= \int_\Delta P(x(\tau_1)=k \mid x(\tau_1-0)=a) \\
&\quad \times P(x(t+\tau_1)=j \mid x(\tau_1-0)=a,\, x(\tau_1)=k) P_{x(\tau_1-0)}(da) \\
&= \int_\Delta \Pi(a, k) P(x(t)=j \mid x(0)=k) P_{x(\tau_1-0)}(da) \\
&= \int_\Delta \Pi(a, k) P_{kj}(t) p_{x(\tau_1-0)}(da).
\end{aligned} \tag{10.2.34}$$

It follows from the arbitrariness of Δ that

$$\begin{aligned}
&P(x(t+\tau_1)=j,\, x(\tau_1)=k \mid x(\tau_1-0)=a) \\
&\qquad = \Pi(a, k) p_{kj}(t) \qquad (a\in(\partial X)_e).
\end{aligned} \tag{10.2:35}$$

Consequently

$$P(x(t+\tau_1)=j|x(\tau_1-0)=a)$$

$$=\sum_{k\in E} P(x(t+\tau_1)=j,\ x(\tau_1)=k|x(\tau_1-0)=a)$$

$$=\sum_{k\in E} \Pi(a,k)p_{kj}(t). \tag{10.2.36}$$

Equations (10.2.33) and (10.2.36) imply (10.2.30).□

§ 10.3 Distribution and moments of the first passage time

Lemma 10.3.1 *Suppose* $\Lambda\in\mathscr{B}_{[0,\tau_1)}$. *Then*

$$P_i(\Lambda,\ x(\tau_1)\in A)=\int_{(\partial X)_e}\Pi(b,A)P_i(\Lambda,\ x(\tau_1-0)\in db), \tag{10.3.1}$$

where $\mathscr{B}_{[0,\tau_1)}=\mathscr{F}\left\{\bigcup_{n=1}^{\infty}\mathscr{B}_{[0,\tau^{(n)}]}\right\}$ and $\mathscr{B}_{[0,\tau^{(n)}]}=\mathscr{F}\{x_t,\ t\leqslant\tau^{(n)}\}$.

Proof. We shall prove this in the following three steps:

(1) To prove

$$P_i(x(\tau_1)\in A)=\int_{(\partial X)_e}\Pi(b,A)h(i,db). \tag{10.3.2}$$

Noticing the statements between formulae (5.9) and (5.10) in [14], and formula (5.9) in [14], we obtain

$$P_i(\tau_1\leqslant t,\ x(\tau_1)\in A)=\int_{(\partial X)_e}\Pi(b,A)P_i(\tau_1\leqslant t,\ x(\tau_1-0)\in db). \tag{10.3.3}$$

In (10.3.3), let $t\uparrow+\infty$; we obtain

$$P_i(\tau_1<+\infty,\ x(\tau_1)\in A)$$

$$=\int_{(\partial X)_e}\Pi(b,A)P_i(\tau_1<+\infty,\ x(\tau_1-0)\in db). \tag{10.3.4}$$

But

$$(x(\tau_1)\in A)\subset(\tau_1<+\infty) \tag{10.3.5}$$

and

$$\left(x(\tau_1-0)\in(\partial X)_e\right)\subset(\tau_1<+\infty). \tag{10.3.6}$$

Equation (10.3.2) now follows from (10.3.4), (10.3.5) and (10.3.6).

(2) To prove

$$P_i(\hat{\Lambda},\ x(\tau_1)\in A)=\int_{(\partial X)_e}\Pi(b,\ A)P_i(\hat{\Lambda},\ x(\tau_1-0)\in db), \tag{10.3.7}$$

where $\hat{\Lambda}\in\bigcup_{n=1}^{\infty}\mathscr{B}_{[0,\,\tau^{(n)}]}$.

We see from the definition of $\hat{\Lambda}$ that there exists $n>0$ such that $\hat{\Lambda}\in\mathscr{B}_{[0,\,\tau^{(n)}]}$. Observe

$$\theta_{\tau^{(n)}}(x(\tau_1)\in A)=(x(\tau_1)\in A), \tag{10.3.8}$$

where $\theta_{\tau^{(n)}}$ is the translation operator defined in [6]. Owing to (10.3.8), the strong Markov property and the homogeneity, we get

$$P_i(x(\tau_1)\in A|\hat{\Lambda},\ x(\tau^{(n)})=j)=P_j(x(\tau_1)\in A), \tag{10.3.9}$$

$$P_i(x(\tau_1-0)\in db|\hat{\Lambda},\ x(\tau^{(n)})=j)=P_j(x(\tau_1-0)\in db). \tag{10.3.10}$$

Equations (10.3.2) and (10.3.3) assure that

$$P_i(\hat{\Lambda},\ x(\tau_1)\in A)=\sum_{j\in E}P_i(\hat{\Lambda},\ x(\tau^{(n)})=j,\ x(\tau_1)\in A)$$

$$=\sum_{j\in E}P_i(\hat{\Lambda},\ x(\tau^{(n)})=j)P_i(x(\tau_1)\in A|\hat{\Lambda},\ x(\tau^{(n)})=j)$$

$$=\sum_{j\in E}P_i(\hat{\Lambda},\ x(\tau^{(n)})=j)P_j(x(\tau_1)\in A)$$

$$=\sum_{j\in E}P_i(\hat{\Lambda},x(\tau^{(n)})=j)\int_{(\partial X)_e}\Pi(b,\ A)P_j(x(\tau_1-0)\in db)$$

$$=\int_{(\partial X)_e}\Pi(b,\ A)\sum_{j\in E}P_i(\hat{\Lambda},\ x(\tau^{(n)})=j)$$

$$\times\ P_i(x(\tau_1-0)\in db|\hat{\Lambda},\ x(\tau^{(n)}=j)$$

$$=\int_{(\partial X)_e}\Pi(b,\ A)\sum_{j\in E}P_i(\hat{\Lambda},\ x(\tau^{(n)}=j),\ x(\tau_1-0)\in db)$$

$$= \int_{(\partial X)_e} \Pi(b, A) P_i(\hat{\Lambda}, x(\tau_1 - 0) \in db). \tag{10.3.11}$$

Consequently (10.3.7) is valid.

(3) To prove (10.3.1).

Let

$$\mathcal{W} = \{\Lambda: \Lambda \text{ such that (10.3.1) holds}\}. \tag{10.3.12}$$

We know from (2) that

$$\mathcal{W} \supseteq \bigcup_{n=1}^{p} \mathcal{B}_{[0, \tau^{(n)}]}. \tag{10.3.13}$$

It is not hard to show that $\mathcal{W}$ is a Λ-system. And from $\mathcal{B}_{[0, \tau^{(n)}]} \subseteq \mathcal{B}_{[0, \tau^{(n+1)}]}$ $(n \geqslant 1)$ we infer that $\bigcup_{n=1}^{\infty} \mathcal{B}_{[0, \tau^{(n)}]}$ is a Π-system. Hence, (10.3.13) and Lemma 1.1 of [21] imply that

$$\mathcal{W} \supseteq \mathcal{F}\left\{\bigcup_{n=1}^{\infty} \mathcal{B}_{[0, \tau^{(n)}]}\right\} = \mathcal{B}_{[0, \tau_1]}. \tag{10.3.14}$$

Therefore (10.3.1) is proved.

This completes the proof of the theorem.□

Lemma 10.3.2 *Suppose* $\Delta \in \mathcal{B}_e$. *Then*

$$P_i(x(\tau_1 - 0) \in \Delta,\ x(\tau_1) \in A) = \int_{\Delta} \Pi(b, A) h(i, db). \tag{10.3.15}$$

Proof. It follows from $(x(\tau_1 - 0)\Delta \in \mathcal{B}_{[0, \tau_1]}$ and Lemma 10.3.1 that

$$P_i(x(\tau_1 - 0) \in \Delta,\ x(\tau_1) \in A)$$

$$= \int_{(\partial X)_e} \Pi(b, A) P_i(x(\tau_1 - 0) \in \Delta,\ x(\tau_1 - 0) \in db)$$

$$= \int_{(\partial X)_e \cap \Delta} \Pi(b, A) P_i(x(\tau_1 - 0) \in db)$$

$$= \int_{\Delta} \Pi(b, A) h(i, db). \ \square \tag{10.3.16}$$

Suppose D is a nonempty subset of E and $B \in B_e$. Let

$$\mathcal{J}_{iD}^{*(B)} = P_i(\text{there exists } s,\ \tau^{(1)} \leqslant s \leqslant \tau_1, \text{ such}$$

$$\text{that } x_s \in D,\ x(\tau_1 - 0) \in B), \tag{10.3.17}$$

$$\tilde{g}_{iD}^{(B)}(t) = P_i(x_s \notin D,\ \tau^{(1)} \leqslant s < \tau_1,\ x(\tau_1 - 0) \in B,\ \tau_1 \leqslant t), \tag{10.3.18}$$

$$\tilde{G}_{iD}^{(B)}(\lambda) = \int_{-0}^{\infty} e^{-\lambda t} d\tilde{g}_{iD}^{(B)}(t) \qquad (\lambda \geqslant 0), \tag{10.3.19}$$

$$\tilde{G}_{iD}^{*(B)} = \tilde{G}_{iD}^{(B)}(0), \tag{10.3.20}$$

$$\tilde{L}_{iD}^{(B,\,p)} = \int_{-0}^{\infty} t^p d\tilde{g}_{iD}^{(B)}(t). \tag{10.3.21}$$

Hence

$$\tilde{L}_{iD}^{(B,\,0)} = \tilde{G}_{iD}^{*(B)} = h(i,\,B) - \tilde{f}_{iD}^{*(B)}. \tag{10.3.22}$$

Lemma 10.3.3 $\{\tilde{f}_{iD}^{*(B)},\ i \in E\}$ *is the minimal nonnegative solution of the pseudo-normal system of equations*

$$x_i = \sum_{j \notin D \cup \{i\}} \frac{q_{ij}}{q_i} x_j + \sum_{j \in D \setminus \{i\}} \frac{q_{ij}}{q_i} h(j,\,B) \qquad (i \in E). \tag{10.3.23}$$

Proof. Observing that $(x(\tau_1 - 0) \in B)$ is a translatable set and letting $v(i) \equiv 1\ (i \in E)$, $\Lambda = (x(\tau_1 - 0) \in B)$, $H \neq \varnothing$, $D = A$ in Theorem 9.7.2, we obtain this lemma.□

Lemma 10.3.4 *For* $\lambda > 0$, $\{h(i,\,B) - \tilde{f}_{iD}^{*(B)} - \tilde{G}_{iD}^{(B)}(\lambda),\ i \in E\}$ *is the minimal nonnegative solution of the pseudo-normal system of equations*

$$x_i = \sum_{j \notin D \cup \{i\}} \frac{q_{ij}}{\lambda + q_i} x_j + \frac{\lambda(h(i,\,B) - \tilde{f}_{iD}^{*(B)})}{\lambda + q_i} \qquad (i \in E). \tag{10.3.24}$$

Proof. In Theorem 9.6.6, letting $v(i) \equiv 1\ (i \in E)$, $A = D$, $\Delta = \Omega$, we obtain this lemma.□

Lemma 10.3.5 $\{\tilde{L}_{iD}^{(B,\,p)},\ i \in E\}\ (p \geqslant 1)$ *is the minimal nonnegative solution of the first-type system of* 1*-bounded equations*

$$x_i = \sum_{j \notin D \cup \{i\}} \frac{q_{ij}}{q_i} x_j + \frac{p}{q_i} \tilde{L}_{iD}^{(B,\,p-1)} \qquad (i \in E). \tag{10.3.25}$$

Proof. This lemma is an immediate consequence of Theorem 9.6.6.□

Lemma 10.3.6

$$p_i(\tau_1 \leqslant t,\ x_s \notin A \cup H,\ \tau^{(1)} \leqslant s < \tau_1,\ x(\tau_1) = j)$$

$$= \int_{(\partial X)_e} \tilde{g}^{(da)}_{i, A \cup H}(t) \Pi(a, j) \qquad (i \in E). \tag{10.3.26}$$

Proof. This lemma follows readily from Lemma 10.3.1.□

In what follows, the notations ${}_H\tilde{f}_{iA}(t)$, ${}_H\tilde{\Phi}_{iA}(\lambda)$, ${}_H\tilde{f}^*_{iA}$ and ${}_H\tilde{m}^{(p)}_{iA}$ introduced in §9.3 are replaced by ${}_Hf_{iA}(t)$, ${}_H\Phi_{iA}(\lambda)$, ${}_Hf^*_{iA}$ and ${}_Hm^{(p)}_{iA}$, respectively, while ${}_H\sigma_A$, ${}_Hf_{iA}(t)$, ${}_H\Phi_{iA}(\lambda)$, ${}_Hf^*_{iA}$ and ${}_Hm^{(p)}_{iA}$ are used for Q processes of order one.

Let $\tau_1(\omega)$ denote the first flying point of $x(\cdot, \omega)$. Let $\tau_2(\omega) = \sigma(\omega)$,if $\tau_1(\omega) = \sigma(\omega)$; otherwise let $\tau_2(\omega)$ denote the first flying point after $\tau_1(\omega)$. Supposing $\tau_{k-1}(\omega)$ has been defined, let $\tau_k(\omega) = \sigma(\omega)$, if $\tau_{k-1}(\omega) = \sigma(\omega)$; otherwise let $\tau_k(\omega)$ denote the first flying point after $\tau_{k-1}(\omega)$. It is an easy matter to show that $\tau_k(\omega)$ $(k = 1, 2, \cdots)$ is well defined, and

$$\lim_{n \to \infty} \tau_k(\omega) = \sigma(\omega) \qquad (\omega \in \Omega). \tag{10.3.27}$$

If $\tau_{n+1}(\omega) > \tau_n(\omega)$, then let $\tau^{(n, m)}(\omega)$ denote the mth jump point of $x(\cdot, \omega)$ after $\tau_n(\omega)$, otherwise, let $\tau^{(n, m)}(\omega) = \tau_n(\omega)$. It is easy to see that $\tau^{(n, m)}(\omega)$ is well defined.

Both A and H are subsets of E, and $A \neq \varnothing$. Let

$$\sigma_A = \begin{cases} \inf\{t: \tau^{(0, 1)} \leqslant t < \sigma, x_t \in A\}, \\ +\infty, \quad \text{if the above set is empty}, \end{cases} \tag{10.3.28}$$

$${}_H\sigma_A = \begin{cases} \sigma_A, & \text{if } \sigma_A \leqslant \sigma_H, \\ +\infty, & \sigma_A > \sigma_H, \end{cases} \tag{10.3.29}$$

$${}_Hf^{(n)}_{iA}(t) = P_i({}_H\sigma_A \leqslant t, \tau^{(n-1, 1)} \leqslant {}_H\sigma_A < \tau^{(n, 1)}) \qquad (n \geqslant 1), \tag{10.3.30}$$

$${}_Hf_{iA}(t) = P_i({}_H\sigma_A \leqslant t), \tag{10.3.31}$$

$${}_H\Phi^{(n)}_{iA}(\lambda) = \int_{-0}^{\infty} e^{-\lambda t} d\,{}_Hf^{(n)}_{iA}(t) \qquad (n \geqslant 1, \lambda \geqslant 0), \tag{10.3.32}$$

$${}_H\Phi_{iA}(\lambda) = \int_{-0}^{\infty} e^{-\lambda t} d\,{}_Hf_{iA}(t) \qquad (\lambda \geqslant 0), \tag{10.3.33}$$

$${}_Hm^{(n, p)}_{iA} = \int_{-0}^{\infty} t^p d\,{}_Hf^{(n)}_{iA}(t) \qquad (n \geqslant 1, p \geqslant 0), \tag{10.3.34}$$

$${}_Hm^{(p)}_{iA} = \int_{-0}^{\infty} t^p d\,{}_Hf_{iA}(t) \qquad (p \geqslant 0). \tag{10.3.35}$$

Obviously we have

$$ {}_Hf_{iA}(t) = \sum_{n=1}^{\infty} {}_Hf_{iA}^{(n)}(t), \tag{10.3.36} $$

$$ {}_H\Phi_{iA}(\lambda) = \sum_{n=1}^{\infty} {}_H\Phi_{iA}^{(n)}(\lambda), \tag{10.3.37} $$

$$ {}_Hm_{iA}^{(p)} = \sum_{n=1}^{\infty} {}_Hm_{iA}^{(n,p)}. \tag{10.3.38} $$

Sometimes we write ${}_Hf_{iA}^*$ and ${}_Hm_{iA}$ for ${}_Hm_{iA}^{(0)}$, and ${}_Hm_{iA}^{(1)}$, respectively.

Lemma 10.3.7 *${}_Hf_{iA}^{(n)}(t)$ is determined uniquely by the following recursion formulae*

$$ {}_Hf_{iA}^{(1)}(t) = {}_H\tilde{f}_{iA}(t) + \sum_{j\in A} \int_{(\partial X)_e} \tilde{g}_{i,\,A\cup H}^{(da)}(t)\Pi(a, j) \qquad (i\in E), \tag{10.3.39} $$

$$ {}_Hf_{iA}^{(n+1)}(t) = \sum_{j\notin A\cup H} \int_{-0}^{t} {}_Hf_{jA}^{(n)}(t-u)d\int_{(\partial X)_e} \tilde{g}_{i,A\cup H}(u)\Pi(a, j) \qquad (i\in E,\ n\geqslant 1). \tag{10.3.40} $$

Proof. Equation (10.3.39) follows from Lemma 10.3.6 and

$$ \begin{aligned} {}_Hf_{iA}^{(1)}(t) &= P_i({}_H\sigma_A \leqslant t,\ \tau^{(0,1)} \leqslant {}_H\sigma_A < \tau^{(1,1)}) \\ &= P_i({}_H\sigma_A \leqslant t,\ \tau^{(0,1)} \leqslant {}_H\sigma_A < \tau_1) + P_i({}_H\sigma_A \leqslant t,\ {}_H\sigma_A = \tau_1) \\ &= P_i({}_H\sigma_A \leqslant t) + \sum_{j\in A} P_i(\tau_1 \leqslant t,\ x\notin A\bigcup H,\ \tau^{(0,1)} \leqslant s < \tau_1,\ x(\tau_1) = j). \end{aligned} $$

It follows from the strong Markov property, homogeneity, Lemmas 10.3.6 and 9.3.2 that

$$ \begin{aligned} {}_Hf_{iA}^{(n+1)}(t) &= \sum_{j\notin A\bigcup H} P_i({}_H\sigma_A \leqslant t,\ \tau^{(n,1)} \leqslant {}_H\sigma_A < \tau^{(n+1,1)},\ x(\tau_1) = j) \\ &= \sum_{j\notin A\bigcup H} P_i(x_s\notin A\bigcup H,\ \tau^{(0,1)} \leqslant s < \tau_1,\ {}_H\sigma_A \leqslant t, \\ &\qquad x(\tau_1) = j,\ \tau^{(n,1)} \leqslant {}_H\sigma_A < \tau^{(n+1,1)}) \\ &= \sum_{j\notin A\bigcup H} P_i(x(\tau_1) = j)P_i(x_s\notin A\bigcup H,\ \tau^{(0,1)} \leqslant s < \tau_1, \end{aligned} $$

$$\left({}_H\sigma_A - \tau_1\right) + \tau_1 \leqslant t,\ \tau^{(n,1)} - \tau_1 \leqslant {}_H\sigma_A$$

$$- \tau_1 < \tau^{(n+1,1)} - \tau_1 | x(\tau) = j)$$

$$= \sum_{j \notin A \bigcup H} P_i(x(\tau_1) = j) \int_{-0}^{t} P_i({}_H\sigma_A - \tau_1 \leqslant t - u,$$

$$\tau^{(n,1)} - \tau_1 \leqslant {}_H\sigma_A - \tau_1 < \tau^{(n+1,1)} - \tau_1 | x(\tau_1) = j)$$

$$\times\, dP_i(\tau_1 \leqslant u,\ x_s \notin A \bigcup H,\ \tau^{(0,1)} \leqslant s < \tau_1 | x(\tau_1) = j)$$

$$= \sum_{j \notin A \bigcup H} \int_{-0}^{t} P_i({}_H\sigma_A \leqslant t - u,\ \tau^{(n-1,1)} \leqslant {}_H\sigma_A < \tau^{(n,1)})$$

$$\times\, dP_i(\tau_1 \leqslant u,\ x_s \notin A \bigcup H,\ \tau^{(0,1)} \leqslant s < \tau_1,\ x(\tau_1) = j)$$

$$= \sum_{j \notin A \bigcup H} \int_{-0}^{t} {}_H f_{jA}^{(n)}(t-u)\, d \int_{(\partial X)_e} \tilde{g}_{i,A \bigcup H}^{(da)}(u) \Pi(a, j).$$

Hence (10.3.40) is valid.□

Lemma 10.3.8 ${}_H\Phi_{iA}^{(n)}(\lambda)$ *is determined uniquely by the following recursion formulae*:

$$ {}_H\Phi_{iA}^{(1)}(\lambda) = {}_H\tilde{\Phi}_{iA}^{(n)}(\lambda) + \sum_{j \in A} \int_{(\partial X)_e} \tilde{G}_{i,A \bigcup H}^{(da)}(\lambda) \Pi(a, j) \qquad (i \in E), \qquad (10.3.41)$$

$$ {}_H\Phi_{iA}^{(n+1)}(\lambda) = \sum_{j \bar{\in} A \bigcup H} \tilde{G}_{i,A \bigcup H}^{(da)}(\lambda) \Pi(a, j)\, {}_H\Phi_{iA}^{(n)}(\lambda) \qquad (i \in E,\ n \geqslant 1). \qquad (10.3.42)$$

Proof. This lemma is a consequence of Lemmas 10.3.7 and 9.3.3.□

Lemma 10.3.9 ${}_H m_{iA}^{(n,p)}$ *is determined uniquely by the following recursion formulae*:

$$ {}_H m_{iA}^{(1,p)} = {}_H \tilde{m}_{iA}^{(p)} + \sum_{j \in A} \int_{(\partial X)_e} \tilde{L}_{i,A \bigcup H}^{(da,p)} \Pi(a, j) \qquad (i \in E), \qquad (10.3.43)$$

$$ {}_H m_{iA}^{(n+1,p)} = \sum_{l=0}^{p} \sum_{i \notin A \bigcup H} \int_{(\partial X)_e} C_p^l \tilde{L}_{i,A \bigcup H}^{(da,l)} \Pi(a, j)\, {}_H m_{jA}^{(n,p-1)} \qquad (i \in E,\ n \geqslant 1). \qquad (10.3.44)$$

Proof. This lemma follows from Lemmas 10.3.7 and 9.3.4.□

Theorem 10.3.1 $\{{}_H\Phi_{iA}(\lambda),\ i \in E\}$ *is the minimal nonnegative solution of the pseudo-normal system of equations*

$$x_i = \sum_{j \notin A \cup H} \int_{(\partial X)_e} G^{(da)}_{i, A \cup H}(\lambda) \Pi(a, j) x_j + {}_H\tilde{\Phi}_{iA}(\lambda)$$

$$+ \sum_{j \in A} \int_{(\partial X)_e} \tilde{G}^{(da)}_{i, A \cup H}(\lambda) \Pi(a, j) \qquad (i \in E). \tag{10.3.45}$$

Proof. A little reflection shows that (10.3.45) is a pseudo-normal system of equations. The theorem is deduced readily from Lemma 10.3.8 and Theorem 3.2.1.□

Theorem 10.3.2

$${}_H\Phi_{iA}(\lambda) = \int_{(\partial X)_e} \tilde{G}^{(da)}_{i, A \cup H}(\lambda) \xi^{(a)}(\lambda)$$

$$+ \int_{(\partial X)_e} \tilde{G}^{(da)}_{i, A \cup H}(\lambda) \xi^{(a)}_0 + {}_H\tilde{\Phi}_{iA}(\lambda) \qquad (i \in E), \tag{10.3.46}$$

where

$$\xi^{(a)}_0 = \sum_{j \notin A} \Pi(a, j) \qquad (a \in (\partial X)_e), \tag{10.3.47}$$

and $\xi^{(a)}_{(\lambda)} (a \in (\partial X)_e)$ *is the minimal nonnegative solution of the integral equation*

$$\xi^{(b)} = \int_{(\partial X)_e} \sum_{i \notin A \cup H} \tilde{G}^{(da)}_{i, A \cup H}(\lambda) \Pi(b, i) \xi^{(a)}$$

$$+ \int_{(\partial X)_e} \sum_{i \notin H} \tilde{G}^{(da)}_{i, A \cup H}(\lambda) \Pi(b, i) \sum_{j \in A} \Pi(a, j) + \sum_{i \notin A \cup H} {}_H\tilde{\Phi}^{(\lambda)}_{iA} \Pi(b, i). \tag{10.3.48}$$

Proof. We see from Theorem 10.3.1 that it suffices to show

$$\sum_{j \notin A \cup H} \Pi(a, j) {}_H\tilde{\Phi}_{jA}(\lambda) \qquad (a \in (\partial X)_e)$$

is the minimal nonnegative solution of the integral equation (10.3.48). Hence let

$$x^{(0)}_i = 0,$$

$$\left.\begin{aligned} x^{(n+1)}_i = {} & \sum_{j \notin A \cup H} \int_{(\partial X)_e} \tilde{G}^{(da)}_{i, A \cup H}(\lambda) \Pi(a, j) x^{(n)}_j + {}_H\tilde{\Phi}_{iA}(\lambda) \\ & + \sum_{j \in A} \int_{(\partial X)_e} \tilde{G}^{(da)}_{i, A \cup H}(\lambda) \Pi(a, j) \qquad (n \geqslant 1, i \in E). \end{aligned}\right\} \tag{10.3.49}$$

$$\left.\begin{aligned}
&\overset{(0)}{\xi}{}^{(b)} \equiv 0,\\
&\overset{(n+1)}{\xi}{}^{(b)} = \int_{(\partial X)_e} \sum_{i\notin A\cup H} \tilde{G}^{(da)}_{i,\,A\cup H}(\lambda)\Pi(b,\, i)\overset{(n)}{\xi}{}^{(a)}\\
&\quad + \int_{(\partial X)_e} \sum_{i\notin A\cup H} \tilde{G}^{(da)}_{i,\,A\cup H}(\lambda)\Pi(b,\, i)\sum_{j\in A}\Pi(a,\, j)\\
&\quad + \sum_{i\notin A\cup H} {}_H\tilde{\Phi}_{iA}(\lambda)\Pi(b,\, i) \qquad (n \geqslant 1,\, b\in(\partial X)_e).
\end{aligned}\right\} \tag{10.3.50}$$

Thus

$$\sum_{i\notin A\cup H} \Pi(b,\, i)x_i^{(0)} = \sum_{i\notin A\cup H} \Pi(b,\, i)\cdot 0 = 0 = \overset{(0)}{\xi}{}^{(b)} \qquad (b\in(\partial X)_e). \tag{10.3.51}$$

We now assume

$$\sum_{i\notin A\cup H} \Pi(b,\, i)x_i^{(n)} = \overset{(n)}{\xi}{}^{(b)} \qquad (b\in(\partial X)_e). \tag{10.3.52}$$

Then

$$\begin{aligned}
\sum_{i\notin A\cup H} \Pi(b,\, i)x_i^{(n+1)} &= \sum_{i\notin A\cup H} \Pi(b,\, i)\Bigg(\sum_{j\notin A\cup H}\Bigg(\int_{(\partial X)_e} \tilde{G}^{(da)}_{i,\,A\cup H}(\lambda)\Pi(a,\, j)x_j^{(n)}\\
&\quad + {}_H\tilde{\Phi}_{iA}(\lambda)\sum_{j\in A}\int_{(\partial X)_e} \tilde{G}^{(da)}_{i,\,A\cup H}(\lambda)\Pi(a,\, j)\Bigg)\Bigg)\\
&= \int_{(\partial X)_e}\Bigg(\sum_{i\notin A\cup H} \tilde{G}^{(da)}_{i,\,A\cup H}(\lambda)\Pi(b,\, i)\Bigg)\sum_{j\notin A\cup H}\Pi(a,\, j)x_j^{(n)}\\
&\quad + \int_{(\partial X)_e}\Bigg(\sum_{i\notin A\cup H} \tilde{G}^{(da)}_{i,\,A\cup H}(\lambda)\Pi(b,\, i)\Bigg)\\
&\quad \times \sum_{j\in A}\Pi(a,\, j) + \sum_{j\notin A\cup H} {}_H\tilde{\Phi}_{iA}(\lambda)\\
&= \int_{(\partial X)_e}\Bigg(\sum_{i\notin A\cup H} \tilde{G}^{(da)}_{i,\,A\cup H}(\lambda)\Pi(b,\, i)\Bigg)\overset{(n)}{\xi}{}^{(a)}\\
&\quad + \int_{(\partial X)_e}\Bigg(\sum_{i\notin A\cup H} \tilde{G}^{(da)}_{i,\,A\cup H}(\lambda)\Pi(b,\, i)\Bigg)
\end{aligned}$$

$$\times \sum_{j\in A} \Pi(a,j) + \sum_{i\notin A\cup H} {}_H\tilde{\Phi}_{iA}(\lambda) \qquad (b\in(\partial X)_e). \tag{10.3.53}$$

It can be seen by induction that

$$\overset{(n)}{\xi}{}^{(b)} = \sum_{i\notin A\cup H} \Pi(b,i)x_i^{(n)} \qquad (b\in(\partial X)_e). \tag{10.3.54}$$

Consequently, it is an easy matter to demonstrate

$$\lim_{n\to\infty} \overset{(n)}{\xi}{}^{(b)} = \sum_{i\notin A\cup H} \Pi(b,i) \lim_{n\to\infty} x_i^{(n)}$$

$$= \sum_{i\notin A\cup H} \Pi(b,i)_H\tilde{\Phi}_{iA}(\lambda) \qquad (b\in(\partial X)_e) \tag{10.3.55}$$

and $\lim_{n\to\infty} \overset{(n)}{\xi}{}^{(b)}$ $(b\in(\partial X)_e)$ (i.e., $\Sigma_{i\notin A\cup H}\Pi(b,i)_H\tilde{\Phi}_{iA}(\lambda)(b\in(\partial X)_e)$) is the minimal nonnegative solution of the integral equation (10.3.48). □

Theorem 10.3.3 $\{{}_Hf_{iA}^*,\ i\in E\}$ *is the minimal nonnegative solution of the pseudo-normal system of equations*

$$x_i = \sum_{i\notin A\cup H} \int_{(\partial X)_e} \tilde{G}^{*(db)}_{i,\,A\cup H}\ \Pi(b,j)x_j + {}_H\tilde{f}_{iA}^*$$

$$+ \sum_{j\in A} \int_{(\partial X)_e} \tilde{G}^{*(db)}_{i,\,A\cup H}\Pi(b,j) \qquad (i\in E). \tag{10.3.56}$$

Proof. This theorem is the special case of Theorem 10.3.1 for $\lambda=0$.□

Theorem 10.3.4

$${}_Hf_{iA}^* = \int_{(\partial X)_e} \tilde{G}^{*(da)}_{i,\,A\cup H}\xi^{*(a)} + \int_{(\partial X)_e} \tilde{G}^{*(da)}_{i,\,A\cup H}\xi_0^{(a)} + {}_H\tilde{f}_{iA}^* \qquad (i\in E), \tag{10.3.57}$$

where

$$\xi^{*(a)} = \xi^{(a)}(0) \qquad (a\in(\partial X)_e). \tag{10.3.58}$$

Proof. This theorem is the special case of the Theorem 10.3.2 for $\lambda=0$. □

Theorem 10.3.5 $\{{}_Hm_{iA}^{(p)},\ i\in E\}\ (p\geqslant 1)$ *is the minimal nonnegative solution of the first-type system of* 1*-bounded equations*

$$x_i = \sum_{j\notin A\cup H} \int_{(\partial X)_e} \tilde{G}^{*(db)}_{i,\,A\cup H}\Pi(b,j)x_j + {}_H\tilde{m}_{iA}^{(p)} + \sum_{j\in A} \int_{(\partial X)_e} \tilde{L}^{(db,\,p)}_{i,\,A\cup H}\Pi(b,j)$$

$$+\sum_{l=1}^{p}\sum_{j\notin A\cup H}\int_{(\partial X)_e} C_p^l \tilde{L}_{i,\,A\cup H}^{(db,\,l)}\Pi(b,\,j)\,{}_Hm_{jA}^{(p-l)} \qquad (i\in E). \tag{10.3.59}$$

Proof. Our theorem is an immediate consequence of Lemma 10.3.9 and Corollary 3.2.2. □

Theorem 10.3.6

$$ {}_Hm_{iA}^{(p)} = \sum_{l=0}^{p} C_p^l \int_{(\partial X)_e} \tilde{L}_{i,\,A\cup H}^{(da,\,l)}\xi^{(a,\,p-l)}$$

$$+\int_{(\partial X)_e} \tilde{L}_{i,\,A\cup H}^{(da,\,p)}\xi_0^{(a)} + {}_H\tilde{m}_{iA}^{(p)} \qquad (i\in E), \tag{10.3.60}$$

where, for any $1\leqslant s\leqslant p$, $\xi^{(a,s)}$ $(a\in(\partial X)_e)$ *is the minimal nonnegative solution of the integral equation*

$$\xi^{(b,s)} = \int_{(\partial X)_e}\sum_{i\notin A\cup H} \tilde{G}^{*(da)}_{i,\,A\cup H}\Pi(b,\,i)\xi^{(a,s)}$$

$$+\sum_{l=1}^{p}\int_{(\partial X)_e}\sum_{i\notin A\cup H} \tilde{L}_{i,\,A\cup H}^{(da,\,l)}\Pi(b,\,i)\xi^{(a,\,s-l)}$$

$$+\int_{(\partial X)_e}\sum_{i\notin A\cup H} \tilde{L}_{i,\,A\cup H}^{(da,\,s)}\Pi(b,\,i)\xi_0^{(a)} \qquad (b\in(\partial X)_e). \tag{10.3.61}$$

Proof. The proof is similar to that of Theorem 10.3.2. □

CHAPTER XI

Arbitrary Q-Processes

§ 11.1 Strengthening of the first construction theorem

In this chapter, we shall investigate arbitrary Q-processes on the basis of the study of minimal Q-processes and Q-processes of order one.

We shall consider three problems: (A) strengthening of the first construction theorem for sample functions; (B) transition probability; (C) the decomposition theorem for excessive measures and excessive functions. The investigation of the first passage time, integral-type functionals and the classification of states are omitted here, since there is nothing new in the methods. In this section we shall consider Problem (A).

Theorem 11.1.1 *Suppose $X(\omega)=\{x(t, \omega), t<\sigma(\omega)\}$ is an arbitrary Q-process, and let*

$$X^{(n)}(\omega) = g_n(X(\omega)). \tag{11.1.1}$$

Then

(1) *$X^{(n)}(\omega)=\{x^{(n)}(t, \omega), t<\sigma^{(n)}(\omega)\}$ $(n\geqslant 1)$ is a $(Q, \Pi_{(\partial X)_e\times D_n})$-process, and we have*

$$X^{(n)}(\omega) = g_n(X^{(n+1)}(\omega)), \tag{11.1.2}$$

where $\Pi_{(\partial X)_e\times D_n}=(\Pi^{(n)}(a, j), a\in(\partial X)_e, j\in D_n)$ is determined uniquely as follows:

$$\Pi^{(n)}(a, j) = P(x(\beta_1^{(n)}) = j|x(\tau_1-0)=a) \qquad (a\in(\partial X)_e). \tag{11.1.3}$$

For the definitions of D_n, g_n and $\beta_1^{(n)}$, see §1.2.

(2) *The matrix sequence $\Pi_{(\partial X)_e\times D_n}=(\Pi^{(n)}(a, j), a\in(\partial X)_e, j\in D_n)$, $(n\geqslant 1)$ has the following properties:*

(i) *For a fixed n and $j\in D_n$, $\Pi^{(n)}(a, j)$ $(a\in(\partial X)_e)$ is a Borel measurable function on β_e.*

(ii) *For any $n\geqslant 1$, we have*

$$\Pi^{(n)}(a, j)\geqslant 0, \qquad (a\in(\partial X)_e, j\in D_n), \tag{11.1.4}$$

$$\sum_{j\in D_n} \Pi^{(n)}(a, j) \leqslant 1, \qquad (a\in(\partial X)_e). \tag{11.1.5}$$

(iii) *For any* $n \geqslant 1$, *we have*

$$\Pi^{(n)}(a, j) = \Pi^{(n+1)}(a, j) + \Pi^{(n+1)}(a, n+1)$$

$$\times \frac{{}_{D_n}\tilde{f}^*_{n+1,j} + \int_{(\partial X)_e} (h(n+1, db) - \tilde{f}^{*(db)}_{n+1, D_n})\Pi^{(n+1)}(b, j)}{-1\int_{(\partial X)_e} (h(n+1, db) - \tilde{f}^{*(db)}_{n+1, D_n})\Pi^{(n+1)}(b, n+1)}$$

$$(a\in(\partial X)_e, j\in D_n). \tag{11.1.6}$$

Here we stipulate $0/0 = 0$, *and the definitions of* ${}_{D_n}\tilde{f}^*_{n+1,j}, \tilde{f}^{*(db)}_{n+1,D_n}$ *are as in* §10.3.

(3) *On* $[0, \sigma(\omega))$, *we have*

$$\lim_{n\to\infty} x^{(n)}(t, \omega) = x(t, \omega) \qquad (\omega\in\Omega). \tag{11.1.7}$$

(4)
$$\lim_{n\to\infty} p_{ij}^{(n)} = p_{ij}(t) \qquad (i, j\in E, t \geqslant 0), \tag{11.1.8}$$

where

$$p_{ij}(t) = P(x(t) = j | x(0) = i) \qquad (i, j\in E, \quad t\geqslant 0), \tag{11.1.9}$$

$$p_{ij}^{(n)}(t) = P(x^{(n)}(t) = j | x^{(n)}(0) = i) \qquad (i, j\in E, t\geqslant 0). \tag{11.1.10}$$

Proof. Steps (1) and (3) hold by Theorem 1.5.1, and we can see that (4) is valid by (3) (and by referring to the proof of [2, Theorem 6.3]). It is clear that (11.1.4) and (11.1.5) hold, so we only need to prove the validity of (11.1.6).

Let

$$\sigma_A^{(n)} = \begin{cases} \inf\{t: \tau^{(1)} \leqslant t < \sigma^{(n)}, x_n(t)\in A\}, \\ +\infty, \quad \text{if the above set is empty,} \end{cases} \tag{11.1.11}$$

$$_H\sigma_A^{(n)} = \begin{cases} \sigma_A^{(n)}, & \text{if } \sigma_A^{(n)} \leqslant \sigma_H^{(n)}, \\ +\infty, & \text{if } \sigma_A^{(n)} > \sigma_H^{(n)}, \end{cases} \tag{11.1.12}$$

$$_Hf_{iA}^{*(n)} = p_i({}_H\sigma_A^{(n)} < +\infty). \tag{11.1.13}$$

By appealing to Theorem 10.3.3 we infer

$${}_{D_n}f^{*(n+1)}_{n+1,j} = \int_{(\partial X)_e} (h(n+1, db) - \tilde{f}^{*(db)}_{n+1,D_n})\Pi^{(n+1)}(b, n+1)\,{}_{D_n}f^{*(n+1)}_{n+1,j} + {}_{D_n}\tilde{f}^*_{n+1,j}$$

$$+\int_{(\partial X)_e}\left(h(n+1, db)-\tilde{f}^{*(db)}_{n+1, D_n}\right)\Pi^{(n+1)}(b, j) \qquad (j\in D_n). \qquad (11.1.14)$$

Hence

$$_{D_n}f^*_{n+1, j}=\frac{_{D_n}\tilde{f}^*_{n+1, j}+\int_{(\partial X)_e}\left(h(n+1, db)-\tilde{f}^{*(db)}_{n+1, D_n}\right)\Pi^{(n+1)}(b, j)}{1-\int_{(\partial X)_e}\left(h(n+1, db)-\tilde{f}^{*(db)}_{n+1, D_n}\right)\Pi^{(n+1)}(b, n+1)} \qquad (j\in D_n). \qquad (11.1.15)$$

Suppose $\Delta\in B_e$ and note that

$$x^{(n)}(t, \omega)=x(t, \omega) \qquad (n=1, 2, \cdots; 0\leqslant t<\tau_1(\omega)). \qquad (11.1.16)$$

For any $j\in D_n$, obviously we have

$$\begin{aligned}&P_i(x^{(n)}(\tau_1)=j,\ x^{(n)}(\tau_1-0)\in\Delta)\\&=P_i(x^{(n+1)}(\tau_1)=j,\ x^{(n+1)}(\tau_1-0)\in\Delta)\\&\quad+P_i(x^{(n+1)}(\tau_1)=j,\ x^{(n+1)}(\tau_1-0)\in\Delta \text{ there}\\&\qquad\text{exists } s,\ \tau^{(1,1)}_{n+1}\leqslant s<\sigma^{(n+1)} \text{ such that } x^{(n+1)}(s)=j\\&\qquad\text{and } x^{(n+1)}(t)\notin D_n,\ \tau^{(1,1)}\leqslant t<s),\end{aligned} \qquad (11.1.17)$$

where $\tau^{(1,1)}_{n+1}(\omega)$ is the first jump point of $X_{n+1}(\omega)$ after $\tau_1(\omega)$ if $\tau_1(\omega)<\sigma^{(n+1)}(\omega)$; otherwise, $\tau^{(1,1)}_{n+1}(\omega)=\sigma^{(n+1)}(\omega)$. By (11.1.16) and Lemma 10.3.2, we see that

$$P_i(x^{(n)}(\tau_1)=j,\ x^{(n)}(\tau_1-0)\in\Delta)=\int_\Delta h(i, da)\Pi^{(n)}(a, j) \qquad (11.1.18)$$

and

$$P_i(x^{(n+1)}(\tau_1)=j,\ x^{(n+1)}(\tau_1-0)\in\Delta)=\int_\Delta h(i, da)\Pi^{(n+1)}(a, j). \qquad (11.1.19)$$

It is easy to obtain from the strong Markov property, the homogeneity and Lemma 10.3.2 that

$$P_i(x^{(n+1)}(\tau_1)=n+1,\ x^{(n+1)}(\tau_1-0)\in\Delta$$

there exists s, $\tau_{n+1}^{(1,1)} \leqslant s < \sigma^{(n+1)}$, such that

$$x^{(n+1)}(s) = j, \text{ and } x^{(n+1)}(t) \notin D_n, \ \tau_{n+1}^{(1,1)} \leqslant t < s)$$

$$= \int_\Delta h(i, da)\Pi^{(n+1)}(a, n+1) \cdot {}_{D_n}f^*_{n+1,j}. \tag{11.1.20}$$

It follows from (11.1.17)—(11.1.20) that

$$\int_\Delta (\Pi^{(n)}(a, j) - \Pi^{(n+1)}(a, j)$$

$$- \Pi^{(n+1)}(a, n+1)_{D_n}f^{*(n+1)}_{n+1,j})h(i, da) = 0, \qquad (i \in E). \tag{11.1.21}$$

By virtue of the arbitrariness of $\Delta \in B_e$, the absolute continuity of the integral and (11.1.21), we get

$$\Pi^{(n)}(a, j) = \Pi^{(n+1)}(a, j) + \Pi^{(n+1)}(a, n+1) \cdot {}_{D_n}f^{*(n+1)}_{n+1,j}. \tag{11.1.22}$$

Equation (11.1.14) follows from (11.1.15) and (11.1.22). Hence (11.1.6) is proved. □

Theorem 11.1.2 *If* (11.1.4) *and* (11.1.5) *hold, then* (11.1.6) *holds iff*

$$\Pi^{(n+1)}(a, j) = \Pi^{(n)}(a, j) - \Pi^{(n+1)}(a, n+1) \cdot {}_{D_n}\mathcal{f}^*_{n+1,j}$$

$$- \Pi^{(n+1)}(a, n+1) \int_{(\partial X)_e} (h(n+1, db)$$

$$- \mathcal{f}^{*(db)}_{n+1, D_n})\Pi^{(n)}(b, j) \qquad (a \in (\partial X)_e, j \in D_n). \tag{11.1.23}$$

Proof. (A) Necessity.
Suppose (11.1.6) holds and we set up the equation

$$X = \int_{(\partial X)_e} (h(n+1, db) - \mathcal{f}^{*(db)}_{n+1, D_n})\Pi^{(n+1)}(b, n+1)X + {}_{D_n}\mathcal{f}^*_{n+1,j}$$

$$+ \int_{(\partial X)_e} (h(n+1, db) - \mathcal{f}^{*(db)}_{n+1, D_n})\Pi^{(n+1)}(b, j). \tag{11.1.24}$$

Hence

$$X = \frac{{}_{Dn}\mathcal{F}^*_{n+1,j} + \int_{(\partial X)_e} (h(n+1, db) - \mathcal{F}^{*(db)}_{n+1, Dn})\Pi^{(n+1)}(b, j)}{1 - \int_{(\partial X)_e} (h(n+1, db) - \mathcal{F}^{*(dp)}_{n+1, Dn})\Pi^{(n+1)}(b, n+1)}. \tag{11.1.25}$$

It follows by (11.1.6) and (11.1.25) that

$$\Pi^{(n)}(a, j) = \Pi^{(n+1)}(a, j) + \Pi^{(n+1)}(a, n+1)X \qquad (j \in D_n). \tag{11.1.26}$$

Substituting (11.1.24) into (11.1.26), we obtain

$$\begin{aligned}\Pi^{(n)}(a, j) = {} & \Pi^{(n+1)}(a, j) + \Pi^{(n+1)}(a, n+1) \\ & \times \int_{(\partial X)_e} (h(n+1, db) - \mathcal{F}^{*(db)}_{n+1, Dn})\Pi^{(n+1)}(b, n+1)X \\ & + \Pi^{(n+1)}(a, n+1)\,{}_{Dn}\mathcal{F}^*_{n+1,j} + \Pi^{(n+1)}(a, n+1) \\ & \times \int_{(\partial X)_e} (h(n+1, db) - \mathcal{F}^{*(db)}_{n+1, Dn})\Pi^{(n+1)}(b, j) \qquad (j \in D_n),\end{aligned} \tag{11.1.27}$$

but

$$\begin{aligned}\Pi^{(n+1)}(a, j) = {} & \Pi^{(n)}(a, j) - \Pi^{(n+1)}(a, n+1)\,{}_{Dn}\mathcal{F}^*_{n+1,j} \\ & - \Pi^{(n+1)}(a, n+1)\Bigg[\int_{(\partial X)_e} (h(n+1, db) - \mathcal{F}^{*(db)}_{n+1, Dn}) \\ & \times (\Pi^{(n+1)}(b, j) + \Pi^{(n+1)}(b, n+1)X)\Bigg] \qquad (j \in D_n)\end{aligned} \tag{11.1.28}$$

and

$$\Pi^{(n)}(b, j) = \Pi^{(n+1)}(b, j) + \Pi^{(n+1)}(b, n+1)X \qquad (j \in D_n). \tag{11.1.29}$$

It follows from (11.1.29) that

$$\begin{aligned}& \int_{(\partial X)_e} (h(n+1, db) - \mathcal{F}^{*(db)}_{(n+1), Dn})\Pi^{(n)}(b, j) \\ & = \int_{(\partial X)_e} (h(n+1, db) - \mathcal{F}^{*(bd)}_{n+1, Dn})(\Pi^{(n+1)}(b, j) \\ & \quad + \Pi^{(n+1)}(b, n+1)X) \qquad (j \in D_n).\end{aligned} \tag{11.1.30}$$

Substituting (11.1.30) into (11.1.28), we obtain (11.1.23).

(B) Sufficiency.

Suppose (11.1.23) holds. Then we have

$$\Pi^{(n)}(a, j) = \Pi^{(n+1)}(a, j) + \Pi^{(n+1)}(a, n+1) \times \left[{}_{Dn}\mathcal{F}^{*}_{n+1,j} + \int_{(\partial X)_e} (h(n+1, db) - \mathcal{F}^{*(db)}_{n+1,Dn}) \Pi^{n}(a, j) \right] \quad (j \in D_n). \tag{11.1.31}$$

It follows from (11.1.31) that

$$\begin{aligned} &\int_{(\partial X)_e} (h(n+1, da) - \mathcal{F}^{*(da)}_{n+1,Dn}) \Pi^{(n)}(a, j) \\ &= \int_{(\partial X)_e} (h(n+1, da) - \mathcal{F}^{*(da)}_{n+1,Dn}) \Pi^{(n+1)}(a, j) \\ &\quad + \int_{(\partial X)_e} (h(n+1, da) - \mathcal{F}^{*(da)}_{n+1,Dn}) \Pi^{(n+1)}(a, n+1) {}_{Dn}\mathcal{F}^{*}_{n+1,j} \\ &\quad + \left[\int_{(\partial X)_e} (h(n+1, da) - \mathcal{F}^{*(da)}_{n+1,Dn}) \Pi^{(n+1)}(a, n+1) \right] \\ &\quad \times \left[\int_{(\partial X)_e} (h(n+1, db) - \mathcal{F}^{*(db)}_{n+1,Dn}) \Pi^{(n)}(b, j) \right] \quad (j \in D_n) \end{aligned} \tag{11.1.32}$$

namely

$$\begin{aligned} &\int_{(\partial X)_e} (h(n+1, db) - \mathcal{F}^{*(db)}_{n+1,Dn}) \Pi^{(n)}(b, j) \\ &= \int_{(\partial X)_e} (h(n+1, db) - \mathcal{F}^{*(db)}_{n+1,Dn}) \Pi^{(n+1)}(b, j) \\ &\quad + \int_{(\partial X)_e} (h(n+1, db) - \mathcal{F}^{*(db)}_{n+1,Dn}) \Pi^{(n+1)}(b, n+1) {}_{Dn}\mathcal{F}^{*}_{n+1,j} \\ &\quad + \left[\int_{(\partial X)_e} (h(n+1, db) - \mathcal{F}^{*(db)}_{n+1,Dn}) \Pi^{(n+1)}(b, n+1) \right] \end{aligned}$$

$$\times\left[\int_{(\partial X)_e}\left(h(n+1, db)-\mathcal{J}^{*(db)}_{n+1, D_n}\right)\Pi^{(n)}(b, j)\right] \quad (j\in D_n). \tag{11.1.33}$$

Therefore

$$\int_{(\partial X)_e}\left(h(n+1, db)-\mathcal{J}^{*(db)}_{n+1, D_n}\right)\Pi^{(n)}(b, j)$$

$$=\left[\int_{(\partial X)_e}\left(h(n+1, db)-\mathcal{J}^{*(db)}_{n+1, D_n}\right)\Pi^{(n+1)}(b, j)\right.$$

$$\left.+\int_{(\partial X)_e}\left(h(n+1, db)-\mathcal{J}^{*(db)}_{n+1, D_n}\right)\Pi^{(n+1)}(b, n+1)_{D_n}\mathcal{J}^{*}_{n+1, j}\right]\Big/$$

$$\left[1-\int_{(\partial X)_e}\left(h(n+1, db)-\mathcal{J}^{*(db)}_{n+1, D_n}\right)\Pi^{(n+1)}(b, n+1)\right] \quad (j\in D_n). \tag{11.1.34}$$

Substituting (11.1.34) into (11.1.31), we obtain (11.1.6). □

§ 11.2 Transition probability

Let

$$\Pi^{(n)}(a, n)=\Pi(a, n) \qquad (n=1, 2, \cdots). \tag{11.2.1}$$

We know from Theorem 11.1.1 that $P(t)=(p_{ij}(t), i, j\in E)$ is determined uniquely by $P^{(n)}(t)=(p_{ij}^{(n)}(t), i, j\in E)$, from Theorem 10.2.1 that $P^{(n)}(t)$ $(n\geqslant 1)$ is determined uniquely by Q and $\Pi_{(\partial X)_e}\times D_n$ $(n\geqslant 1)$. Based on Theorem 11.1.2, $\Pi_{(\partial X)_e}\times D_n$ $(n=1, 2, \cdots)$ is determined uniquely by Q and $\Pi_{(\partial X)_e\times E}=(\Pi(a, n), a\in(\partial X)_e, n\in E)$. Hence $P(t)$ is determined uniquely by Q and $\Pi_{(\partial X)_e\times E}$. So we introduce the following terminology.

Definition 11.2.1 *Suppose $X(\omega)=\{x(t, \omega), t<\sigma(\omega)\}$ is a Q-process. If*

$$\Pi(a, n)=P(x(\beta_1^{(n)})=n|X(\tau-0)=a), \tag{11.2.2}$$

then we say $\Pi_{(\partial X)_e\times E}=(\Pi(a, n), a\in(\partial X)_e, n\in E)$ is the characteristic matrix of $X(\omega)$, and $X(\omega)$ is a $[Q, \Pi_{(\partial X)_e\times E}]$-process.

Let

$$p_{ij}(\lambda)=\int_0^\infty e^{-\lambda t}p_{ij}(t)dt, \tag{11.2.3}$$

$$p_{ij}^{(n)}(\lambda) = \int_0^\infty e^{-\lambda t} p_{ij}^{(n)}(t)dt. \tag{11.2.4}$$

Theorem 11.2.1 *Suppose $X(\omega)$ is a $[Q, \Pi_{(\partial X)_e \times E}]$-process. Then*

$$p_{ij}(\lambda) = p_{ij}^{\min}(\lambda) + \int_{(\partial X)_e} h_\lambda(i, da)\xi_j^{(a)}(\lambda) \qquad (i, j \in E), \tag{11.2.5}$$

where

$$\xi_j^{(a)}(\lambda) = \int_0^\infty e^{-\lambda t} P(x(t+\tau) = j | x(\tau - 0) = a)dt$$

$$= \lim_{n\to\infty} \xi_j^{(n,a)}(\lambda) \qquad (a \in (\partial X)_e, j \in E), \tag{11.2.6}$$

and $\{\xi_j^{(n,a)}(\lambda), a \in (\partial X)_e\}$ is the minimal nonnegative solution of the integral equation

$$\xi^{(a)} = \int_{(\partial X)_e} \left(\sum_{i \in D_n} \Pi^{(n)}(a, i) h_\lambda(i, db) \right) \xi^{(b)}$$

$$+ \sum_{i \in D_n} \Pi^{(n)}(a, i) p_{ij}^{\min}(\lambda) \qquad (a \in (\partial X)_e), \tag{11.2.7}$$

and

$$\Pi^{(n)}(a, j) = P(x(\beta_1^{(n)}) = j | x(\tau - 0) = a) \qquad (j \in D_n) \tag{11.2.8}$$

is determined uniquely by the Q-matrix and the characteristic matrix $\Pi_{(\partial X)_e \times E} = (\Pi(a, n), a \in (\partial X)_e, n \in E)$.

Proof. In the light of Theorem 10.2.3 we see that

$$p_{ij}^{(n)}(\lambda) = p_{ij}^{\min}(\lambda) + \int_{(\partial X)_e} h_\lambda(i, da)\xi_j^{(n,a)}(\lambda), \tag{11.2.9}$$

where $\xi_j^{(n,a)}(\lambda)$ $(a \in (\partial X)_e)$ is the minimal nonnegative solution of the integral equation (11.2.7), and

$$\xi_j^{(n,a)}(\lambda) = \int_0^\infty e^{-\lambda t} P(x^{(n)}(t+\tau) = j | x(\tau - 0) = a)dt \qquad (a \in (\partial X)_e). \tag{11.2.10}$$

From Theorem 11.1.1,

$$\lim_{n\to\infty} \{\omega: x^{(n)}(t+\tau) = j\} = \{\omega: x(t+\tau) = j\}. \tag{11.2.11}$$

Hence by the dominated convergence theorem we obtain

$$\lim_{n\to\infty} P(x^{(n)}(t)=j|x(\tau-0)=a)$$

$$=P(x(t+\tau)=j|x(\tau-0)=a) \qquad (a\in(\partial X)_e). \tag{11.2.12}$$

Therefore

$$\lim_{n\to\infty} \xi_j^{(n,\,a)}(\lambda)=\int_0^\infty e^{-\lambda t}P(x(t+\tau)=j|x(\tau-0)=a)dt \qquad (a\in(\partial X)_e). \tag{11.2.13}$$

Taking limits from both sides of (11.2.9) when $n\to\infty$ and noting (11.2.13) and

$$\int_{(\partial X)_e} h_\lambda(i,\,da)\leqslant 1, \tag{11.2.14}$$

we get

$$p_{ij}(\lambda)=\int_{(\partial X)_e} h_\lambda(i,\,da)\int_0^\infty e^{-\lambda t}P(x(t+\tau)$$

$$=j|x(\tau-0)=a)dt \qquad (a\in(\partial X)_e). \tag{11.2.15}$$

Consequently the theorem follows from Theorem 11.1.2. □

§ 11.3 Decomposition theorems for excessive measures and excessive functions

A matrix function Π on E is a family $(\Pi(t))_{t\in[0,\,\infty)}$ of $E\times E$ matrices, and we shall denote by $p_{ij}(t)$ the $(i,\,j)$ coefficient of $\Pi(t)$. Π is called a *general transition matrix* (abbreviated GTM), if it satisfies the following properties:

(i) $\Pi(t)\geqslant 0$ for all t(i.e., $p_{ij}(t)\geqslant 0$ for all $t,\ i,\ j$).

(ii) $\Pi(s+t)=\Pi(s)\Pi(t)$ for all $s,\ t$(i.e., we have $p_{ij}(s+t)=\sum_k p'_{ik}(s)p_{kj}(t)$ for all $i,\ j,\ s,\ t$).

(iii) $\Pi(0)=I=\lim_{t\to 0}\Pi(t)$ (i.e., $p_{ij}(0)=\delta_{ij}=\lim_{t\to 0}p_{ij}(t)$ for all $i,\ j$).

For example, a transition probability matrix of a homogeneous denumerable Markov process is a GTM.

An entrance law is a family $(G(t))_{t>0}$ of positive row vectors (measure) on E such that $G(s)\Pi(t)=G(s+t)$ $(s>0,\ t\geqslant 0)$. In the same way, exit laws are families $(F(t))_{t>0}$ of positive column vectors (functions) such that $\Pi(t)F(s)=F(s+t)$ $(s>0,\ t\geqslant 0)$.

Definition 11.3.1 *A row vector* $\mathbf{H}=(h_i)$ *is called an excessive measure, if*

$$0 \leqslant \mathbf{H} < +\infty, \tag{11.3.1}$$

$$\mathbf{H}\Pi(t) \leqslant \mathbf{H} \quad \text{for all } t \geqslant 0. \tag{11.3.2}$$

If, in addition, we have

$$\lim_{t \to \infty} \mathbf{H}\Pi(t) = 0, \tag{11.3.3}$$

then **H** *is said to be a strictly excessive measure. An excessive measure* **H** *is said to be a harmonic measure, if equality holds in* (11.3.2). *Here* $\Pi(t)$ *is a GTM.*

Definition 11.3.2 *A column vector* $\mathbf{D} = (d_i)$ *is called an excessive function if*

$$0 \leqslant \mathbf{D} < +\infty, \tag{11.3.4}$$

$$\Pi(t)\mathbf{D} \leqslant \mathbf{D} \quad \text{for all } t \geqslant 0. \tag{11.3.5}$$

If, in addition,

$$\lim_{t \to \infty} \Pi(t)\mathbf{D} = 0, \tag{11.3.6}$$

then we say **D** *is a strictly excessive function, and an excessive function* **D** *is said to be a harmonic function if equality holds in* (11.3.5).

Definition 11.3.3 *An entrance law* $\mathbf{F}(\cdot)$ *is said to be integrable if*

$$\int_0^\infty \mathbf{F}(t)dt < +\infty. \tag{11.3.7}$$

An exit law $\mathbf{W}(\cdot)$ *is said to be integrable if*

$$\int_0^\infty \mathbf{W}(t)dt < +\infty. \tag{11.3.8}$$

Remark 11.3.1 Not all the entrance laws or exit laws are integrable. For instance, Let $P(t)$ be a irreducible recurrent Q-process. So $f_j(t) = p_{ij}(t)$ (i fixed) is a nonintegrable entrance law; and $W_i(t) = p_{ij}(t)$ (j fixed) is a nonintegrable exit law.

Proposition 3 of [23, Chapter I] says: if **H** is a strictly excessive measure, then there exists an entrance law $\mathbf{F}(\cdot)$ such that

$$\mathbf{H} = \int_0^\infty \mathbf{F}(t)dt \tag{11.3.9}$$

and $\mathbf{F}(\cdot)$ is unique.

The purpose of this chapter is to answer the following three deep related problems:

(A) If $\mathbf{F}(\cdot)$ is an entrance law, is $\mathbf{H}$ as defined by (11.3.9) necessarily a strictly excessive measure? If it is not, what additional condition should be imposed?

(B) What are the corresponding results for excessive measures?

(C) What are the corresponding results for excessive functions?

Lemma 11.3.1 *If* $\mathbf{H} \geqslant 0$ *and*

$$\overline{\lim_{t\to\infty}}\, \mathbf{H}\Pi(t) < +\infty, \tag{11.3.10}$$

then

$$\mathbf{H} < +\infty. \tag{11.3.11}$$

Proof. If $h_i = +\infty$ for some $i \in \underset{\sim}{I}$, then $p_{ii}(t) > 0\ (0 \leqslant t < +\infty)$ implies that

$$h_i p_{ii}(t) = +\infty \qquad (0 \leqslant t < +\infty). \tag{11.3.12}$$

Hence

$$h_i p_{ii}(t) \to +\infty \qquad (t \to \infty). \tag{11.3.13}$$

But from $\mathbf{H} \geqslant 0$ and (11.3.10) we have

$$\lim_{t\to\infty} h_i p_{ii}(t) \leqslant \overline{\lim_{t\to\infty}} \sum_j h_j p_{ji}(t) < +\infty, \tag{11.3.14}$$

which contradicts (11.3.13), so (11.3.11) must hold.□

Remark 11.3.2 Lemma 11.3.1 shows that strictly excessive measures can be defined as nonnegative row vectors satisfying the conditions (11.3.2) and (11.3.3).

Lemma 11.3.2 *If* $\mathbf{H}$ *is an excessive measure, then the limit*

$$\lim_{t\to\infty} \mathbf{H}\Pi(t) \tag{11.3.15}$$

exists and is nonnegative and finite. If we let

$$\hat{\mathbf{H}} = \lim_{t\to\infty} \mathbf{H}\Pi(t), \tag{11.3.16}$$

then $\hat{\mathbf{H}}$ *is a harmonic measure.*

Proof. We know from (11.3.1) and (11.3.2) that

$$+\infty > \mathbf{H} \geqslant \mathbf{H}\Pi(t) \geqslant \mathbf{H}\Pi(t+s) \geqslant 0. \tag{11.3.17}$$

So the limit $\hat{\mathbf{H}} = \lim_{t\to\infty} \mathbf{H}\Pi(t)$ exists and is nonnegative and finite. From

$$\sum_j h_j p_{ji}(s) p_{il}(t) \leqslant h_i p_{il}(t) \tag{11.3.18}$$

and

$$\sum_i h_i p_{il}(t) \leqslant h_l, \tag{11.3.19}$$

we see

$$\begin{aligned}\sum_j \hat{h}_i p_{il}(t) &= \sum_i \left(\lim_{s\to\infty} \sum_j h_j p_{ji}(s) \right) p_{il}(t) \\ &= \sum_i \lim_{s\to\infty} \sum_j h_j p_{ji}(s) p_{il}(t) \\ &= \lim_{s\to\infty} \sum_i \sum_j h_j p_{ji}(s) p_{il}(t) \\ &= \lim_{s\to\infty} \sum_j h_j \sum_i p_{ji}(s) p_{il}(t) \\ &= \lim_{s\to\infty} \sum_j h_j p_{jl}(t+s) \\ &= \lim_{t\to\infty} \sum_j h_j p_{jl}(t) = \hat{h}_l. \end{aligned} \tag{11.3.20}$$

Hence $\hat{\mathbf{H}}$ is a harmonic measure. □

Theorem 11.3.1 *$\mathbf{H}$ is an excessive measure iff it can be expressed in the following form:*

$$\mathbf{H} = \hat{\mathbf{H}} + \int_0^\infty \mathbf{F}(t)\,dt, \tag{11.3.21}$$

where $\hat{\mathbf{H}}$ is a harmonic measure, $\mathbf{F}(\cdot)$ is an integrable entrance law, and

$$\hat{\mathbf{H}} = \lim_{t\to\infty} \mathbf{H}\Pi(t), \tag{11.3.22}$$

$$\mathbf{F}(t) = -\frac{d\mathbf{H}\Pi(t)}{dt}. \tag{11.3.23}$$

Hence the expression (11.3.21) *of an excessive measure is unique.*

Proof. Suppose $\mathbf{H}$ is an excessive measure. Let

$$\mathbf{G}(s) = \mathbf{H} - \mathbf{H}\Pi(s). \tag{11.3.24}$$

In the light of [23, Chapter I, Proposition 2], there exists an entrance law $F(\cdot)$ such that

$$\mathbf{H}\Pi(t) - \mathbf{H}\Pi(t+s) = \mathbf{G}(t+s) - \mathbf{G}(t) = \int_t^{s+t} \mathbf{F}(u)du. \tag{11.3.25}$$

According to [23, Chapter I, Proposition 1, (b)], we see that

$$\lim_{t\to 0} \mathbf{H}\Pi(t) = \mathbf{H}. \tag{11.3.26}$$

By Lemma 11.3.2,

$$\lim_{t\to\infty} \mathbf{H}\Pi(t+s) = \lim_{t\to\infty} \mathbf{H}\Pi(t) = \hat{\mathbf{H}} \tag{11.3.27}$$

is a harmonic measure. Hence it follows from (11.3.25) that

$$+\infty > \mathbf{H} - \hat{\mathbf{H}} = \int_0^{\infty} \mathbf{F}(u)du. \tag{11.3.28}$$

Consequently, we know that $\mathbf{F}(\cdot)$ is integrable and that (11.3.21) holds.

Conversely, suppose $\hat{\mathbf{H}}$ is a harmonic measure and $\mathbf{F}(\cdot)$ is an integrable entrance law. Let

$$\mathbf{H} = \hat{\mathbf{H}} + \int_0^{\infty} \mathbf{F}(u)du. \tag{11.3.29}$$

Then

$$\mathbf{H}\Pi(t) = \hat{\mathbf{H}}\Pi(t) + \int_t^{\infty} \mathbf{F}(u)du. \tag{11.3.30}$$

By (11.3.29), (11.3.30) and since $\hat{\mathbf{H}}$ is a harmonic measure and $\mathbf{F}(t)$ is nonnegative and integrable, we obtain

$$0 \leqslant \mathbf{H} < +\infty, \qquad \mathbf{H}\Pi(t) \leqslant \hat{\mathbf{H}} + \int_0^{\infty} \mathbf{F}(u)du = \mathbf{H}. \tag{11.3.31}$$

Therefore $\mathbf{H}$ is an excessive measure.

Suppose (11.3.21) holds. Then we have

$$\mathbf{H}\Pi(t) = \hat{\mathbf{H}}\Pi(t) + \int_t^{\infty} \mathbf{F}(u)du = \hat{\mathbf{H}} + \int_t^{\infty} \mathbf{F}(u)du. \tag{11.3.32}$$

Equations (11.3.22) and (11.3.23) follow readily from (11.3.32) since $\mathbf{F}(\cdot)$ is integrable.□

Corollary 11.3.1 $\mathbf{H}$ *is a strictly excessive measure iff it can be expressed in the following form:*

$$\mathbf{H} = \int_0^\infty \mathbf{F}(t)dt, \tag{11.3.33}$$

where $\mathbf{F}(\cdot)$ *is an integrable entrance law, and*

$$\mathbf{F}(t) = -\frac{d\mathbf{H}\Pi(t)}{dt}. \tag{11.3.34}$$

Hence the expression (11.3.33) *for a strictly excessive measure is unique.*□

According to the symmetry of the entrance law and exit law with respect to GTM pointed out in [23, Chapter I], and observing Theorem 11.3.1, we conclude the following.

Theorem 11.3.2 **D** *is an excessive function iff it can be expressed in the following form*:

$$\mathbf{D} = \hat{\mathbf{D}} + \int_0^\infty \mathbf{W}(t)dt, \tag{11.3.35}$$

where $\hat{\mathbf{D}}$ *is a harmonic function,* $\mathbf{W}(\cdot)$ *is an integrable exit law, and*

$$\hat{\mathbf{D}} = \lim_{t\to\infty} \Pi(t)\mathbf{D}, \tag{11.3.36}$$

$$\mathbf{W}(t) = -\frac{d\Pi(t)\mathbf{D}}{dt}. \tag{11.3.37}$$

Hence the expression (11.3.35) *for an excessive function is unique.*

Corollary 11.3.2 **D** *is a strictly excessive function iff it can be expressed in the following form*:

$$\mathbf{D} = \int_0^\infty \mathbf{W}(t)dt, \tag{11.3.38}$$

where $\mathbf{W}(\cdot)$ *is an integrable exit law, and*

$$\mathbf{W}(t) = -\frac{d\Pi(t)\mathbf{D}}{dt}. \tag{11.3.39}$$

Hence the expression (11.3.38) *for a strictly excessive function is unique.*□

PART V

CONSTRUCTION THEORY OF HOMOGENEOUS DENUMERABLE MARKOV PROCESSES

CHAPTER XII

Criteria for the Uniqueness of Q-Processes

§ 12.1 Introduction

Suppose $X = \{x(t, \omega), t < \sigma(\omega)\}$ is a homogeneous denumerable Markov process defined on a complete probability space $(\Omega, \mathscr{F}, P)$, with phase space $E = \{1, 2, \cdots\}$, and transition probabilities $p_{ij}(t)$, $i, j \in E$, $t \geqslant 0$, which are real-valued functions satisfying the following conditions:

$$p_{ij}(t) \geqslant 0, \tag{12.1.1}$$

$$\sum_{j \in E} p_{ij}(t) \leqslant 1, \tag{12.1.2}$$

$$\sum_{k \in E} p_{ik}(t) p_{kj}(s) = p_{ij}(t+s), \tag{12.1.3}$$

$$\lim_{t \to 0} p_{ij}(t) = p_{ij}(0) = \delta_{ij}, \tag{12.1.4}$$

where $\delta_{ii} = 1$, $\delta_{ij} = 0$ $(i \neq j)$. It is well known that the limit

$$\lim_{t \to 0} \frac{p_{ij}(t) - \delta_{ij}}{t} = q_{ij} \tag{12.1.5}$$

exists, and $0 \leqslant q_{ij} < +\infty$ $(i \neq j)$, $0 \leqslant q_i \equiv -q_{ii} \leqslant +\infty$, $\Sigma_{j \in E}\, q_{ij} \leqslant 0$. The process is called differentiable if $q_i < +\infty$ $(i \in E)$. $Q = (q_{ij})$ is simply called the density matrix of the process, and the process $X = \{x(t, \omega), t < \sigma(\omega)\}$ is simply called a Q-process so as to indicate its relation with Q by (12.1.5). If two Q-processes have the same $p_{ij}(t)$, we identify them as one Q-process. So a matrix $(p_{ij}(t))$ satisfying (12.1.1)–(12.1.5) or its Laplace transform is also called a Q-process.

Definition 12.1.1 *A matrix $Q = (q_{ij})$ defined on $E \times E$ is called a Q-matrix if Q satisfies*

$$0 \leqslant q_{ij} < +\infty \quad (i \neq j), \qquad 0 \leqslant q_i \equiv -q_{ii} < +\infty,$$

$$\sum_{j\in E} q_{ij} \leqslant 0 \qquad (i\in E) \tag{12.1.6}$$

and Q is called a conservative Q-matrix if we have in addition

$$\sum_{j\in E} q_{ij} = 0 \qquad (i\in E). \tag{12.1.7}$$

Obviously, the density matrix of every differentiable process is a Q-matrix. Conversely, supposing a Q-matrix is given, the following three fundamental problems should be answered. (A) Does there exist a corresponding Q-process? (B) If one does exist, under what conditions is it unique? (C) If we know that such a Q-process is not unique, then how can all these Q-processes be constructed?

Feller [25] solved Problem (A) in 1940. He proved that for any given Q-matrix there always exist corresponding Q-processes. Problem (B) i.e., the uniqueness of the Q-process, for the conservative case, was solved by Feller [25] early in 1940. As for problem (C) i.e., to construct all Q-processes, it has attracted wide-spread interest, but it remains far from solved.

The aim of this chapter is to find necessary and sufficient condition for the uniqueness of a Q-process for any given Q-matrix (not necessarily be conservative).

Theorem 12.1.1 *Suppose a Q-matrix is given, then it is necessary and sufficient for the existence of a unique Q-process that the following two conditions hold simultaneously:*

(i)

$$\inf_{i\in E} \lambda \sum_{j\in E} p_{ij}^{\min}(\lambda) = \eta_\lambda > 0, \qquad 0 < \lambda < +\infty, \tag{12.1.8}$$

where $P_\lambda^{\min} = (p_{ij}^{\min}(\lambda),\ i,\ j\in E)$ is a minimal Q-process, i.e.,

$$p_{ij}^{\min}(\lambda) = \int_0^\infty e^{-\lambda t} f_{ij}(t)\,dt \qquad (i,\ j\in E,\ 0<\lambda<+\infty), \tag{12.1.9}$$

and

$$f_{ij}(t) = \sum_{n=0}^{\infty} f_{ij}^{(n)}(t) \qquad (i,\ j\in E), \tag{12.1.10}$$

$$\left.\begin{aligned} f_{ij}^{(0)}(t) &= \delta_{ij} e^{-q_i t} \qquad (i,\ j\in E), \\ f_{ij}^{(n+1)}(t) &= \sum_{k\neq i} \int_0^t e^{-q_i(t-s)} q_{ik} f_{kj}^{(n)}(s)\,ds \qquad (n\geqslant 0,\ i,\ j\in E). \end{aligned}\right\} \tag{12.1.11}$$

(ii) *The Q-process satisfying Kolmogorov's system of forward differential equations is unique, i.e., its minimal Q-process is honest (i.e., $\lambda\ \Sigma_{j\in E} p_{ij}^{\min}(\lambda) =$*

$1(i\in E)$) *or the equation*

$$\begin{cases}\lambda\mathbf{n}-\mathbf{n}Q=\mathbf{0}_-, & \lambda>0,\\ \mathbf{0}_-\leqslant\mathbf{n}, & \sum_i n(i)<+\infty\end{cases} \tag{12.1.12}$$

has only the zero solution, where $\mathbf{n}=(n(1),\ n(2),\ \cdots)$, $\mathbf{0}_-=(0,\ 0,\ \cdots)$. *And the minimal Q-process is honest iff Q is conservative and the equation*

$$\begin{cases}\lambda\mathbf{U}-Q\mathbf{U}=\mathbf{0}_1, & \lambda>0,\\ \mathbf{0}_1\leqslant\mathbf{U}\leqslant 1\end{cases} \tag{12.1.13}$$

has only the zero solution, where

$$\mathbf{U}=\begin{pmatrix}u(1)\\ u(2)\\ \vdots\end{pmatrix},\qquad \mathbf{0}_1=\begin{pmatrix}0\\ 0\\ \vdots\end{pmatrix},\qquad \mathbf{I}=\begin{pmatrix}1\\ 1\\ \vdots\end{pmatrix}.$$

We first give the proof of Theorem 12.1.1, then apply it to some special cases and give another equivalent form, finally we shall prove the independence of conditions (i) and (ii) and present the probability interpretation of condition(i).

§ 12.2 Lemmas

Lemma 12.2.1 *If Condition* (i) *in Theorem* 12.1.1 *holds, then the equation*

$$\begin{cases}\lambda\mathbf{U}-Q\mathbf{U}=\mathbf{0}_1, & \lambda>0,\\ \mathbf{0}_1\leqslant\mathbf{U}\leqslant \mathbf{1}\end{cases} \tag{12.2.1}$$

has only the zero solution.

Proof. By (12.1.9)–(12.1.11), it is easy to prove that $\{p_{ij}^{\min}(\lambda),\ i\in E\}$ is the minimal nonnegative solution of the first-type system of 1-bounded equations

$$u(i)=\sum_{k\neq i}\frac{q_{ik}}{\lambda+q_i}u(k)+\frac{\delta_{ij}}{\lambda+q_i}\qquad (i\in E). \tag{12.2.2}$$

Hence we see by Theorem 3.3.2 that $\{\lambda\Sigma_{j\in E}\, q_{ij}^{\min}(\lambda),\ i\in E\}$ is the minimal nonnegative solution of the pseudo-normal system of equations

$$u(i)=\sum_{k\neq i}\frac{q_{ik}}{\lambda+q_i}u(k)+\frac{\lambda}{\lambda+q_i}\qquad (i\in E). \tag{12.2.3}$$

Hence from $\lambda\,\Sigma_{j\in E}\,p_{ij}^{\min}(\lambda)\leqslant 1$, Condition (i) in Theorem 12.1.1 and Corollary 3.2.4

we deduce that the equation

$$\left.\begin{aligned} &u(i) = \sum_{k \neq i} \frac{q_{ik}}{\lambda + q_i} u(k), \qquad \lambda > 0, \\ &0 \leqslant u(i) \leqslant 1 \qquad (i \in E) \end{aligned}\right\} \tag{12.2.4}$$

has only the zero solution. But (12.2.4) is a equivalent form of (12.2.1), so Equation (12.2.1) has only the zero solution. □

Remark 12.2.1 In view of §12.7, the converse of Lemma 12.2.1 is not true in general. But in the finite nonconservative case, the converse of the lemma is true, i.e., it is equivalent to Condition (i) in Theorem 12.2.1 that the Equation (12.2.1) has only zero solution.This is an immediate consequence of Lemmas 12.2.1 and 12.8.2.

Lemma 12.2.2 *Let*

$$u_\lambda(i) = 1 - \lambda \sum_{j \in E} p_{ij}^{\min}(\lambda) \qquad (i \in E). \tag{12.2.5}$$

Then $u_\lambda(i)$ does not increase as λ increases, i.e., $\lambda \Sigma_{j \in E}\, p_{ij}^{\min}(\lambda)$ does not decrease as λ increases.

Proof. $\{\lambda \Sigma_{j \in E}\, p_{ij}^{\min}(\lambda),\ i \in E\}$ is the minimal nonnegative solution of (12.2.3), and $0 \leqslant \lambda \Sigma_{j \in E}\, p_{ij}^{\min}(\lambda) \leqslant 1$, so $\{u_\lambda(i), i \in E\}$ is the maximal solution of the system of nonnegative linear equations

$$u(i) = \sum_{k \neq i} \frac{q_{ik}}{\lambda + q_i} u(k) + \frac{-\sum_{j \in E} q_{ij}}{\lambda + q_i}, \qquad i \in E, \qquad 0 \leqslant u(i) \leqslant 1 \tag{12.2.6}$$

and the maximal solution may be obtained as follows. Let

$$\left.\begin{aligned} &u^{(0)}(i) \equiv 1 \qquad (i \in E), \\ &u^{(n+1)}(i) = \sum_{k \neq i} \frac{q_{ik}}{\lambda + q_i} u^{(n)}(k) + \frac{-\sum_{j \in E} q_{ij}}{\lambda + q_i} \\ &(i \in E,\ n \geqslant 1). \end{aligned}\right\} \tag{12.2.7}$$

Then

$$u^{(n)}(i) \downarrow u(i) \qquad (n \uparrow +\infty). \tag{12.2.8}$$

Hence, by virtue of (12.2.7) and (12.2.8), $u_\lambda(i)$ does not increase as λ increases. □

Lemma 12.2.3 *Condition* (i) *in Theorem* 12.1.1 *is equivalent to the following condition:*

There exists a certain constant $0 < \lambda_0 < +\infty$, *such that*

$$\inf_{i\in E} \lambda_0 \sum_{j\in E} p_{ij}^{\min}(\lambda_0) = \eta_{\lambda_0} > 0. \tag{12.2.9}$$

Proof. Obviously, it suffices to prove that Condition (i) in Theorem 12.1.1 can be derived from the above condition. This may be done in two cases:

(1) $0 < \lambda < \lambda_0$,

$$\eta_\lambda \geqslant \inf_{i\in E} \lambda \sum_{j\in E} p_{ij}^{\min}(\lambda_0) = \frac{\lambda}{\lambda_0} \eta_{\lambda_0} > 0. \tag{12.2.10}$$

(2) $\lambda_0 \leqslant \lambda < +\infty$.

By Lemma 12.2.2 we see that $\eta_\lambda \geqslant \eta_{\lambda_0} > 0$, proving this lemma.□

Remark 12.2.2 It is well known that Condition (ii) in Theorem 12.1.1 is equivalent to the fact that the minimal Q-process is honest, or that (12.2.12), for a certain constant $0 < \lambda = \lambda_0 < +\infty$, has only the zero solution. Therefore we know by Lemma 12.2.3 that the two conditions in Theorem 12.1.1 can be replaced by formally weaker but in essence, equivalent conditions.

Lemma 12.2.4 *There exists a row vector* $\alpha = (\alpha(1), \alpha(2), \cdots) > \mathbf{0}_-$, *which is independent of* λ *such that* $\alpha P_\lambda^{\min}$ *is summable*[1] *but* α *is not summable, i.e.,*

$$\lambda \sum_{j\in E} \sum_{i\in E} \alpha(i) p_{ij}^{\min}(\lambda) = \sum_{i\in E} \alpha(i) \lambda \sum_{j\in E} p_{ij}^{\min}(\lambda) < +\infty \qquad (0 < \lambda < +\infty) \tag{12.2.11}$$

and

$$\sum_{i\in E} \alpha(i) = +\infty, \tag{12.2.12}$$

if and only if

$$\inf_{i\in E} \lambda \sum_{j\in E} p_{ij}^{\min}(\lambda) = \eta = 0 \qquad (0 < \lambda < +\infty), \tag{12.2.13}$$

or equivalently there exists a certain constant $0 < \lambda_0 < +\infty$, *such that*

$$\inf_{i\in E} \lambda_0 \sum_{j\in E} p_{ij}^{\min}(\lambda_0) = \eta_{\lambda_0} = 0. \tag{12.2.14}$$

Proof. Sufficiency. Since $0 < \lambda \sum_{j\in E} p_{ij}^{\min}(\lambda) \leqslant 1$, there exists a row vector $\alpha > \mathbf{0}_-$ which satisfies (12.2.11), for example, choose $\alpha(i) = 2^{-i}$ $(i \in E)$. If (12.2.12)

1 Generally, we say the fact that a row vector $\rho = (\rho(1), \rho(2), \cdots)$ is summable means $\Sigma_j \rho(j) < \infty$.

holds in addition, then the sufficiency is proved. In what follows therefore, we assume that we choose $\alpha > \mathbf{0}_-$, such that (12.2.11) holds, but (12.2.12) does not hold. Choosing an arbitrarily fixed number $0 < \bar{\lambda}_0 < +\infty$, by (12.2.13) we may choose an infinite subset $\hat{E}$ of E such that

$$\sum_{i\in\hat{E}} \bar{\lambda}_0 \sum_{j\in E} p_{ij}^{\min}(\bar{\lambda}_0) < +\infty. \tag{12.2.15}$$

Let

$$\hat{E}_1 = \{i:\ i\in\hat{E},\ \alpha(i) > 1\}, \tag{12.2.16}$$

$$\hat{E}_2 = \{i:\ i\in\hat{E},\ \alpha(i) \leqslant 1\}, \tag{12.2.17}$$

$$\hat{\alpha}(i) = \begin{cases} \alpha(i), & i\in(E\setminus\hat{E}_2), \\ 1, & i\in\hat{E}_2. \end{cases} \tag{12.2.18}$$

Then

$$\hat{\alpha}(i) \geqslant \alpha(i) > 0 \qquad (i\in E), \tag{12.2.19}$$

$$\begin{aligned} \sum_{i\in E} \hat{\alpha}(i)\bar{\lambda}_0 \sum_{j\in E} p_{ij}^{\min}(\bar{\lambda}_0) &= \sum_{i\in E\setminus\hat{E}} \hat{\alpha}(i)\bar{\lambda}_0 \sum_{j\in E} p_{ij}^{\min}(\bar{\lambda}_0) \\ &\quad + \sum_{i\in E} \lambda_0 \sum_{j\in E} p_{ij}^{\min}(\lambda_0) \leqslant \sum_{i\in E} \alpha(i)\bar{\lambda}_0 \sum_{j\in E} p_{ij}^{\min}(\lambda_0) \\ &\quad + \sum_{i\in\hat{E}} \bar{\lambda}_0 \sum_{j\in E} p_{ij}^{\min}(\bar{\lambda}_0) < +\infty, \end{aligned} \tag{12.2.20}$$

and

$$\sum_{\in E} \hat{\alpha}(i) \geqslant \sum_{i\in\hat{E}} \hat{\alpha}(i) \geqslant \sum_{i\in\hat{E}} 1 = +\infty. \tag{12.2.21}$$

Evidently, by referring to the proof of Lemma 12.2.2, we have

$$\sum_{i\in E} \hat{\alpha}(i)\lambda \sum_{j\in E} p_{ij}^{\min}(\lambda) < +\infty \tag{12.2.22}$$

for any $0 < \lambda < +\infty$. This proves the sufficiency in Lemma 12.2.4.

Necessity. If (12.2.14) does not hold, then it follows from Lemma 12.2.3 that

$$\inf_{i\in E} \lambda \sum_{j\in E} p_{ij}^{\min}(\lambda) = \eta_\lambda > 0 \qquad (0 < \lambda < +\infty). \tag{12.2.23}$$

If there exists $\alpha > \mathbf{0}_-$ such that

$$\sum_{i\in E}\alpha(i)\lambda\sum_{j\in E}p_{ij}^{\min}(\lambda)<+\infty \qquad (0<\lambda<+\infty), \tag{12.2.24}$$

then, by (12.2.23), we have

$$\sum_{i\in E}\alpha(i)\lambda\sum_{j\in E}p_{ij}^{\min}(\lambda)\geqslant\eta_\lambda\sum_{i\in E}\alpha(i) \qquad (0<\lambda<+\infty). \tag{12.2.25}$$

From (12.2.24) and (12.2.25), we deduce

$$\eta_\lambda\sum_{i\in E}\alpha(i)<+\infty. \tag{12.2.26}$$

From $\eta_\lambda>0$ and (12.2.26) it follows that

$$\sum_{i\in E}\alpha(i)<+\infty. \tag{12.2.27}$$

This proves the necessity in Lemma 12.2.4. □

Lemma 12.2.5 *Let* $\Phi(t)=(\phi_{ij}(t);\ i,j\in E,\ t\geqslant 0)$ *be a Q-process satisfying Kolmogorov's backward differential equation*

$$\Phi'(t)=Q\Phi(t). \tag{12.2.28}$$

If we set

$$p_{ij}(\lambda)=\int_0^\infty e^{-\lambda t}\phi_{ij}(t)\,dt \qquad (i,j\in E;\ 0<\lambda<+\infty), \tag{12.2.29}$$

$$\mathbf{P}_\lambda^{(j)}=\begin{pmatrix}p_{1j}(\lambda)\\ p_{2j}(\lambda)\\ \vdots\end{pmatrix} \qquad (j\in E), \tag{12.2.30}$$

then the column vectors $\mathbf{P}_\lambda^{(j)}$ $(j\in E)$ *are linearly independent, i.e., if there exists a collection of real numbers* α_j $(j\in E)$ *such that* $\sum_{j\in E}\alpha_j\mathbf{P}_\lambda^{(j)}$ *absolutely converges to* $\mathbf{0}_1$ *for some* $\lambda>0$, *then*

$$\alpha_j=0 \qquad (j\in E). \tag{12.2.31}$$

In particular, the column vectors

$$\mathbf{P}_\lambda^{\min(j)}=\begin{pmatrix}p_{1j}^{\min}(\lambda)\\ p_{2j}^{\min}(\lambda)\\ \vdots\end{pmatrix} \qquad (j\in E) \tag{12.2.32}$$

are linearly independent.

Proof. From (12.2.29) and the fact that $\Phi(t)$ is a Q-process satisfying Equation (12.2.28) we have

$$p_{ij}(\lambda)=\sum_{k\neq i}\frac{q_{ik}}{\lambda+q_i}p_{kj}(\lambda)+\frac{\delta_{ij}}{\lambda+q_i}\qquad (i,\ j\in E). \tag{12.2.33}$$

Hence

$$0=\sum_{j\in E}\alpha_j p_{ij}(\lambda)=\sum_{j\in E}\sum_{k\neq i}\frac{q_{ik}}{\lambda+q_i}\alpha_j p_{kj}(\lambda)+\sum_{j\in E}\frac{\alpha_j\delta_{ij}}{\lambda+q_i}$$

$$=\sum_{k\neq i}\frac{q_{ik}}{\lambda+q_i}\sum_{j\in E}\alpha_j p_{kj}(\lambda)+\frac{\alpha_i}{\lambda+q_i}=0+\frac{\alpha_i}{\lambda+q_i}\qquad (i\in E). \tag{12.2.34}$$

Consequently (12.2.31) holds. The last part of the lemma may be immediately established by the fact that the minimal Q-process $(f_{ij}(t);\ i,\ j\in E,\ t\geqslant 0)$ satisfies (12.2.28). □

§ 12.3 Proof of the main theorem

The proof of Theorem 12.1.1.

First, we prove the necessity.

If Condition (ii) in Theorem 12.1.1 does not hold, then, in view of [27], there is more than one Q-process satisfying the Kolmogorov forward differential equations, so the Q-processes is not unique.

If Condition (i) in Theorem 12.1.1 does not hold, then, by Lemma 12.2.4, there exists a row vector $\alpha=(\alpha(1),\alpha(2),\cdots)>\mathbf{0}_-$, independent of λ, such that $\alpha p_\lambda^{\min}$ $(0<\lambda<+\infty)$ is summable, but α is not. Since α is not summable, we have

$$\frac{\left(-\sum_{k\in E}q_{ik}\right)\alpha_j}{\sum_{k\in E}\alpha_k}=0\qquad (i,\ j\in E). \tag{12.3.1}$$

Hence, by [28], we see

$$P_\lambda=P_\lambda^{\min}+\frac{(\mathbf{1}-\lambda_\lambda^{\min}\mathbf{1})}{\lambda\alpha P_\lambda^{\min}\mathbf{1}}\cdot\alpha P_\lambda^{\min} \tag{12.3.2}$$

is a Q-process and is honest. Therefore

$$\lambda P_\lambda\mathbf{1}=\mathbf{1} \tag{12.3.3}$$

Since Condition (i) does not hold,

$$\lambda P_\lambda^{\min}\mathbf{1} \not\equiv \mathbf{1}. \tag{12.3.4}$$

Hence

$$P_\lambda \not\equiv P_\lambda^{\min}. \tag{12.3.5}$$

Consequently, the Q-processes are not unique. Thus we have proved the necessity.

Now we turn to prove the sufficiency.

Suppose Conditions (i) and (ii) in Theorem 12.1.1 hold, we now prove that the Q-process is unique. We may furthermore assume that the minimal Q-process is not honest, otherwise the Q-process is unique.

Let $P_\lambda = \{p_{ij}(\lambda);\ i, j \in E\}$ denote any Q-process. It is well known that $\{p_{ij}(\lambda), i \in E\}$, for every $j \in E$, satisfies the system of inequalities

$$x_i \geqslant \sum_{k \neq i} \frac{q_{ik}}{\lambda + q_i} x_k + \frac{\delta_{ij}}{\lambda + q_i} \qquad (i \in E) \tag{12.3.6}$$

and $\{p_{ij}^{\min}(\lambda);\ i \in E\}$ is the minimal (nonnegative) solution of the system of equations

$$x_i = \sum_{k \neq i} \frac{q_{ik}}{\lambda + q_i} x_k + \frac{\delta_{ij}}{\lambda + q_i} \qquad (i \in E). \tag{12.3.7}$$

Therefore it is easy to verify that

$$p_{ij}(\lambda) - p_{ij}^{\min}(\lambda) \geqslant 0 \qquad (i,\ j \in E), \tag{12.3.8}$$

and $\{p_{ij}(\lambda) - p_{ij}^{\min}(\lambda);\ i \in E\}$ satisfies the system

$$x_i \geqslant \sum_{k \neq i} \frac{q_{ik}}{\lambda + q_i} x_k \qquad (i \in E). \tag{12.3.9}$$

Let

$$r_{ij}(\lambda) = \begin{cases} \dfrac{q_{ij}}{\lambda + q_i}, & i \neq j, \\ 0, & i = j, \end{cases} \tag{12.3.10}$$

$$R_\lambda = (r_{ij}(\lambda);\ i,\ j \in E). \tag{12.3.11}$$

Hence $\{p_{ij}(\lambda) - p_{ij}^{\min}(\lambda);\ i \in E\}$ is the excessive function of a Markov chain with R_λ as its transition probability matrix. The chain is evidently nonrecurrent. From Lemma 12.2.1, [13] and

$$\sum_{i\in E} 2^{-(i+1)}(p_{ij}(\lambda) - p_{ij}^{\min}(\lambda)) \leqslant \sum_{i\in E} 2^{-(i+1)} < +\infty,$$

we deduce

$$p_{ij}(\lambda) - p_{ij}^{\min}(\lambda) = \sum_{a\in E} k_\lambda(i, a) \tilde{f}_\lambda^{(a)}(j) \geqslant 0 \qquad (i, j, a\in E), \tag{12.3.12}$$

where

$$k_\lambda(i,a) = \frac{p_{ia}^{\min}(\lambda)}{\sum_{k\in E} 2^{-(k+1)} p_{ka}^{\min}(\lambda)} \qquad (a, j\in E). \tag{12.3.13}$$

Let

$$f_\lambda^{(a)}(j) = \frac{\tilde{f}_\lambda^{(a)}(j)}{\sum_{k\in E} 2^{-(k+1)} p_{ka}^{\min}(\lambda)} \qquad (a, j\in E). \tag{12.3.14}$$

Substituting (12.3.13) and (12.3.14) into (12.3.12) we obtain

$$p_{ij}(\lambda) = p_{ij}^{\min}(\lambda) + \sum_{a\in E} p_{ia}^{\min}(\lambda) f_\lambda^{(a)}(j),$$

$$f_\lambda^{(a)}(j) \geqslant 0 \qquad (\lambda > 0;\ i, j, a\in E). \tag{12.3.15}$$

From [29] we know that $P_\lambda = \{p_{ij}(\lambda);\ i, j\in E\}$ must satisfy

$$P_\lambda \geqslant 0 \qquad (\lambda > 0), \tag{12.3.16}$$

$$\lambda P\mathbf{1} \leqslant \mathbf{1} \qquad (\lambda > 0), \tag{12.3.17}$$

$$P_\lambda - P_\mu + (\lambda - \mu) P_\lambda P_\mu = \mathbf{0} \qquad (\lambda, \mu > 0), \tag{12.3.18}$$

$$\lim_{\lambda\to\infty} \lambda(\lambda P_\lambda - \mathbf{I}) = Q, \tag{12.3.19}$$

where $\mathbf{0}$ and $\mathbf{I}$ denote the zero and the unit matrix respectively. It follows from (12.3.15) and (12.3.17) that

$$\mathbf{F}_\lambda^{(a)} = (f_\lambda^{(a)}(1), f_\lambda^{(a)}(2), \cdots) \geqslant \mathbf{0}_- \qquad (\lambda > 0,\ a\in E) \tag{12.3.20}$$

and

$$\lambda\sum_{j\in E} p_{ij}^{\min}(\lambda) + \lambda\sum_{a\in E} p_{ia}^{\min}(\lambda)\cdot[\mathbf{F}_\lambda^{(a)}, \mathbf{1}\ ] \leqslant 1 \qquad (i\in E), \tag{12.3.21}$$

using (12.1.4), we have

$$\lambda p_{aa}^{\min}(\lambda) > 0 \qquad (\lambda > 0,\ a \in E). \tag{12.3.22}$$

Equations (12.3.21) and (12.3.22) imply that

$$[\mathbf{F}_\lambda^{(a)},\ \mathbf{1}] < +\infty \qquad (\lambda > 0,\ a \in E). \tag{12.3.23}$$

Here and henceforth we use

$$[\rho,\ \mathbf{C}] = \rho\mathbf{C} = \sum_{i \in E} \rho(i)\, C(i) \tag{12.3.24}$$

to denote the inner product of the row vector $\rho = (\rho(1), \rho(2), \cdots)$ and the column vector

$$\mathbf{C} = \begin{pmatrix} C(1) \\ C(2) \\ \vdots \end{pmatrix}.$$

Since $P_\lambda^{\min}$ satisfies (12.3.18), we get

$$\mathbf{P}_\lambda^{\min(a)} - \mathbf{P}_\mu^{\min(a)} + (\lambda - \mu) P_\lambda^{\min} \mathbf{P}_\mu^{\min(a)} = \mathbf{0}_1 \quad (\lambda,\ \mu > 0,\ a \in E). \tag{12.3.25}$$

Let

$$A(\mu,\ \lambda) = I + (\mu - \lambda) P_\lambda^{\min} \qquad (\lambda,\ \mu > 0). \tag{12.3.26}$$

Since $P_\lambda^{\min}$ satisfies (12.3.18), we obtain

$$A(\mu,\ \nu) A(\nu,\ \lambda) = A(\mu,\ \lambda), \tag{12.3.27}$$

$$A(\mu,\lambda) P_\mu^{\min} = P_\lambda^{\min}, \tag{12.3.28}$$

i.e.,

$$A(\mu,\lambda) \mathbf{P}_\mu^{\min(a)} = \mathbf{P}_\lambda^{\min(a)} \qquad (a \in E). \tag{12.3.29}$$

By substituting (12.3.15) into (12.3.18), taking account of the fact that $P_\lambda^{\min}$ also satisfies (12.3.18) and Lemma 12.2.5, we get

$$\mathbf{F}_\lambda^{(a)} A(\lambda,\mu) = \mathbf{F}_\mu^{(a)} + (\mu - \lambda) \sum_{t \in E} [\mathbf{F}_\lambda^{(a)}, \mathbf{P}_\mu^{\min(t)}]\, \mathbf{F}_\mu^{(t)}. \tag{12.3.30}$$

It follows from (12.3.20) and (12.3.26) that $\mathbf{F}_\lambda^{(a)} A(\lambda,\mu) \geqslant \mathbf{0}_-$ when $\lambda \geqslant \mu > 0$. By (12.3.30), $\mathbf{F}_\lambda^{(a)} A(\lambda,\mu) \geqslant \mathbf{0}_-$ when $\mu \geqslant \lambda > 0$. So $\mathbf{F}_\lambda^{(a)} A(\lambda,\mu) \geqslant \mathbf{0}_-$ always holds for any $\lambda,\ \mu > 0$. It follows by (12.3.30) that

$$[\mathbf{F}_\lambda^{(a)} A(\lambda,\mu), \mathbf{1}] = [\mathbf{F}_\mu^{(a)}, \mathbf{1}] + (\mu-\lambda) \sum_{t\in E} [\mathbf{F}_\lambda^{(a)}, \mathbf{P}_\mu^{\min(t)}][\mathbf{F}_\mu^{(t)}, \mathbf{1}], \tag{12.3.31}$$

and by (12.3.15) that

$$P_\mu \mathbf{1} = P_\mu^{\min}\mathbf{1} + \sum_{t\in E} \mathbf{P}_\mu^{\min(t)} [\mathbf{F}_\mu^{(t)}, \mathbf{1}]. \tag{12.3.32}$$

Hence

$$[\mathbf{F}_\lambda^{(a)}, P_\mu \mathbf{1}] = [\mathbf{F}_\lambda^{(a)}, \mathbf{P}_\mu^{\min}\mathbf{1}] + \sum_{t\in E} [\mathbf{F}_\lambda^{(a)}, \mathbf{P}_\mu^{\min(t)}][\mathbf{F}_\mu^{(t)}, \mathbf{1}]. \tag{12.3.33}$$

But it is known by (12.3.17) and (12.3.23) that

$$[\mathbf{F}_\lambda^{(a)}, P_\mu \mathbf{1}] = \frac{1}{\mu}[\mathbf{F}_\lambda^{(a)}, \mu P_\mu \mathbf{1}] \leqslant \frac{1}{\mu}[\mathbf{F}_\lambda^{(a)}, \mathbf{1}] < +\infty \qquad (\mu > 0), \tag{12.3.34}$$

by (12.3.23) and (12.3.34) that

$$\sum_{t\in E} [\mathbf{F}_\lambda^{(a)}, \mathbf{P}_\mu^{\min(t)}][\mathbf{F}_\mu^{(t)}, \mathbf{1}] < +\infty, \tag{12.3.35}$$

and by (12.3.23), (12.3.31) and (12.3.35) that

$$[\mathbf{F}_\lambda^{(a)} A(\lambda,\mu), \mathbf{1}] < +\infty, \tag{12.3.36}$$

i.e., $\mathbf{F}_\lambda^{(a)} A(\lambda,\mu)$ is summable. Temporarily, we fix a and $\lambda > 0$, and put

$$\rho_\mu = \mathbf{F}_\lambda^{(a)} A(\lambda,\mu) \qquad (\mu > 0). \tag{12.3.37}$$

Hence $\rho_\mu \geqslant \mathbf{0}_-$ and is summable, and it follows by (12.3.27) that

$$\rho_\mu A(\mu,\nu) = \rho_\nu \qquad (\mu,\nu > 0). \tag{12.3.38}$$

According to [29, Lemma 2.2] and Condition (ii) in Theorem 12.1.1, there exists a row vector $\beta_\lambda^{(a)} \geqslant \mathbf{0}_-$ independent of μ (but dependent on a and λ), such that $\beta_\lambda^{(a)} P_\mu^{\min}$ is summable,

$$\mu \mathbf{F}_\lambda^{(a)} A(\lambda,\mu) - \mathbf{F}_\lambda^{(a)} A(\lambda,\mu) Q = \beta_\lambda^{(a)} \qquad (\mu > 0), \tag{12.3.39}$$

and

$$\mathbf{F}_\lambda^{(a)} A(\lambda,\mu) = \beta_\lambda^{(a)} P_\mu^{\min} \qquad (\mu > 0). \tag{12.3.40}$$

By (12.3.39) and (12.3.40), particularly when $\mu = \lambda$, we can write

$$\lambda \mathbf{F}_\lambda^{(a)} - \mathbf{F}_\lambda^{(a)} Q = \beta_\lambda^{(a)} \qquad (\lambda > 0), \tag{12.3.41}$$

$$\mathbf{F}_\lambda^{(a)} = \beta_\lambda^{(a)} P_\lambda^{\min} \qquad (\lambda > 0). \tag{12.3.42}$$

Multiplying both sides of (12.3.30) from the right by $(\mu I - Q)$, we get

$$\beta_\lambda^{(a)} = \beta_\mu^{(a)} + (\mu - \lambda) \sum_{t \in E} [\mathbf{F}_\lambda^{(a)}, \mathbf{P}_\mu^{\min(t)}]\, \beta_\mu^{(t)}. \tag{12.3.43}$$

Hence we have

$$[\beta_\lambda^{(a)}, \mathbf{1}] = [\beta_\mu^{(a)}, \mathbf{1}] + (\mu - \lambda) \sum_{t \in E} [\mathbf{F}_\lambda^{(a)}, \mathbf{P}_\mu^{\min(t)}]\, [\beta_\mu^{(t)}, \mathbf{1}]. \tag{12.3.44}$$

Since both P_λ and $P_\lambda^{\min}$ satisfy (12.3.19) and (12.3.15), we get

$$\lim_{\lambda \to \infty} \lambda p_{ia}^{\min}(\lambda)\, \lambda f_\lambda^{(a)}(j) = 0 \qquad (i, j, a \in E), \tag{12.3.45}$$

in particular,

$$\lim_{\lambda \to \infty} \lambda p_{aa}^{\min}(\lambda)\, \lambda f_\lambda^{(a)}(j) = 0. \tag{12.3.46}$$

But

$$\lim_{\lambda \to \infty} \lambda p_{aa}^{\min}(\lambda) = 1. \tag{12.3.47}$$

Hence

$$\lim_{\lambda \to \infty} \lambda f_\lambda^{(a)}(j) = 0, \tag{12.3.48}$$

and furthermore

$$\lim_{\lambda \to \infty} f_\lambda^{(a)}(j) = 0. \tag{12.3.49}$$

It follows from (12.3.41) that

$$\lambda f_\lambda^{(a)}(j) + q_j f_\lambda^{(a)}(j) = \sum_{i \neq j} f_\lambda^{(a)}(i)\, q_{ij} + \beta_\lambda^{(a)}(j). \tag{12.3.50}$$

It can be seen from (12.3.43) that $\beta_\mu^{(a)}$ does not increase as μ increases. Moreover, (12.3.48), (12.3.49) and (12.3.50) imply that

$$\beta_\lambda^{(a)} \downarrow \mathbf{0}_- \qquad (\lambda \uparrow +\infty). \tag{12.3.51}$$

We shall complete the proof of the theorem in the following steps:
(1) To prove

$$[\beta_\lambda^{(a)}, \mathbf{1}] \leqslant \frac{\lambda}{\eta_\lambda}[\mathbf{F}_\lambda^{(a)}, \mathbf{1}] < +\infty \qquad (\lambda > 0, a \in E). \tag{12.3.52}$$

Proof. By (12.3.23) and Condition (i) in Theorem 12.1.1 it is enough to prove the left part of (12.3.52). It can be deduced from (12.3.42) that

$$\lambda[\mathbf{F}_\lambda^{(a)}, \mathbf{1}] = \lambda[\beta_\lambda^{(a)} P_\lambda^{\min}, \mathbf{1}]$$

$$= [\beta_\lambda^{(a)}, \lambda P_\lambda^{\min} \mathbf{1}] \geqslant \eta_\lambda[\beta_\lambda^{(a)}, \mathbf{1}]. \tag{12.3.53}$$

Hence

$$[\beta_\lambda^{(a)}, \mathbf{1}] \leqslant \frac{\lambda}{\eta_\lambda}[\mathbf{F}_\lambda^{(a)}, \mathbf{1}], \tag{12.3.54}$$

and (12.3.52) is proved.
(2) To prove

$$(\mu - \lambda)[\mathbf{F}_\lambda^{(a)}, \mathbf{P}_\mu^{\min(t)}]$$

$$= [\beta_\lambda^{(a)}, \mathbf{P}_\lambda^{\min(t)}] - [\beta_\lambda^{(a)}, \mathbf{P}_\mu^{\min(t)}] < +\infty \quad (\lambda, \mu > 0, a, t \in E). \tag{12.3.55}$$

Proof. Note that

$$[\beta_\lambda^{(a)}, \mathbf{P}_\mu^{\min(t)} \leqslant \frac{1}{\mu}[\beta_\lambda^{(a)}, \mathbf{1}] < +\infty \qquad (\lambda, \mu > 0), \tag{12.3.56}$$

$$\sup_{i \in E} |p_{it}^{\min}(\lambda) - p_{it}^{\min}(\mu)| \leqslant \frac{1}{\lambda} + \frac{1}{\mu} < +\infty \qquad (\lambda, \mu > 0), \tag{12.3.57}$$

and

$$(\mu - \lambda)[\mathbf{F}_\lambda^{(a)}, \mathbf{P}_\mu^{\min(t)}] = (\mu - \lambda)[\beta_\lambda^{(a)}, P_\lambda^{\min}, \mathbf{P}_\mu^{\min(t)}]$$

$$= (\mu - \lambda)[\beta_\lambda^{(a)}, P_\lambda^{\min}, \mathbf{P}_\mu^{\min(t)}] = [\beta_\lambda^{(a)}, (\mu - \lambda) P_\lambda^{\min}, \mathbf{P}_\mu^{\min(t)}]$$

$$= [\beta_\lambda^{(a)}, \mathbf{P}_\lambda^{\min(t)} -- \mathbf{P}_\mu^{\min(t)}]$$

$$= [\beta_\lambda^{(a)}, \mathbf{P}_\lambda^{\min(t)}] - [\beta_\lambda^{(a)}, \mathbf{P}_\mu^{\min(t)}], \tag{12.3.58}$$

as desired.

(3) To prove

$$\sum_{t\in E} [\beta_\mu^{(t)}, \mathbf{1}]\,\mathbf{P}_\mu^{\min(t)} \leqslant \left(\frac{1}{\eta_\mu} - 1\right)\mathbf{1} \qquad (\mu > 0). \tag{12.3.59}$$

Proof. It follows by (12.3.15), (12.3.17) and (12.3.52) that

$$\begin{aligned}\mathbf{1} \geqslant \mu P_\mu \mathbf{1} &= \mu P_\mu^{\min}\mathbf{1} + \sum_{t\in E} \mu[\mathbf{P}_\mu^{(t)}, \mathbf{1}]\,\mathbf{P}_\mu^{\min(t)} \\ &\geqslant \eta_\mu \mathbf{1} + \sum_{t\in E} \eta_\mu [\beta_\mu^{(t)},\ \mathbf{1}]\,\mathbf{P}_\mu^{\min(t)} \\ &= \eta_\mu\left(\mathbf{1} + \sum_{t\in E} [\beta_\mu^{(t)}, \mathbf{1}]\,\mathbf{P}_\mu^{\min(t)}\right).\end{aligned} \tag{12.3.60}$$

Hence (12.3.59) holds.

(4) To prove

$$\sum_{t\in E} [\beta_\lambda^{(a)}, [\beta_\mu^{(t)}, \mathbf{1}]\,\mathbf{P}_\mu^{\min(t)}] \leqslant \left(\frac{1}{\eta_\mu} - 1\right)[\beta_\lambda^{(a)}, \mathbf{1}] < +\infty \qquad (\lambda, \mu > 0, a \in E). \tag{12.3.61}$$

Proof. It follows from (12.3.52) and (12.3.60) that

$$\begin{aligned}\sum_{t\in E} [\beta_\lambda^{(a)}, [\beta_\mu^{(t)}, \mathbf{1}]\,\mathbf{P}_\mu^{\min(t)}] &= [\beta_\lambda^{(a)}, \sum_{t\in E} [\beta_\mu^{(t)}, \mathbf{1}]\,\mathbf{P}_\mu^{\min(t)}] \\ &\leqslant \left[\beta_\lambda^{(a)}, \left(\frac{1}{\eta_\mu} - 1\right)\mathbf{1}\right] = \left(\frac{1}{\eta_\mu} - 1\right)[\beta_\lambda^{(a)}, \mathbf{1}] < +\infty.\end{aligned} \tag{12.3.62}$$

(5) To prove

$$\begin{aligned}[\beta_\lambda^{(a)}, \mathbf{1}] &= [\beta_\mu^{(a)}, \mathbf{1}] + \sum_{t\in E} [\beta_\lambda^{(a)}, [\beta_\mu^{(t)}, \mathbf{1}]\,\mathbf{P}_\lambda^{\min(t)}] \\ &\quad - \sum_{t\in E} [\beta_\lambda^{(a)}, [\beta_\mu^{(t)}, \mathbf{1}]\,\mathbf{P}_\mu^{\min(t)}] < +\infty \quad (\lambda, \mu > 0, a \in E).\end{aligned} \tag{12.3.63}$$

Proof. From (12.3.52), it is evident that we need only prove the first half of (12.3.63). It follows from (12.3.44), (12.3.55) and (12.3.61) that

$$[\beta_\lambda^{(a)}, \mathbf{1}] = [\beta_\mu^{(a)}, \mathbf{1}] + \sum_{t\in E} (\mu - \lambda)\,[\mathbf{F}_\lambda^{(a)}, \mathbf{P}_\mu^{\min(t)}],\ [\beta_\mu^{(t)}, \mathbf{1}]$$

$$= [\beta_\mu^{(a)}, \mathbf{1}] + \sum_{t\in E} \{ [\beta_\lambda^{(a)}, \mathbf{P}_\lambda^{\min(t)}]$$

$$- [\beta_\lambda^{(a)}, \mathbf{P}_\mu^{\min(t)}] \} [\beta_\mu^{(t)}, \mathbf{1}]$$

$$= [\beta_\mu^{(a)}, \mathbf{1}] + \sum_{t\in E} [\beta_\lambda^{(a)}, [\beta_\mu^{(t)}, \mathbf{1}] \mathbf{P}_\lambda^{\min(t)}]$$

$$- \sum_{t\in E} [\beta_\lambda^{(a)}, [\beta_\mu^{(t)}, \mathbf{1}] \mathbf{P}_\mu^{\min(t)}], \qquad (12.3.64)$$

as desired.

(6) To prove

$$[\beta_\mu^{(a)}, \mathbf{1}] \downarrow 0 \qquad (\mu \uparrow +\infty), \qquad (12.3.65)$$

$$[\beta_\lambda^{(a)}, [\beta_\mu^{(t)}, \mathbf{1}] \mathbf{P}_\mu^{\min(t)}] \downarrow 0 \qquad (\mu \uparrow +\infty) \qquad (12.3.66)$$

and

$$[\beta_\lambda^{(a)}, [\beta_\mu^{(t)}, \mathbf{1}] \mathbf{P}_\lambda^{\min(t)}] \downarrow 0 \qquad (\mu \uparrow +\infty). \qquad (12.3.67)$$

Proof. Equations (12.3.65) and (12.3.67) follow immediately from (12.3.51); (12.3.66) follows immediately from (12.3.51) and

$$p_{it}^{\min}(\mu) \downarrow 0 \qquad (\mu \uparrow +\infty). \qquad (12.3.68)$$

(7) To prove

$$[\beta_\lambda^{(a)}, \mathbf{1}] \equiv 0 \qquad (\lambda > 0, a \in E). \qquad (12.3.69)$$

Proof. Taking limits on both sides of (12.3.64) as $\mu \uparrow +\infty$, we obtain (12.3.69) from (12.3.65), (12.3.66) and (12.3.67) immediately.

(8) To prove the uniqueness of the Q-process.

It follows from (12.3.69) that

$$\beta_\lambda^{(a)} \equiv \mathbf{0}_- \qquad (\lambda > 0, a \in E), \qquad (12.3.70)$$

from (12.3.42) and (12.3.70) that

$$\mathbf{F}_\lambda^{(a)} \equiv \mathbf{0}_- \qquad (\lambda > 0, a \in E), \qquad (12.3.71)$$

and from (12.3.15) and (12.3.71) that

$$P_\lambda \equiv P_\lambda^{\min} \qquad (\lambda > 0), \qquad (12.3.72)$$

i.e., the Q-process is unique.

Now sufficiency in Theorem 12.1.1 has been proved. □

§ 12.4 The case of diagonal type[1]

Definition 12.4.1 *The Q-matrix is called a Q-matrix of diagonal type if it has the following properties:*

$$q_{ij} = 0 \quad (i \neq j), \qquad q_{ii} \not\equiv 0 \quad (i \in E). \tag{12.4.1}$$

Theorem 12.4.1 *If a Q-matrix is of diagonal type, then a necessary and sufficient condition for the uniqueness of the Q-process is that*

$$\sup_{i \in E} q_i = C < +\infty. \tag{12.4.2}$$

Proof.

(1) Sufficiency.

If (12.4.2) holds, since $\{\lambda \sum_{j \in E} p_{ij}^{\min}(\lambda), i \in E\}$ is the minimal (nonnegative) solution of the system of equations

$$x_i = \frac{\lambda}{\lambda + q_i} \qquad (i \in E), \tag{12.4.3}$$

we have

$$\lambda \sum_{j \in E} p_{ij}^{\min}(\lambda) = \frac{\lambda}{\lambda + q_i} \geqslant \frac{\lambda}{\lambda + c} > 0 \qquad (i \in E). \tag{12.4.4}$$

Thus Condition (i) in Theorem 12.1.1 holds, and Equation (12.1.12) transforms into

$$\begin{cases} \lambda n_i = q_i n_i, & \lambda > 0, \\ 0 \leqslant n_i, & \sum_i n_i < +\infty. \end{cases} \tag{12.4.5}$$

Hence $n_i \equiv 0$ $(i \in E)$, and Condition (ii) in Theorem 12.1.1 is also satisfied. Consequently the Q-process is unique.

(2) Necessity.

If (12.4.2) does not hold, then

$$\inf_{i \in E} \lambda \sum_{j \in E} p_{ij}^{\min}(\lambda) = \inf_{i \in E} \frac{\lambda}{\lambda + q_i} = 0. \tag{12.4.6}$$

---

1 The results of this section have been got by G. E. H. Reuter[28] with a different approach.

So Condition (i) in Theorem 12.1.1 is not satisfied. Consequently the Q-process is not unique. □

§ 12.5 The bounded case[1]

Lemma 12.5.1 *If* $n(j)$ $(j\in E)$ *is an arbitrary nonzero solution of Equation* (12.1.12), *then*

$$\sum_{j\in E}\left(\sum_{k\neq j} q_{jk}\right)n(j) = +\infty. \tag{12.5.1}$$

Proof. It can be seen from (12.1.12) that

$$(\lambda + q_j)n(j) = \sum_{k\neq j} q_{kj}n(k) \qquad (j\in E). \tag{12.5.2}$$

Hence we have

$$\lambda\sum_{j\in E} n(j) + \sum_{j\in E} q_j n(j) = \sum_{j\in E}\Big(\sum_{k\neq j} q_{kj}n(k)\Big)$$

$$= \sum_{k\in E}\left(\sum_{j\neq k} q_{kj}\right)n(k) = \sum_{j\in E}\left(\sum_{k\neq j} q_{jk}\right)n(j). \tag{12.5.3}$$

If

$$\sum_{j\in E}\left(\sum_{k\neq j} q_{jk}\right)n(j) < +\infty, \tag{12.5.4}$$

then from $\lambda > 0$, (12.5.4) and

$$q_j \geqslant \sum_{k\neq j} q_{ik} \quad (j\in E), \qquad \sum_{j\in E} n(j) < +\infty, \tag{12.5.5}$$

we find that

$$\sum_{j\in E} n(j) = 0. \tag{12.5.6}$$

Using (12.5.6) and

$$n(j) \geqslant 0 \qquad (j\in E), \tag{12.5.7}$$

we obtain that

1 The result of this section was reached independently by G. E. H. Reuter.

$$n(j) \equiv 0 \qquad (j \in E). \tag{12.5.8}$$

But this contradicts the fact that $n(j)$ $(j \in E)$ is a nonzero solution of (12.2.12). So (12.5.4) does not hold. Consequently (12.5.1) holds. □

Definition 12.5.1 *If there exists a constant* $0 \leqslant c < +\infty$, *such that*

$$-q_{ii} \leqslant c \qquad (i \in E), \tag{12.5.9}$$

then the Q-matrix is said to be bounded.

Lemma 12.5.2 *If*

$$\sum_{j \neq i} q_{ij} \leqslant c < +\infty \qquad (i \in E), \tag{12.5.10}$$

then Equation (12.1.12) *has only the zero solution.*

Proof. (12.5.10) shows that we have

$$\sum_{j \in E} \left(\sum_{k \neq j} q_{jk} \right) n(j) \leqslant c \sum_{j \in E} n(j) < +\infty \tag{12.5.11}$$

for any solution $n(j)$ $(j \in E)$ of (12.1.12). Hence we get what should be proved by Lemma 12.5.1. □

Theorem 12.5.1 *If a Q-matrix is bounded, then the Q-process is unique.*

Proof. Evidently, we need to prove this theorem only for the nonconservative case of Q. The proof of Lemma 12.2.1 shows that

$$\left\{ \lambda \sum_{j \in E} p_{ij}^{\min}(\lambda),\ i \in E \right\} \tag{12.5.12}$$

is the minimal nonnegative solution of (12.2.3), so we have

$$\lambda \sum_{j \in E} p_{ij}^{\min}(\lambda) \geqslant \frac{\lambda}{\lambda + q_i} \geqslant \frac{\lambda}{\lambda + c} > 0 \qquad (i \in E). \tag{12.5.13}$$

Hence Condition (i) in Theorem 12.1.1 is satisfied. It is known from (12.5.9) that

$$\sum_{j \neq i} q_{ij} \leqslant q_i \leqslant c < +\infty \qquad (i \in E). \tag{12.5.14}$$

By (12.5.14) and Lemma 12.5.2 we see that the Equation (12.1.12) has only the zero solution. Therefore Condition (ii) is satisfied. By Theorem 12.1.1 we get the theorem. □

§ 12.6 The case when E is finite

Theorem 12.6.1 *If E is a finite set, then the Q-process is unique.* *Proof.* Evidently Theorem 12.1.1 is also valid for finite E. It is easy to complete this proof similarly to the proof of Theorem 12.5.1. □

§ 12.7 The case of a branch Q-matrix

Definition 12.7.1 *If $E=(0, 1, 2, \cdots)$ and the Q-matrix satisfies*

$$q_{ij}=\begin{cases} iq_{1,j-i+1}, & j \geqslant i-1 \\ 0, & j<i-1 \end{cases} \qquad (i, j \in E), \tag{12.7.1}$$

then Q is called a branch Q-matrix.

Definition 12.7.2 *If the Q-matrix of the Q-process $P(t)=(p_{ij}(t),\ i, j \in E)$ is a branch Q-matrix, and*

$$p_{ij}(t)=\sum_{\sum_{s=1}^{i} j_s=j} \prod_{s=1}^{i} p_{1j_s}(t) \qquad (i, j \in E, t \geqslant 0), \tag{12.7.2}$$

then $P(t)$ is called a branch Q-process.

Lemma 12.7.1 *For any given branch Q-matrix, there is one and only one branch Q-process, and this is just the minimal Q-process.*

Proof. In the case of Q being conservative, the proof of this lemma is contained in [30, Chapter V, Appendix 1]. In fact, the method used there is also effective for the case of Q being nonconservative, so we shall not go into it further. □

Lemma 12.7.2 *For any given branch Q-matrix, Condition* (ii) *in Theorem* 12.1.1 *is satisfied, hence the Q-process satisfying Kolmogorov's system of forward differential equations is unique.*

Proof.[31] We consider the following equation

$$(n(0), n(1), n(2), \cdots)\begin{Bmatrix} \lambda & 0 & 0 & \cdots \\ -q_{10} & \lambda-q_{11} & -q_{12} & \cdots \\ 0 & -2q_{10} & \lambda-2q_{11} & \cdots \\ \vdots & \vdots & \vdots & \cdots \end{Bmatrix}=0. \tag{$*$}$$

(1) Clearly, Equation $(*)$ has only the zero solution if $q_{10}=0$.

(2) If $q_{10}>0$, since the demension of the solution space of Equation (12.1.12)

is independent of $\lambda > 0$ [27, Lemma 6], we assume that $\lambda = 2q_{10} > 0$. Suppose that $\mathbf{n} = (n(0), n(1), n(2), \cdots)$ satisfies(*) and $\mathbf{n} \neq \mathbf{0}_1$, then

$$n(k+1) = \frac{1}{(k+1)q_{10}}[kq_1 n(k) - (k-1)q_{12}n(k-1)$$

$$- (k-2)q_{13}n(k-2) - \cdots - q_{1k}n(1) + \lambda n(k)],$$

$$n(k) = \frac{1}{kq_{10}}[(k-1)q_1 n(k-1)$$

$$- (k-2)q_{12}n(k-2) - \cdots - q_{1,k-1}n(1) + \lambda n(k-1)],$$

$$n(k-1) = \frac{1}{(k-1)q_{10}}[(k-2)q_1 n(k-2) - \cdots - q_{1,k-2}n(1) + \lambda n(k-1)],$$

$$\vdots$$

$$n(2) = \frac{1}{2q_{10}}[q_1 n(1) + \lambda n(1)],$$

hence

$$\sum_{j=2}^{k-1} n(j) = \frac{\lambda}{q_{10}} \sum_{j=1}^{k} \frac{n(j)}{j+1} + \frac{1}{q_{10}}\left[\frac{k}{k+1} q_1 n(k) + (k-1)\left(\frac{q_1}{k} - \frac{q_{12}}{k+1}\right)n(k-1)\right.$$

$$+ (k-2)\left(\frac{q_1}{k-1} - \frac{q_{12}}{k} - \frac{q_{12}}{k+1}\right)n(k-2)$$

$$\left. + \cdots + 1\cdot\left(\frac{q_1}{2} - \frac{q_{12}}{3} - \cdots - \frac{q_{1k}}{k+1}\right)n(1)\right]$$

$$\geqslant \frac{\lambda}{q_{10}} \sum_{j=1}^{k} \frac{n(j)}{j+1} + \frac{1}{q_{10}}\left[\frac{k}{k+1} q_1 n(k) + (k-1)\left(\frac{q_1 - q_{12}}{k}\right)n(k-1)\right.$$

$$+ (k-2)\left(\frac{q_1 - q_{12} - q_{13}}{k-1}\right)n(k-2)$$

$$\left. + \cdots + 1\cdot\left(\frac{q_1 - q_{12} - \cdots - q_{1k}}{2}\right)n(1)\right]$$

$$\geqslant \frac{\lambda}{q_{10}} \sum_{j=1}^{k} \frac{n(j)}{j+1} + \frac{1}{q_{10}}\left[\frac{k}{k+1} q_{10} n(k)\right.$$

$$+\frac{k-1}{k}q_{10}n(k-1)+\cdots+\frac{1}{2}q_{10}n(1)\Bigg]$$

$$=2\sum_{j=1}^{k}\frac{n(j)}{j+1}+\sum_{j=1}^{k}\frac{jn(j)}{j+1}=\sum_{j=1}^{k}\frac{n(j)}{j+1}+\sum_{j=1}^{k}n(j).$$

Letting $k\to\infty$, we obtain

$$\sum_{j=2}^{\infty}n(j)\geqslant\sum_{j=1}^{\infty}\frac{n(j)}{j+1}+\sum_{j=1}^{\infty}n(j),$$

and so, from $\mathbf{n}\geqslant 0$ and $\mathbf{n}\neq 0$, we have

$$\sum_{j=1}^{\infty}n(j)=\infty.$$

This implies the Condition (ii) in Theorem 12.1.1. □

Lemma 12.7.3 *Let Q be a conservative branch Q-matrix, then Equation* (12.2.1) *has only the zero solution, i.e., necessary and sufficient condition for the Q-process satisfying Kolmogorov's system of backward differential equations to be unique is that, for any $\varepsilon>0$, the integral*

$$\int_{1-\varepsilon}^{1}\frac{dx}{f(x)}\tag{12.7.3}$$

diverges, where

$$f(x)=\sum_{j=0}^{\infty}q_{1j}x^{j}\qquad(|x|\leqslant 1).\tag{12.7.4}$$

Proof. This lemma follows from [30, Chapter V, Theorems 4.2 and 9.1]. □

Lemma 12.7.4 *If Q is a non-conservative Q-matrix, then Equation* (12.2.1) *has only the zero solution, i.e., the Q-process satisfying Kolmogorov system of backward differential equations is unique.*

Proof. We prove this for two cases:

(1)
$$\sum_{j\neq 1}q_{1j}=0.\tag{12.7.5}$$

Then the equation

$$(\lambda\mathbf{1}-Q)\mathbf{U}=\mathbf{0}_1\tag{12.7.6}$$

yields

$$\begin{aligned} \lambda u_0 &= 0, \\ (\lambda + q_1) u_1 &= 0, \\ (\lambda + 2q_1) u_2 &= 0, \\ &\cdots\cdots\cdots \\ (\lambda + kq_1) u_k &= 0, \\ &\cdots\cdots\cdots. \end{aligned} \tag{12.7.7}$$

Thus $u_k = 0$ $(k = 0, 1, \cdots)$. Consequently Equation (12.1.1) has only the zero solution.

(2)
$$\sum_{j \neq 1} q_{1j} > 0. \tag{12.7.8}$$

Let

$$\alpha = \frac{q_1}{\sum_{j \neq 1} q_{1j}} > 0. \tag{12.7.9}$$

Since Q is nonconservative, we see that

$$\alpha > 1. \tag{12.7.10}$$

And Equation (12.7.6) is just

$$\begin{aligned} \lambda u_0 &= 0, \\ -q_{10}u_0 + (\lambda + q_1) u_1 - q_{12}u_2 - q_{13}u_3 - \cdots &= 0, \\ -2q_{10}u_1 + (\lambda + 2q_1) u_2 - q_{12}u_3 - \cdots &= 0, \\ -3q_{10}u_2 + (\lambda + 3q_1) u_3 - \cdots &= 0, \\ \cdots\cdots\cdots\cdots\cdots\cdots&. \end{aligned} \tag{12.7.11}$$

If the above system of equations has solutions $\mathbf{U} \neq \mathbf{0}_1$, $\mathbf{U} \geqslant \mathbf{0}_1$, then, of course, $u_0 = 0$. Let u_{k_1} be the first among $u_1, u_2, \cdots$ which is greater than zero. It can be seen by (12.7.10) and (12.7.11) that there must exist $n > k_1$ such that

$$u_n \geqslant \frac{\alpha(\lambda + k_1 q_1)}{k_1 q_1} u_{k_1}. \tag{12.7.12}$$

Let

$$k_2 = \min\left\{n \mid u_n \geqslant \frac{\alpha(\lambda + k_1 q_1)}{k_1 q_1} u_{k_1},\ n > k_1\right\}. \tag{12.7.13}$$

In view of (12.7.10), (12.7.11) and the fact that $u_{k_2 - 1} < u_{k_2}$, we see that there exists $n > k_2$ such that

$$u_n \geqslant \frac{\alpha(\lambda + k_2 q_1)}{k_2 q_1} u_{k_2}. \tag{12.7.14}$$

Let

$$k_3 = \min\left\{n \mid u_n \geqslant \frac{\alpha(\lambda + k_2 q_1)}{k_2 q_1} u_{k_2},\ n > k_2\right\}. \tag{12.7.15}$$

Proceeding in this way, we obtain a sequence of strictly increasing positive integers $\{k_m\}$, and

$$u_{k_{m+1}} \geqslant \alpha^m \left(\frac{\lambda + k_m q_1}{k_m q_1}\right)\left(\frac{\lambda + k_{m-1} q_1}{k_{m-1} q_1}\right)\cdots\left(\frac{\lambda + k_1 q_1}{k_1 q_1}\right) u_{k_1}$$

$$\geqslant u_{k_1}\alpha^m \uparrow +\infty \qquad (m \uparrow +\infty). \tag{12.7.16}$$

So $\{u_{k_m}\}$ is unbounded. Hence $\{u_n\}$ is unbounded. Consequently Equation (12.2.1) has only the zero solution. □

Lemma 12.7.5 *If Q is a nonconservative branch Q-matrix, then Condition* (i) *in Theorem* 12.1.1 *does not hold.*

Proof. We first prove that if $P(t) = (p_{ij}(t),\ i, j \in E)$ is a minimal Q-process, then

$$\sum_{j\in E} p_{1j}(t) < 1 \qquad (t > 0). \tag{12.7.17}$$

It is known from Lemma 12.7.1 and [31, Lemma 6.2] that

$$\sum_{j\in E} p_{ij}(t) = \left(\sum_{j\in E} p_{1j}(t)\right)^i \qquad (i \in E, t \geqslant 0). \tag{12.7.18}$$

Owing to (12.7.18) and the fact that Q is nonconservative, there exists $t_0 > 0$ such that

$$\sum_{j\in E} p_{1j}(t_0) < 1. \tag{12.7.19}$$

It follows from (12.1.1)—(12.1.3) and (12.7.19) that

$$\sum_{j\in E} p_{1j}(t_0 + s) = \sum_{j\in E}\sum_{k\in E} p_{1k}(t_0) p_{kj}(s) = \sum_{k\in E} p_{1k}(t_0) \sum_{j\in E} p_{kj}(s)$$

$$\leqslant \sum_{k\in E} p_{1k}(t_0) < 1 \qquad (s \geqslant 0). \tag{12.7.20}$$

Let

$$\varepsilon = \inf\left\{\left(t\colon \sum_{j\in E} p_{1j}(t) < 1\right)\right\}. \tag{12.7.21}$$

From (12.7.18), (12.7.20) and (12.7.21) we deduce that

$$\sum_{j\in E} p_{ij}(t) = 1 \qquad (t < \varepsilon) \tag{12.7.22}$$

and

$$\sum_{j\in E} p_{ij}(t) < 1 \qquad (t > \varepsilon). \tag{12.7.23}$$

Hence

$$\begin{aligned} \sum_{j\in E} p_{1j}\left(\frac{3}{2}\varepsilon\right) &= \sum_{j\in E}\sum_{k\in E} p_{1k}\left(\frac{3}{4}\varepsilon\right) p_{kj}\left(\frac{3}{4}\varepsilon\right) \\ &= \sum_{k\in E} p_{1k}\left(\frac{3}{4}\varepsilon\right) \sum_{j\in E} p_{kj}\left(\frac{3}{4}\varepsilon\right) \\ &= \sum_{k\in E} p_{1k}\left(\frac{3}{4}\varepsilon\right) = 1. \end{aligned} \tag{12.7.24}$$

Equations (12.7.22) ~ (12.7.24) imply that $\varepsilon = 0$. Therefore (12.7.17) holds. It can be seen by (12.7.17) that

$$\left(\sum_{j\in E} p_{1j}(t)\right)^i \downarrow 0 \qquad (i \uparrow +\infty, t > 0). \tag{12.7.25}$$

Using (12.7.18) and (12.7.25), we have that

$$\begin{aligned} \lim_{i\to\infty} \lambda \sum_{j\in E} \int_0^\infty e^{-\lambda t} p_{ij}(t)\,dt &= \lim_{i\to\infty} \lambda \int_0^\infty e^{-\lambda t} \sum_{j\in E} p_{ij}(t)\,dt \\ &= \lim_{i\to\infty} \lambda \int_0^\infty e^{-\lambda t} \left(\sum_{j\in E} p_{1j}(t)\right)^i dt \\ &= \lambda \int_0^\infty e^{-\lambda t} \cdot 0\,dt = 0. \square \end{aligned} \tag{12.7.26}$$

By Theorem 12.1.1 and the above lemmas, we obtain the following

Theorem 12.7.1 *Let Q be a branch Q-matrix. If Q is conservative, then the Q-process is unique iff for any $\varepsilon > 0$ the integral*

$$\int_{1-\varepsilon}^{1} \frac{dx}{f(x)} \tag{12.7.27}$$

diverges. If Q is nonconservative, then the Q-process is not unique. □

§ 12.8 Another criterion and the finite and nonconservative case

Lemma 12.8.1 *Let x_i^* $(i \in E)$ be the minimal nonnegative solution of the first-type system of* 1*-bounded equations*

$$x_i = \sum_{k \in E} a_{ik} x_k + b_i \qquad (i \in E). \tag{12.8.1}$$

Let $D = \{i\colon\ b_i > 0\}$. Then

$$x_i^* \leqslant \sup_{j \in D} x_j^* \qquad (i \in E), \tag{12.8.2}$$

i.e.,

$$\sup_{i \in E} x_i^* = \sup_{j \in D} x_j^*. \tag{12.8.3}$$

Proof. Let

$$\left.\begin{aligned} x_i^{(1)} &= b_1 \qquad (i \in E), \\ x_i^{(n+1)} &= \sum_{k \in E} a_{ik} x_k^{(n)} + b_i \qquad (i \in E,\ n \geqslant 1). \end{aligned}\right\} \tag{12.8.4}$$

Then we have

$$x_i^{(1)} = \begin{cases} b_i > 0 & (i \in D), \\ 0 & (i \in E \setminus D). \end{cases} \tag{12.8.5}$$

Consequently

$$x_i^{(1)} \leqslant \sup_{j \in D} x_j^{(1)} \qquad (i \in E). \tag{12.8.6}$$

Suppose

$$x_i^{(n)} \leqslant \sup_{j \in D} x_j^{(n)} \qquad (i \in E). \tag{12.8.7}$$

Noting the fact that $x_i^{(n)}$ $(i \in E)$ does not decrease as n increases and that (12.8.1) is a first-type system of 1-bounded equations, we get

$$x_i^{(n+1)} = \sum_{k \in E} a_{ik} x_k^{(n)} + b_i = \sum_{k \in E} a_{ik} x_k^{(n)} \leqslant \sum_{k \in E} a_{ik} \cdot \sup_{j \in D} x_j^{(n)}$$

$$= \sup_{j\in D} x_j^{(n)} \cdot \sum_{k\in E} a_{ik} \leqslant \sup_{j\in D} x_j^{(n)} \leqslant \sup_{j\in D} x_j^{(n+1)} \qquad (i\in E\setminus D). \tag{12.8.8}$$

Hence we have

$$x_i^{(n+1)} \leqslant \sup_{j\in D} x_j^{(n+1)} \qquad (i\in E). \tag{12.8.9}$$

It follows by induction that for all natural numbers n, we have

$$x_i^{(n)} \leqslant \sup_{j\in D} x_j^{(n)} \qquad (i\in E). \tag{12.8.10}$$

Equation (12.8.2) can be deduced from the fact that $x_i^{(n)}\uparrow (n\uparrow +\infty)$, $\sup_{j\in D} x_j^{(n)}\uparrow (n\uparrow +\infty)$ and (12.8.10). □

Lemma 12.8.2 *If Equation* (12.2.1) *has only the zero solution, then*

$$\inf_{i\in E} \lambda \sum_{j\in E} p_{ij}^{\min}(\lambda) = \inf_{i\in \dot{E}} \lambda \sum_{j\in E} p_{ij}^{\min}(\lambda), \tag{12.8.11}$$

where $\dot{E} = (i\colon \sum_{j\in E} q_{ij} < 0)$ *is called a nonconservative state set.*

Proof. Since Equation (12.2.1) has only the zero solution, we see by the proof of Lemma 12.2.1 that

$$u(i) = 1 - \lambda \sum_{j\in E} p_{ij}^{\min}(\lambda) \qquad (i\in E) \tag{12.8.12}$$

is the minimal nonnegative solution of the pseudo-normal system of equations

$$x_i = \sum_{k\neq i} \frac{q_{ik}}{\lambda + q_i} x_k + \frac{-\sum_{j\in E} q_{ij}}{\lambda + q_i} \qquad (i\in E). \tag{12.8.13}$$

Thus it follows by Lemma 12.8.1 that

$$\sup_{i\in E} \left(1 - \lambda \sum_{j\in E} p_{ij}^{\min}(\lambda)\right) = \sup_{i\in \dot{E}} \left(1 - \lambda \sum_{j\in E} p_{ij}^{\min}(\lambda)\right). \tag{12.8.14}$$

Consequently (12.8.11) holds. □

By Theorem 12.1.1, Lemmas 12.2.1 and 12.8.2, we obtain the following.

Theorem 12.8.1 *For a given Q-matrix, a unique Q-process exist iff the following three conditions hold simultaneously.*

(i)′ $$\inf_{i\in \dot{E}} \lambda \sum_{j\in E} p_{ij}^{\min}(\lambda) > 0, \qquad 0 < \lambda < +\infty. \tag{12.8.15}$$

(i)″ *Equation* (12.2.1) *has only the zero solution.*

(ii) *The minimal Q-process is honest or Equation* (12.1.12) *has only the zero solution.* □

Remark 12.8.1 The criterion for the uniqueness of Q-process given in Theorem 12.8.1 is better in form than that in Theorem 12.1.1 because the meanings and roles of Condition (i)′, (i)″ and (ii) are very explicit. Conditions (i)″ and (ii) are the restrictions imposed on the Q-processes satisfying Kolmogorov backward and forward differential equation systems respectively, which ensure the uniqueness of the Q-process when Q is conservative. But it will not ensure the uniqueness of the Q-process when Q is nonconservative. Therefore a plain and intuitive idea might arise: in general, besides Conditions (i)″ and (ii), what restriction should be added in the nonconservative case so that it will ensure the uniqueness of the Q-process? Theorem 12.8.1 gives the answer: The restriction added is just Condition (i)′, so the above idea is very fruitful. In this way, the restrictions(i)″ and (ii) imposed on the Q-process satisfying Kolmogorov's systems of backward and forward differential equations are symmetrical.

Definition 12.8.1 *If $\dot{E}$ is a finite set, then Q is said to be finite nonconservative.*

Theorem 12.8.2 *If a Q-matrix is finite and nonconservative, then a necessary and sufficient condition for the Q-process to be unique is that the Q-processes satisfying the Kolmogorov's systems of backward and forward differential equations are both unique, i.e., the following two conditions hold simultaneously.*

(A) *Equation* (12.2.1) *has only the zero solution.*

(B) *The minimal Q-process is honest or Equation* (12.1.2) *has only the zero solution.*

Proof. Since

$$\lambda\sum_{j\in E} p_{ij}^{\min}(\lambda) > 0 \qquad (i\in E,\ \lambda>0) \tag{12.8.16}$$

and $\dot{E}$ is finite, we see that Condition (i)′ in Theorem 12.8.1 holds, as desired. □

§ 12.9 Independence of the two conditions in Theorem 12.1.1

A birth and death process is given with density matrix

$$Q=\begin{pmatrix} -(\delta_0+\beta_0) & \beta_0 & 0 & 0 & \cdots \\ \delta_1 & -(\delta_1+\beta_1) & \beta_1 & 0 & \cdots \\ 0 & \delta_2 & -(\delta_2+\beta_2) & \beta_2 & \cdots \\ \cdots\cdots\cdots & \cdots\cdots\cdots & \cdots\cdots\cdots & \cdots\cdots & \cdots \end{pmatrix} \tag{12.9.1}$$

where $\delta_0>0$. Let

$$
\begin{aligned}
x_0 &= \frac{1}{\delta_0}, \\
x_1 &= x_0 + \frac{1}{\beta_0} \\
&\cdots\cdots \\
x_n &= x_0 + \frac{1}{\beta_0} + \cdots + \frac{\delta_1\delta_2\cdots\delta_{n-1}}{\beta_0\beta_1\cdots\beta_{n-1}} \qquad (n = 2,\ 3,\ \cdots),
\end{aligned}
\tag{12.9.2}
$$

$$x_\infty = \lim_{n\to\infty} x_n, \tag{12.9.3}$$

$$\mu_0 = 1, \qquad \mu_n = \frac{\beta_0\beta_1\cdots\beta_{n-1}}{\delta_1\delta_2\cdots\delta_n} \qquad (n = 1,\ 2,\ \cdots), \tag{12.9.4}$$

$$R = \sum_{i=0}^{\infty} (x_\infty - x_i)\,\mu_i, \qquad S = \sum_{i=1}^{\infty} x_i\mu_i. \tag{12.9.5}$$

(1) An example in which both Conditions (i) and (ii) in Theorem 12.1.1 hold. Choose

$$\delta_n = \beta_n = 1 \qquad (n = 0,\ 1,\ 2,\ \cdots). \tag{12.9.6}$$

Then

$$R = S = +\infty. \tag{12.9.7}$$

Since Q is finite and nonconservative, it follows from Lemma 12.8.2, [4] and [32, Theorem 1] that both Conditions (i) and (ii) hold.

(2) An example in which Condition (i) in Theorem 12.1.1 holds but Condition (ii) does not. Choose

$$\delta_n = \beta_n = (n+1)2^n \qquad (n = 0,\ 1,\ \cdots). \tag{12.9.8}$$

Then

$$x_n = n + 1 \qquad (n = 0,\ 1,\ \cdots), \tag{12.9.9}$$

$$\mu_n = \frac{1}{(n+1)2^n} \qquad (n = 0,\ 1,\ \cdots), \tag{12.9.10}$$

$$x_\infty = +\infty, \tag{12.9.11}$$

$$R = +\infty, \qquad S = 1 < +\infty. \tag{12.9.12}$$

By [4] and [32, Theorems 1 and 4], Condition (i) holds but (ii) does not.

(3) An example in which Condition (i) in Theorem 12.1.1 does not hold but Condition (ii) holds. Choose

$$\delta_n = \frac{1}{2}\beta_n = 2^n \qquad (n = 0, 1, \cdots). \tag{12.9.13}$$

Then

$$x_n = 2\left(1 - 2^{-(n+1)}\right) \qquad (n = 0, 1, \cdots), \tag{12.9.14}$$

$$x_\infty = 2, \qquad x_\infty - x_i = 2^{-i} \qquad (i = 0, 1, \cdots), \tag{12.9.15}$$

$$\mu_n = 1 \qquad (n = 0, 1, \cdots), \tag{12.9.16}$$

$$R = 2 < +\infty, \qquad S = +\infty. \tag{12.9.17}$$

It can be seen by [4] and [32, Theorems 1 and 2], that Condition (i) does not hold but (ii) holds.

(4) An example in which neither Condition (i) nor (ii) in Theorem 12.1.1 holds. Choose

$$\delta_n = \frac{1}{2}\beta_n = 2^{2n} \qquad (n = 0, 1, \cdots). \tag{12.9.18}$$

Then

$$x_n = 2\left(1 - 2^{-(n+1)}\right) \qquad (n = 0, 1, \cdots), \tag{12.9.19}$$

$$x_\infty = 2, \qquad x_\infty - x_i = 2^{-i} \qquad (i = 0, 1, \cdots), \tag{12.9.20}$$

$$\mu_n = 2^{-n} \qquad (n = 0, 1, \cdots), \tag{12.9.21}$$

$$R = \frac{4}{3} < +\infty, \qquad S = \frac{5}{3} < +\infty. \tag{12.9.22}$$

We can deduce from [4] and [32, Theorem 2 and 4] that neither condition holds.

We see from (1)—(4) that Conditions (i) and (ii) in Theorem 12.1.1 are mutually independent.

The results of this section also prove that all the four types (of nature, entrance, exit and regularity) of birth and death processes can be realized.

§ 12.10 Probability interpretation of Condition (i) in Theorem 12.1.1[1]

---

1 The problem discussed in this section was proposed by Professor K. L. Chung and the result was also obtained independently by G. E. H. Reuter.

Lemma 12.10.1 *Let $J \subset E$. Then the following two conditions are equivalent:*

$$(\alpha_1) \qquad \inf_{i\in J} \lambda \sum_{j\in E} p_{ij}^{\min}(\lambda) > 0, \qquad 0 < \lambda + \infty. \tag{12.10.1}$$

$$(\beta_1) \quad \inf_{i\in J} P_i(\tau > t_0) = \zeta_t > 0, \quad \textit{for some fixed} \quad 0 < t_0 < +\infty, \tag{12.10.2}$$

where τ stands for the first infinity of the minimal Q-process. For its meaning, see [1, II, § 19].

Proof. If (12.10.2) holds, then

$$P_i(\tau > t) \geqslant \zeta_{t_0} \qquad (i \in J,\ 0 \leqslant t \leqslant t_0). \tag{12.10.3}$$

It follows by (12.10.3) that

$$\inf_{i\in J} \lambda \sum_{i\in E} p_{ij}^{\min}(\lambda) = \inf_{i\in J} \lambda \int_0^\infty e^{-\lambda t} P_i(\tau > t)\,dt$$

$$\geqslant \inf_{i\in J} \lambda \int_0^{t_0} e^{-\lambda t} \zeta_{t_0} dt = \lambda \zeta_{t_0} \int_0^{t_0} e^{-\lambda t} dt > 0. \tag{12.10.4}$$

Therefore (12.10.1) holds. Conversely, if (12.10.2) does not hold. i.e.,

$$\inf_{i\in J} P_i(\tau > t) = 0 \qquad (0 < t < +\infty), \tag{12.10.5}$$

then for any fixed $\delta > 0$ and $\varepsilon > 0$, there exists $s \in J$, such that

$$P_s(\tau > \delta) < \varepsilon. \tag{12.10.6}$$

Thus

$$P_s(\tau > t) < \varepsilon \qquad (t \geqslant \delta). \tag{12.10.7}$$

Hence

$$\inf_{i\in J} \lambda \sum_{j\in E} p_{ij}^{\min}(\lambda) = \inf_{i\in J} \lambda \int_0^\infty e^{-\lambda t} p_i(\tau > t)\,dt$$

$$\leqslant \lambda \int_0^\infty e^{-\lambda t} P_s(\tau > t)\,dt = \lambda\left[\int_0^\delta e^{-\lambda t} P_s(\tau > t)\,dt\right.$$

$$\left. + \int_\delta^\infty e^{-\lambda t} P_s(\tau > t)\,dt\right] \leqslant \lambda\left[\int_0^\delta dt + \int_\delta^\infty e^{-\lambda t}\varepsilon\,dt\right]$$

$$\leqslant \lambda\left[\delta + \varepsilon \int_0^\infty e^{-\lambda t}\, dt\right] = \lambda\delta + \varepsilon. \tag{12.10.8}$$

Since δ and ε are chosen arbitrarily, we have

$$\inf_{i \in J} \lambda \sum_{j \in E} p_{ij}^{\min}(\lambda) = 0. \tag{12.10.9}$$

Consequently (12.10.1) does not hold. Therefore (α_1) and (β_1) are equivalent, as claimed. □

Let

$$\hat{\Omega} = \{\omega\colon \tau(\omega) = +\infty \text{ or for any } \varepsilon > 0, \text{ there are infinitely many points of discontinuity of the sample function of the minimal } Q\text{-process in } (\tau(\omega) - \varepsilon,\ \tau(\omega))\}. \tag{12.10.10}$$

The following lemma is well known.

Lemma 12.10.2 *The following two conditions are equivalent:*
(α_2) Equation

$$\begin{cases} \lambda \mathbf{U} - Q\mathbf{U} = \mathbf{0}_1, & \lambda > 0, \\ \mathbf{0}_1 \leqslant \mathbf{U} \leqslant \mathbf{1} \end{cases} \tag{12.10.11}$$

has only the zero solution.
(β_2) $\tau(\omega) = +\infty$ almost everywhere on $\hat{\Omega}$. □

Lemma 12.10.3 *If the equation*

$$\begin{cases} \lambda \mathbf{U} - Q\mathbf{U} = \mathbf{0}_1, & \lambda > 0, \\ \mathbf{0}_1 \leqslant \mathbf{U} \leqslant \mathbf{1}. \end{cases} \tag{12.10.12}$$

has only the zero solution, then

$$P_i(\tau > t) \geqslant \inf_{j \in \dot{E}} P_j(\tau > t) \qquad (t > 0,\ i \in E), \tag{12.10.13}$$

i.e.,

$$\inf_{i \in E} P_i(\tau > t) = \inf_{i \in \dot{E}} P_i(\tau > t). \tag{12.10.14}$$

Proof. Let

$$F(t_i) = P_i(\tau \leqslant t) \qquad (i \in E). \tag{12.10.15}$$

It is easy to prove that, under the hypothesis that (12.10.12) possesses only the zero solution, $F_i(t)$ $(i \in E)$ is the unique solution of the system of the integral equation

$$F_i(t) = \sum_{j \neq i} \int_0^t e^{-q_i s} q_{ij} F_j(t-s)\,ds + \frac{-\sum_{k \in E} q_{ik}}{q_i}(1 - e^{-q_i t}) \qquad (i \in E), \tag{12.10.16}$$

(where $0/0 = 0$ by convention) and it can be found in the following way:
Let

$$\left.\begin{aligned} F_i^{(1)}(t) &= \frac{-\sum_{k \in E} q_{ik}}{q_i}(1 - e^{-q_i t}) \qquad (i \in E), \\ F_i^{(n+1)}(t) &= \sum_{j \neq i} \int_0^t e^{-q_i s} q_{ij} F_j^{(n)}(t-s)\,ds \\ &\quad + \frac{-\sum_{k \in E} q_{ik}}{q_i}(1 - e^{-q_i t}) \qquad (i \in E,\ n \geqslant 1). \end{aligned}\right\} \tag{12.10.17}$$

Then

$$F_i^{(n)}(t) \uparrow F_i(t) \qquad (n \uparrow +\infty). \tag{12.10.18}$$

Now we are going to prove

$$F_i(t) \leqslant \sup_{j \in \dot{E}} F_j(t) \qquad (i \in E). \tag{12.10.19}$$

By (12.10.17) and the definition of $\dot{E}$,

$$F_i^{(1)}(t) = \begin{cases} \dfrac{-\sum_{k \in E} q_{ik}}{q_i}(1 - e^{-q_i t}) > 0 & (i \in E), \\ 0 & (i \in E \setminus \dot{E}). \end{cases} \tag{12.10.20}$$

So

$$F_i^{(1)}(t) \leqslant \sup_{j \in \dot{E}} F_j^{(1)}(t) \qquad (i \in E). \tag{12.10.21}$$

Now suppose

$$F_i^{(n)}(t) \leqslant \sup_{j \in \dot{E}} F_j^{(n)}(t). \tag{12.10.22}$$

Then $F_i^{(n)}(t)$ does not decrease as n and/or t increase and

$$\sum_{j\neq i}\int_0^t e^{-q_is}q_{ij}ds = \frac{-\sum_{j\neq i} q_{ij}}{q_i}(1-e^{-q_it}) \leqslant 1 \qquad (i\in E). \qquad (12.10.23)$$

It follows immediately that

$$F_i^{(n+1)}(t) = \sum_{j\neq i}\int_0^t e^{-q_is}q_{ij}F_j^{(n)}(t-s)\,ds + \frac{-\sum_{k\in E} q_{ik}}{q_i}(1-e^{-q_it})$$

$$= \sum_{j\neq i}\int_0^t e^{-q_is}q_{ij}F_j^{(n)}(t-s)\,ds \leqslant \sum_{j\neq i}\int_0^t e^{-q_is}q_{ij}F_j^{(n)}(t)\,ds$$

$$= \sum_{j\neq i} F_j^{(n)}(t)\int_0^t e^{-q_is}q_{ij}ds \leqslant \sum_{j\neq i}\left(\sup_{k\in \dot{E}} F_k^{(n)}(t)\right)\int_0^t e^{-q_is}q_{ij}ds$$

$$= \sup_{k\in\dot{E}} F_k^{(n)}(t)\sum_{j\neq i}\int_0^t e^{-q_is}q_{ij}ds$$

$$\leqslant \sup_{j\in\dot{E}} F_j^{(n)}(t) \leqslant \sup_{j\in\dot{E}} F_j^{(n+1)}(t) \qquad (i\in E\setminus\dot{E}). \qquad (12.10.24)$$

Therefore we have

$$F_i^{(n+1)}(t) \leqslant \sup_{j\in\dot{E}} F_j^{(n+1)}(t) \qquad (i\in E). \qquad (12.10.25)$$

It can be seen by induction that for all natural numbers n, we have

$$F_i^{(n)}(t) \leqslant \sup_{j\in\dot{E}} F_j^{(n)}(t) \qquad (i\in E). \qquad (12.10.26)$$

Equation (12.10.19) follows from $F_i^{(n)}(t)\uparrow(n\uparrow+\infty)$, $\sup_{j\in\dot{E}} F_j^{(n)}\uparrow(n\uparrow+\infty)$ and (12.10.26).

Equation (12.10.13) follows from (12.10.19) and

$$P_i(\tau > t) = 1 - F_i(t) \qquad (i\in E), \qquad (12.10.27)$$

and the lemma is proved. □

Theorem 12.10.1 *The following conditions are equivalent*

(i) $$\inf_{i\in E}\lambda\sum_{j\in E} P_{ij}^{\min}(\lambda) = \eta_\lambda > 0, \qquad 0<\lambda<+\infty. \qquad (12.10.28)$$

(ii) *The equation*

$$\begin{cases} \lambda \mathbf{U} - Q\mathbf{U} = \mathbf{0}_1, & \lambda > 0, \\ \mathbf{0}_1 \leqslant \mathbf{U} \leqslant \mathbf{1} \end{cases} \tag{12.10.29}$$

has only the zero solution, and

$$\inf_{i \in \dot{E}} \lambda \sum_{j \in E} p_{ij}^{\min}(\lambda) = \dot{\eta}_\lambda > 0, \qquad 0 < \lambda < +\infty. \tag{12.10.30}$$

(iii)
$$\inf_{i \in E} P_i(\tau > t) = \zeta_t > 0, \qquad 0 < t < +\infty. \tag{12.10.31}$$

(iv) $\tau(\omega) = +\infty$ *almost everywhere on* $\hat{\Omega}$, *and*

$$\inf_{i \in \dot{E}} p_i(\tau > t) = \zeta_t > 0, \qquad 0 < t < +\infty. \tag{12.10.32}$$

The above conditions remain equivalent to one another if λ *and* t *are replaced in the conditions by arbitrarily fixed* $0 < \lambda_0 < +\infty$ *and* $0 < t_0 < +\infty$, *respectively.*

Proof. It is known from §12.8 that (i) and (ii) are equivalent.
From Lemma 12.10.1 we deduce that (i) is equivalent to

$$\inf_{i \in E} P_i(\tau > t_0) = \zeta_{t_0} > 0 \quad \text{for some fixed } t_0 \qquad (0 < t_0 < +\infty). \tag{12.10.33}$$

Therefore, in order to show the equivalence of (i) and (iii) it suffices to prove that (12.10.33) implies (12.10.31).

Suppose (12.10.33) holds. Let

$$t' = \sup\left(t\colon \inf_{i \in E} P_i(\tau > t) = \zeta_t > 0\right). \tag{12.10.34}$$

By (12.10.33), we see that

$$t' \geqslant t_0 > 0. \tag{12.10.35}$$

Since $P_i(\tau > t)$ is a nonincreasing function of t, we know

$$\zeta_t > 0 \qquad (0 \leqslant t < t'). \tag{12.10.36}$$

If $t' < +\infty$, then

$$\inf_{i \in E} P_i\left(\tau > \frac{3}{2}t'\right) = \inf_{i \in E} \sum_{k \in E} p_{ik}^{\min}\left(\frac{3}{4}t'\right) P_k\left(\tau > \frac{3}{4}t'\right)$$

$$\geqslant \inf_{i\in E} \sum_{k\in E} p_{ik}\left(\frac{3}{4}t'\right)\zeta_{\frac{3}{4}t'}$$

$$= \zeta_{\frac{3}{4}t'} \inf_{i\in E} P_i\left(\tau > \frac{3}{4}t'\right) = (\zeta_{\frac{3}{4}t'})^2 > 0. \tag{12.10.37}$$

But this contradicts (12.10.34). Consequently, by (12.10.35),

$$t' = +\infty. \tag{12.10.38}$$

Thus by (12.10.36), (12.10.31) holds. Hence (i) and (iii) are equivalent.

In view of Lemmas 12.10.1 and 12.10.2, we know that in order to show (ii) and (iv) are equivalent it suffices to prove that if (12.10.29) has only the zero solution and

$$\inf_{i\in \dot{E}} p_i(\tau > \dot{t}_0) = \dot{\zeta}_{\dot{t}_0} > 0 \quad \text{for a certain fixed } 0 < \dot{t}_0 < +\infty, \tag{12.10.39}$$

then (12.10.32) holds.

Let

$$\hat{t} = \sup\left(t\colon \inf_{i\in \dot{E}} p_i(\tau > t) = \zeta_t > 0\right). \tag{12.10.40}$$

Then

$$\hat{t} \geqslant \dot{t}_0 > 0, \tag{12.10.41}$$

$$\zeta_t > 0 \qquad (0 \leqslant t < \dot{t}_0). \tag{12.10.42}$$

If $\hat{t} < +\infty$, then by Lemma 12.10.3, we see that

$$\inf_{i\in \dot{E}} P_i\left(\tau > \frac{3}{2}\hat{t}\right) = \inf_{i\in \dot{E}} \sum_{k\in E} P_{ik}^{\min}\left(\frac{3}{4}\hat{t}\right) P_k\left(\tau > \frac{3}{4}\hat{t}\right)$$

$$\geqslant \inf_{i\in \dot{E}} \sum_{k\in E} P_{ik}^{\min}\left(\frac{3}{4}\hat{t}\right)\zeta_{\frac{3}{4}\hat{t}}$$

$$= \zeta_{\frac{3}{4}\hat{t}} \inf_{i\in \dot{E}} P_i\left(\tau > \frac{3}{4}\hat{t}\right) = (\zeta_{\frac{3}{4}\hat{t}})^2 > 0. \tag{12.10.43}$$

But this contradicts (12.10.40), so we must have

$$\hat{t} = +\infty.$$

Consequently (12.10.32) holds, so (ii) and (iv) are equivalent.

The validity of the last assertion of this theorem is easy to establish. And this completes the proof. □

CHAPTER XIII
Construction of Q-Processes

§ 13.1 Construction theorem

First we stipulate that all the Q-processes that occurred in §§13.1—13.3 possess the Property (D) introduced in §1.1, and their Q-matrices satisfy (9.1.1).

Definition 13.1.1 *If the mapping $\hat{\Pi}(a, j)$ giving the entries of the matrix $\hat{\Pi}_{(\partial X)_e \times E} = (\Pi(a, j),\ a \in (\partial X)_e,\ j \in E)$, for an arbitrary fixed $j \in E$, is a Borel measurable function on $(\partial X)_e$ and satisfies the inequality*

$$0 \leqslant \hat{\Pi}(a, j) \leqslant 1 \qquad (a \in (\partial X)_e,\ j \in E), \tag{13.1.1}$$

then this matrix is called a 1*-bounded matrix.*

Suppose $\hat{\Pi}_{(\partial X)_e \times E} = (\hat{\Pi}(a, j),\ a \in (\partial X)_e,\ j \in E)$ is a 1-bounded matrix. Let

$$\hat{\Pi}^{(1)}(a, 1) = \hat{\Pi}(a, 1) \qquad (a \in (\partial X)_e). \tag{13.1.2}$$

We now obtain a matrix $\hat{\Pi}_{(\partial X)_e \times D_1} = (\hat{\Pi}^{(1)}(a, j),\ a \in (\partial X)_e,\ j \in D_1)$. Suppose the matrix $\Pi_{(\partial X)_e \times D_n} = (\hat{\Pi}^{(n)}(a, j),\ a \in (\partial X)_e,\ j \in D_n)$ has been constructed, then let

$$\hat{\Pi}^{(n+1)}(a, n+1) = \hat{\Pi}(a, n+1) \qquad (a \in (\partial X)_e), \tag{13.1.3}$$

$$\begin{aligned} \hat{\Pi}^{(n+1)}(a, j) = {} & \hat{\Pi}^{(n)}(a, j) - \hat{\Pi}(a, n+1) \\ & \times \left({}_{D_n}\tilde{f}^{*}_{n+1, j} + \int_{(\partial X)_e} \left(h(n+1, db) - \tilde{f}^{*(db)}_{n+1, D_n}\right) \hat{\Pi}^{(n)}(b, j) \right) \\ & (a \in (\partial X)_e,\ j \in D_n); \end{aligned} \tag{13.1.4}$$

we obtain a matrix $\hat{\Pi}_{(\partial X)_e \times D_{n+1}} = (\hat{\Pi}^{(n+1)}(a, j),\ a \in (\partial X)_e,\ j \in D_{n+1})$. Proceeding like this, we will construct a sequence of matrices

$$\hat{\Pi}_{(\partial X)_e \times D_n} = (\hat{\Pi}^{(n)}(a, j),\ a \in (\partial X)_e,\ j \in D_n) \qquad (n = 1, 2, \cdots). \tag{13.1.5}$$

It is easy to prove the following:

Lemma 13.1.1 *For any $n \geqslant 1$, $\hat{\Pi}^{(n)}(a, j)$ can be defined as above, and*

$$|\hat{\Pi}^{(n)}(a, j)| \leqslant n! \qquad (a \in (\partial X)_e,\ j \in D_n). \ \square \tag{13.1.6}$$

Lemma 13.1.1 ensures the workability of the above procedure of constructing the matrix sequence $\hat{\Pi}_{(\partial X)_e \times D_n}$ $(n = 1, 2, \cdots)$ from the 1-bounded matrix $\hat{\Pi}_{(\partial X)_e \times E}$.

Definition 13.1.2 *The matrix sequence $\hat{\Pi}_{(\partial X)_e \times D_n}$ $(n = 1, 2, \cdots)$ is called a Q-derived-matrix sequence of the 1-bounded matrix $\hat{\Pi}_{(\partial X)_e \times E}$.*

Definition 13.1.3 *The 1-bounded matrix $\hat{\Pi}_{(\partial X)_e \times E} = (\hat{\Pi}(a, j);\ a \in (\partial X)_e, j \in E)$ is called a generating matrix, if its Q-derived-matrix sequence $\hat{\Pi}_{(\partial X)_e \times D_n} = (\hat{\Pi}^{(n)}(a, j);\ a \in (\partial X)_e,\ j \in D_n)$ satisfies:*

$$\text{(i)} \quad \left.\begin{array}{ll} \hat{\Pi}^{(n)}(a, j) \geqslant 0 & (a \in (\partial X)_e,\ j \in D_n), \\ \sum_{j \in D_n} \hat{\Pi}^{(n)}(a, j) \leqslant 1 & (a \in (\partial X)_e). \end{array}\right\} \tag{13.1.7}$$

$$\text{(ii)} \quad \lim_{n \to \infty} \lim_{\lambda \to 0} \Phi_{ij}^{(n)}(\lambda) = \lim_{\lambda \to 0} \lim_{n \to \infty} \Phi_{ij}^{(n)}(\lambda), \qquad i, j \in E, \tag{13.1.8}$$

where $\{\Phi_{ij}^{(n)}(\lambda),\ i \in E\}$ is the minimal nonnegative solution of (10.3.45) *for every $j \in E$.*

Theorem 13.1.1 *Suppose $X(\omega) = \{x(t, \omega),\ t < \sigma(\omega)\}$ is a Q-process. For any $n \geqslant 1$, let*

$$X^{(n)}(\omega) = g_n(X(\omega)), \tag{13.1.9}$$

$$\Pi^{(n)}(a, j) = P(x(\beta_1^{(n)}) = j \,|\, x(\tau - 0) = a)$$

$$(a \in (\partial X)_e,\ j \in D_n), \tag{13.1.10}$$

$$\Pi(a, n) = \Pi^{(n)}(a, n) = P(x(\beta_1^{(n)}) = n \,|\, x(\tau - 0) = a)$$

$$(a \in (\partial X)_e), \tag{13.1.11}$$

where $\beta_1^{(n)}$ is defined as in §1.2. *Then*

(i) *$\Pi_{(\partial X)_e \times E} = (\Pi(a, n),\ a \in (\partial X)_e,\ n \in E)$ is a Q-generating matrix and $\Pi_{(\partial X)_e \times D_n} = (\Pi^{(n)}(a, j),\ a \in (\partial X)_e,\ j \in D_n)$ $(n = 1, 2, \cdots)$ is a Q-derived-matrix sequence of $\Pi_{(\partial X)_e \times E}$.*

(ii) *$X^{(n)}(\omega) = \{x^{(n)}(t, \omega),\ t < \sigma^{(n)}(\omega)\}$ is a $(Q, \Pi_{(\partial X)_e \times D_n})$-process, and*

$$X^{(n)}(\omega) = g_n(X^{(n+1)}(\omega)). \tag{13.1.12}$$

(iii) *For any point t on $[0, \sigma(\omega))$ there exists*

$$\lim_{n\to\infty} x^{(n)}(t, \omega) = x(t, \omega) \qquad (\omega\in\Omega). \tag{13.1.13}$$

(iv)
$$\lim_{n\to\infty} p_{ij}^{(n)}(t) = p_{ij}(t), \tag{13.1.14}$$

where

$$p_{ij}^{(n)}(t) = P(x^{(n)}(t) = j \mid x^{(n)}(0) = i), \tag{13.1.15}$$

$$p_{ij}(t) = P(x(t) = j \mid x(0) = i). \tag{13.1.16}$$

Conversely, suppose $\Pi_{(\partial X)_e\times E} = (\Pi(a, n); a\in(\partial X)_e, n\in E)$ *is a Q-generating matrix and* $\Pi_{(\partial X)_e\times D_n} = (\Pi^{(n)}(a, j)$: $a\in(\partial X)_e, j\in D_n)$ $(n = 1, 2, \cdots)$ *is its Q-derived-matrix sequence, then there exists a complete probability space* $(\Omega, \mathcal{F}, P)$, *on which we can define a sequence of Q-processes* $X^{(n)}(\omega) = \{x^{(n)}(t, \omega), t < \sigma^{(n)}(\omega)\}$ $(n = 1, 2, \cdots)$ *such that*

(A) $X^{(n)}(\omega)$ *is a* $(Q, \Pi_{(\partial X)_e\times D_n})$*-process, and*

$$X^{(n)}(\omega) = g_n(X^{(n+1)}(\omega)). \tag{13.1.17}$$

(B) *For any point t on* $[0, \sigma(\omega))$, *there exists*

$$\lim_{n\to\infty} x^{(n)}(t, \omega) = x(t, \omega) \qquad (\omega\in\Omega) \tag{13.1.18}$$

and $X(\omega) = \{x(t, \omega), t < \sigma(\omega)\}$ *is a Q-process, where*

$$\sigma(\omega) = \lim_{n\to\infty} \sigma^{(n)}(\omega) \qquad (\omega\in\Omega). \tag{13.1.19}$$

(C)
$$\lim_{n\to\infty} p_{ij}^{(n)}(t) = p_{ij}(t), \tag{13.1.20}$$

where

$$p_{ij}^{(n)}(t) = P(x^{(n)}(t) = j \mid x^{(n)}(0) = i), \tag{13.1.21}$$

$$p_{ij}(t) = P(x(t) = j \mid x(0) = i). \tag{13.1.22}$$

Proof. It is easy to complete the proof of this theorem by Theorems 11.1.1 and 11.1.2, Lemma 10.1.1 and by referring to the proof of Lemma 7.7 of [2]. □

§ 13.2 Specifications of all the Q-processes

Suppose $\Pi_{(\partial X)_e\times E} = \Pi(a, n)$, $(a\in(\partial X)_e, n\in E)$ is a Q-generating matrix, and $\Pi_{(\partial X)_e\times D_n} = (\Pi^{(n)}(a, j), a\in(\partial X)_e, j\in D_n)$ $(n = 1, 2, \cdots)$ is a Q-derived-matrix sequence of $\Pi_{(\partial X)_e\times E}$. We deduce from Theorem 13.1.1 that for any $n \geqslant 1$, there exists a $(Q, \Pi_{(\partial X)_e \times D_n})$-process $(p_{ij}^{(n)}(t), i, j \in E)_{t \geqslant 0}$, and that the limit

$$\lim_{n \to \infty} p_{ij}^{(n)}(t) = p_{ij}(t) \tag{13.2.1}$$

exists and $(p_{ij}(t), i, j \in E)_{t \geqslant 0}$ is a Q-process. Since $(p_{ij}(t), i, j \in E)_{t \geqslant 0}$ is determined uniquely by Q *and* $\Pi_{(\partial X)_e \times E}$, we also call $(p_{ij}(t); i, j \in E)_{t \geqslant 0}$ a $\{Q, \Pi_{(\partial X)_e \times E}\}$-process. Hence using Theorem 13.1.1 we obtain

Theorem 13.2.1 *The collection of all $\{Q, \Pi_{(\partial X)_e \times E}\}$-processes is just the collection of all Q-processes.* □

§ 13.3 Expression of $\{Q, \Pi_{(\partial X)_e \times E}\}$-processes

Theorem 13.3.1 *Suppose $\Pi_{(\partial X)_e \times E} = (\Pi(a, n), a \in (\partial X)_e, n \in E)$ is a Q-generating matrix, $(p_{ij}(t), i, j \in E)_{t \geqslant 0}$ is a $\{Q, \Pi_{(\partial X)_e \times E}\}$-process. If*

$$p_{ij}(\lambda) = \int_0^\infty e^{-\lambda t} p_{ij}(t) dt, \tag{13.3.1}$$

then

$$p_{ij}(\lambda) = \int_{(\partial X)_e} h_\lambda(i, da)\ \xi_j^{(a)}(\lambda), \tag{13.3.2}$$

where

$$\xi_j^{(a)}(\lambda) = \lim_{n \to \infty} \xi_j^{(n,a)}(\lambda) \qquad (a \in (\partial X)_e) \tag{13.3.3}$$

and $\xi_j^{(n,a)}$ $(a \in (\partial X)_e)$ is the minimal nonnegative solution of the integral equation

$$\begin{aligned} \xi^{(a)} = &\int_{(\partial X)_e} \left(\sum_{i \in E} \Pi^{(n)}(a, i)\ h_\lambda\ (i, db) \right) \xi^{(b)} \\ &+ \sum_{i \in E} \Pi^{(n)}(a, i)\ p_{ij}^{\min}(\lambda) \qquad (a \in (\partial X)_e), \end{aligned} \tag{13.3.4}$$

while $\Pi_{(\partial X)_e \times D_n} = (\Pi^{(n)}(a, j), a \in (\partial X)_e, j \in D_n)$ $(n = 1, 2, \cdots)$ is a Q-derived-matrix sequence of $\Pi_{(\partial X) \times E}$.

Proof. It is similar to the proof of Theorem 11.2.1.□

§ 13.4 Discussion

In §9.1, we pointed out that it was only for convenience in the proof that we assumed $q_i > 0$ and $\sum_{j \in E} q_{ij} = 0$ $(i \in E)$. Otherwise, we may still obtain these results, the difference being that we then obtain not all the Q-processes, but all the Q-processes satisfying Kolmogorov system of backward differential equations. Now let us remove the above restriction only for the contents of this chapter, so as to get all the Q-processes satisfying Kolmogorov system of backward differential equation.

Definition 13.4.1 *A Q-matrix is said to be double-conservative if $q_i > 0$, $\sum_{j\in E} q_{ij} = 0$ $(i \in E)$.*

If an arbitrary conservative Q-matrix is given, let

$$E^0 = \{i\colon\ i \in E,\ q_i = 0\}, \tag{13.4.1}$$

$$E^+ = \{i\colon\ i \in E,\ q_i \neq 0\}, \tag{13.4.2}$$

$$E_0 = \{-i\colon\ i \in E^0\}, \tag{13.4.3}$$

$$\hat{E} = E^0 \bigcup E^+ \bigcup E_0, \tag{13.4.4}$$

we define a matrix $\hat{Q} = (\hat{q}_{ij};\ i,\ j \in \hat{E})$ from Q in the following way:

$$\hat{q}_{ij} = \begin{cases} q_{ij}, & (i,\ j \in E^+), \\ \frac{1}{2} q_{i|j|}, & (i \in E^+,\ j \in E^0 \bigcup E_0), \\ -1, & (i \in E^0 \cup E_0,\ j = i), \\ 1, & (i \in E^0 \cup E_0,\ j = -i), \\ 0, & (i \in E^0 \cup E_0,\ j \neq \pm i). \end{cases} \tag{13.4.5}$$

Obviously we have

Lemma 13.4.1 *The matrix $\hat{Q}$ is a double-conservative Q-matrix.* □

We denote by L the transformation from Q to $\hat{Q}$, i.e.,

$$LQ = \hat{Q}. \tag{13.4.6}$$

Suppose $\hat{P}(t) = (\hat{p}_{ij}(t),\ i,\ j \in \hat{E})$ is a $\hat{Q}$-process. We define a matrix $P(t) = (p_{ij}(t),\ i,\ j \in E)$ from $\hat{P}(t)$ in the following way:

$$p_{ij}(t) = \begin{cases} \hat{p}_{ij}(t) & (i \in E,\ j \in E^+), \\ \hat{p}_{ij}(t) + \hat{p}_{i-j}(t) & (i \in E,\ j \in E^0). \end{cases} \tag{13.4.7}$$

We denote by G the transformation from $\hat{P}(t)$ to $P(t)$, i.e.,

$$G\hat{P}(t) = P(t). \tag{13.4.8}$$

It is easy to prove the following:

Theorem 13.4.1 *If $\hat{P}(t)$ is a $\hat{Q}$-process, then*

$$P(t) = G\hat{P}(t) \tag{13.4.9}$$

is a Q-process; conversely, if $P(t)$ is a Q-process, then there exists a $\hat{Q}$-process $\hat{P}(t)$ such that (13.4.9) *holds.* □

In view of Theorem 13.4.1 and the following Proposition 14.3.1 we see that in this book we obtain the following result: for an arbitrary Q-matrix, we can construct all the Q-processes satisfying Kolmogorov backward differential equations.

CHAPTER XIV
Qualitative Theory

§ 14.1 Introduction

As indicated in §12.1, with respect to any given Q-matrix, the following three fundamental problems should be answered: (A) Whether there exists a Q-process with Q as its density matrix. (B) If one does exist then, what are necessary and sufficient conditions for the Q-process to be unique? (C) If we know that such a Q-process is not unique, then how can we construct all such Q-processes? If we distinguish qualitative and quantitative points of view, the above Problems (A) and (B) are qualitative problems, while Problem (C) is quantitative. In order to elucidate the fact that there are far more qualitative problems than just these two, we first introduce some terminology:

Definiton 14.1.1 *For any given Q-matrix, the Q-process satisfying Kolmogorov's system of backward differential equations*

$$P'(t) = QP(t) \tag{14.1.1}$$

is called a B-type Q-process; the Q-process satisfying Kolmogorov's system of forward differential equations

$$P'(t) = P(t)Q \tag{14.1.2}$$

is called an F-type Q-process; the Q-processes satisfying both (14.1.1) *and* (14.1.2), *neither* (14.1.2) *nor* (14.1.1), *not satisfying* (14.1.1), *not satisfying* (14.1.2), *not satisfying* (14.1.1) *but* (14.1.2), *satisfying* (14.1.1) *but not* (14.1.2) *and satisfying at least one of* (14.1.1) *and* (14.1.2) *are called $B\bigcap F$-type, $\overline{B\bigcup F}$-type, $\bar{B}$-type, $\bar{F}$-type, $\bar{B}\bigcap F$-type, $B\bigcap\bar{F}$-type and $B\bigcup F$-type Q-processes respectively. For convenience, an arbitrary Q-process is called a O-type Q-process; the honest B-type, honest F-type, honest $B\bigcap F$-type, honest $\overline{B\bigcup F}$-type, honest O-type, honest $\bar{B}$-type, honest $\bar{F}$-type, honest $\bar{B}\bigcap F$-type, honest $B\bigcap\bar{F}$-type and honest $B\bigcup F$-type Q-processes are called respectively N-B-type, N-F-type, N-$B\bigcap F$-type, N-$\overline{B\bigcup F}$-type, N-O-type, N-$\bar{B}$-type, N-$\bar{F}$-type, N-$\bar{B}\bigcap F$-type, N-$B\bigcap\bar{F}$*-type and N-*$B\bigcup F$-type Q-processes. There are 20 types all together.*

For a given Q-matrix, there would be none, exactly one, more than one but finite, or infinitely many corresponding Q-processes. If we know this Q-matrix is of a certain type, we have to decide which of the four cases definitely occurs, definitely does not occur, or possibly occurs, and if it possibly occurs, what is a necessary and sufficient condition for its occurence.

The up-to-date situation of the study concerning the qualitative problems is as follows:

(1) Feller proved in [25] that O-type Q-processes always exist, Doob proved that there are only two possible cases, either there exists only one or there are infinitely many. We have given, in Chapter XII, a necessary and suffcient condition for the existence of a unique O-type Q-process.

(2) Doob proved in [24] that there always exist B-type and F-type Q-processes. Reuter in [27] gave a necessary and sufficient condition for the existence of unique B-type (F-type) Q-process and also proved that if B-type (F-type) Q-processes are not unique, then, there must exist infinitely many B-type (F-type) Q-processes. Actually in §12.5 we have given another necessary and sufficient condition for the existence of a unique F-type Q-process.

(3) Reuter proved in [27] that with respect to N-F-type Q-processes, there are only three possible cases: nonexistence, unique existence or existence of infinitely many. He also gave a necessary and sufficient condition for the occurrence of each of the three possibilities.

There are in addition some fragmentary results.

On the basis of the above-mentioned work, the authors have made a comparatively comprehensive study of these qualitative problems. The aim of this chapter is to give a complete answer to these problems. For completeness (without occupying too much space), the above-men tioned results and the new results obtained recently by the authors are summarized together into 20 theorems. Each of the theorems specially deals with one of the twenty types of Q-processes. In the course of investigation below we shall refer to all the existing results with references but without proofs.

§ 14.2 Statement of results

Theorem 14.2.1 *Suppose a Q-matrix is given. Then*

(1) *B-type Q-processes always exist and there are only two possibilities, either there exists only one or there exist infinitely many.*

(2) *A necessary and sufficient condition for a B-type Q-process to be unique is, that the equation*

$$\left.\begin{aligned} &\lambda \mathbf{U} - Q\mathbf{U} = \mathbf{0}_1, \qquad \lambda > 0, \\ &\mathbf{0}_1 \leqslant \mathbf{U} \leqslant \mathbf{1} \end{aligned}\right\} \tag{14.2.1}$$

has only the zero solution.

Theorem 14.2.2 *Suppose a Q-matrix is given. Then*

(1) *F-type Q-processes always exist and there are only two possibilities, either there exists only one or there exist infinitely many.*

(2) *A necessary and sufficient condition for F-type Q-processes to be unique is, that either the minimal Q-process is honest or the equation*

$$\left.\begin{array}{ll}\lambda\mathbf{n} - \mathbf{n}Q = \mathbf{0}_-, & \lambda > 0,\\ \mathbf{0}_- \leqslant \mathbf{n}, & \mathbf{n}\mathbf{l} < +\infty\end{array}\right\} \tag{14.2.2}$$

has only the zero solution. Or equivalently, that either the minimal Q-process is honest or for any solution $\mathbf{n}_\lambda = (n_\lambda(1), n_\lambda(2) \cdots)$ of (14.2.2) there must be $\sum_{j\in E}(\sum_{k\neq j} q_{ik})\ n_\lambda(j) < +\infty$.

Theorem 14.2.3 *Suppose a Q-matrix is given. Then*

(1) *$B\bigcap F$-type Q-processes always exist and there are only two possibilities, either there exists only one or there exist infinitely many.*

(2) *A necessary and sufficient condition for $B\bigcup F$-type Q-process to be unique is, that at most one of the equations*

$$\left.\begin{array}{ll}\lambda\mathbf{U} - Q\mathbf{U} = \mathbf{0}_1, & \lambda > 0,\\ \mathbf{0}_1 \leqslant \mathbf{U} \leqslant \mathbf{1}, & \end{array}\right\} \tag{14.2.3}$$

$$\left.\begin{array}{ll}\lambda\mathbf{n} - \mathbf{n}Q = \mathbf{0}_-, & \lambda > 0,\\ \mathbf{0}_- \leqslant \mathbf{n}, & \mathbf{n}\mathbf{l} < +\infty,\\ \sum\limits_{j\in E}(q_j - \sum\limits_{k\neq j} q_{ik})n(j) < +\infty & \end{array}\right\} \tag{14.2.4}$$

has a nonzero solution.

Theorem 14.2.4 *Suppose a Q-matrix is given. Then*

(1) *There are only two possibilities for $\overline{B\bigcup F}$-type Q-processes, either there exist none or there exist infinitely many.*

(2) *If Q is conservative, then no $\overline{B\bigcup F}$-type Q-process exists.*

(3) *If Q is nonconservative, then a necessary and sufficient condition for the nonexistence of $\overline{B\bigcup F}$-type Q-process is that the following two conditions hold simultaneously:*

$$\text{(1)} \qquad \inf_{i\in E} \lambda \sum_{j\in E} p_{ij}^{\min}(\lambda) = n_\lambda > 0, \qquad 0 < \lambda < +\infty; \tag{14.2.5}$$

$$\text{(ii)} \qquad \sum_{j\in E} (q_j - \sum_{k\neq j} q_{jk})\ n_\lambda(j) < +\infty, \qquad 0 < \lambda < \infty, \tag{14.2.6}$$

where $\mathbf{n}_\lambda = (n_\lambda(1), n_\lambda(2), \cdots)$ is an arbitrary solution of the system of equations

$$\left.\begin{array}{l}\lambda\mathbf{n}_\lambda - \mathbf{n}_\lambda Q = \mathbf{0}_-, \qquad \lambda > 0,\\ \mathbf{0}_- \leqslant \mathbf{n}_\lambda, \qquad \mathbf{n}_\lambda\mathbf{1} < +\infty,\\ \mathbf{n}_\lambda - \mathbf{n}_\mu + (\lambda-\mu)\cdot\mathbf{n}_\mu P_\lambda^{\min} = \mathbf{0}_-.\end{array}\right\} \tag{14.2.7}$$

(4) *A necessary and sufficient condition for condition* (i) *in* (3) *to hold is, that the following two conditions hold simultaneously,*

(i)′

$$\inf_{i\in\dot{E}} \textstyle\sum_{j\in E} p_{ij}^{\min}(\lambda) > 0 \qquad (0<\lambda<+\infty), \tag{14.2.8}$$

where $\dot{E} = (i\colon \sum_{j\in E} q_{ij} < 0)$.

(i)″ *Equation* (14.2.1) *has only the zero solution.*

Condition (ii) *in* (3) *can be replaced by the following condition:*

(ii)′

$$\lim_{\lambda\to\infty} \lambda \sum_{j\in E} n_\lambda\,(j) < +\infty. \tag{14.2.9}$$

Theorem 14.2.5 *Suppose a Q-matrix is given. Then*

(1) *O-type Q-processes always exist and there are only two possibilities: either there exists only one or there exist infinitely many.*

(2) *A necessary and sufficient condition for O-type Q-processes to be unique is, that the following two conditions hold simultaneously:*

(i) *Condition* (i) *in Theorem* 14.2.4, *or equivalently Conditions* (i)′ *and* (i)″ *in Theorem* 14.2.4 *hold.*

(ii) *Either the minimal Q-process is honest or Equation* (14.2.2) *has only the zero solution, or equivalently, either the minimal Q-process is honest or for an arbitrary solution* $\mathbf{n}_\lambda = (n_\lambda(1), n_\lambda(2), \cdots)$ *of* (14.2.2) *we must have*

$$\sum_{j\in E}\left(\sum_{k\neq j} q_{ik}\right) n_\lambda(j) < +\infty. \tag{14.2.10}$$

Theorem 14.2.6 *Suppose a Q-matrix is given. Then*

(1) *There are only three possibilities for N-B-type Q-processes: none exist, exactly one exists, or infinitely many exist.*

(2) *If Q is nonconservative, then no N-B-type Q-process exists.*

(3) *If Q is conservative and the Q-process is unique, then the N-B-type Q-process is unique.*

(4) *If Q is conservative and the Q-process is not unique, then there exist infinitely many N-B-type Q-processes.*

Theorem 14.2.7 *Suppose a Q-matrix is given. Then*

(1) *There are only three possibilities for N-F-type Q-processes: none exist, exactly one exists, or there exist infinitely many.*

(2) *If Q is conservative, the O-type Q-process is not unique and Equation*

(14.2.2) *has only the zero solution, or if Q is nonconservative and* (14.2.2) *has only the zero solution, then no N-F-type Q-process exists.*

(3) *If Q is conservative and the O-type Q-process is unique or* (14.2.2) *has just one linearly independent solution, then there exists only one N-F-type Q-process.*

(4) *If the minimal Q-process is not honest and* (14.2.2) *has more than one linearly independent solution, then there exist infinitely many N-F-type Q-processes.*

Theorem 14.2.8 *Suppose a Q-matrix is given. Then*

(1) *There are three possibilities for* N-$B\bigcap F$-*type Q-processes; none exist, only one exists, or there exist infinitely many.*

(2) *If Q is nonconservative, or if Q is conservative and the O-type Q-process is not unique but the F-type Q-process is unique, then no* N-$B\bigcap F$-*type Q-process exists.*

(3) *If Q is conservative and the O-type Q-process is unique, or Q is conservative and the O-type Q-process is not unique but Equation* (14.2.2) *has just one linearly independent solution, then the* N-$B\bigcap F$-*type Q-process is unique.*

(4) *If Q is conservative and the O-type Q-process is not unique but Equation* (14.2.2) *has more than one linearly independent solution, then there exist infinitely many* N-$B\bigcap F$-*type Q-processes.*

Theorem 14.2.9 *Suppose a Q-matrix is given. Then*

(1) *There are noly two possibilities for* N-$\overline{B\bigcup F}$-*type Q-processes: either there exist none or there exist infinitely many.*

(2) *A necessary and sufficient condition for the nonexistence of* N-$\overline{B\bigcup F}$-*type Q-process is, that no* $\overline{B\bigcup F}$-*type Q-process exists.*

Theorem 14.2.10 *Suppose a Q-matrix is given. Then*

(1) *There are only three possibilities for N-O-type Q-processes: none exist, only one exists, or there exist infinitely many.*

(2) *If Q is nonconservative and the Q-process is unique, then no N-O-type Q-process exists.*

(3) *If Q is conservative and the Q-process is unique, or if Q is nonconservative and the following two conditions hold simultaneously:*

(a) (14.2.5) *holds,*

(b) *Equation* (14.2.2) *has just one linearly independent solution* $\mathbf{n}_\lambda = (n_\lambda(1), n_\lambda(2), \cdots)$, *and*

$$\lim_{\lambda \to \infty} \lambda \sum_{j \in E} n_\lambda(j) < +\infty, \tag{14.2.11}$$

then the N-O-type Q-process is unique.

(4) *If Q is conservative and the Q-process is not unique, or if Q is nonconservative and* (14.2.5) *does not hold, or if Q is nonconservative and the*

following two conditions hold simultaneously:

(α) (14.2.5) *holds.*

(β) *Equation* (14.2.2) *has more than one linearly independent solution, or Equation* (14.2.2) *has just one linearly independent solution* $\mathbf{n}_\lambda = (n_\lambda(1), n_\lambda(2) \cdots)$ *and*

$$\lim_{\lambda\to\infty} \lambda \sum_{j\in E} n_\lambda(j) = +\infty, \tag{14.2.12}$$

then there exist infinitely many N-O-type Q-processes.

Theorem 14.2.11 *Suppose a Q-matrix is given. Then*

(1) *There are only two possibilities for* $\bar{B}$*-type Q-processes: either none exist or there exist infinitely many.*

(2) *A necessary and sufficient condition for the nonexistence of* $\bar{B}$*-type Q-processes is, that Q is conservative, or that Q is nonconservative and the Q-process is unique.*

Theorem 14.2.12 *Suppose a Q-matrix is given. Then*

(1) *There are only two possibilities for* $\bar{F}$*-type Q-processes: either none exist or there exist infinitely many.*

(2) *A necessary and sufficient condition for the non-existence of* $\bar{F}$*-type Q-processes is, that Q is conservative and O-type Q-process is unique, or that Q is nonconservative and no* $\overline{B \cup F}$*-type Q-process exists.*

Theorem 14.2.13 *Suppose a Q-matrix is given. Then*

(1) *There are only two possibilities for* $\bar{B} \cap F$*-type Q-processes: either none exist or there exist infinitely many.*

(2) *A necessary and sufficient condition for the existence of* $\bar{B} \cap F$*-type Q-processes is, that the following two conditions hold simultaneously:*

(i) *Q is nonconservative.*

(ii) *Equation* (14.2.2) *has a nonzero solution.*

Theorem 14.2.14 *Suppose a Q-matrix is given. Then*

(1) *There are only two possibilities for* $B \cap \bar{F}$*-type Q-processes: either none exist or there exist infinitely many.*

(2) *A necessary and sufficient condition for the nonexistence of* $B \cap \bar{F}$*-type Q-process is, that Equation* (14.2.1) *has only the zero solution.*

Theorem 14.2.15 *Suppose a Q-matrix is given. Then*

(1) $B \cup F$*-type Q-processes always exist and there are only two possibilities: either there exists only one or there exist infinitely many.*

(2) *A necessary and sufficient condition for the* $B \cup F$*-type Q-process to be unique is, that both B-type and F-type Q-processes are unique.*

Theorem 14.2.16 *Suppose a Q-matrix is given. Then*

(1) *There are three possibilities for N-*$\bar{B}$*-type Q-processes: none exist, only one exists, or there exist infinitely many.*

(2) *If Q is conservative, or if Q is nonconservative and the Q-process is unique, then no N-$\bar{B}$-type Q-process exists; If Q is nonconservative and*

(a) (14.2.5) *holds; and*

(b) *Equation* (14.2.2) *has just one linearly independent solution* $\mathbf{n}_\lambda = (n_\lambda(1), n_\lambda(2), \cdots)$, *and*

$$\sum_{j\in E}\left(q_j - \sum_{k\neq j} q_{jk}\right)n_\lambda(j) < +\infty, \qquad 0 < \lambda < +\infty, \tag{14.2.13}$$

then there is a unique N-$\bar{B}$-type Q-process. In all other cases there exist infinitely many N-$\bar{B}$-type Q-processes.

Theorem 14.2.17 *Suppose a Q-matrix is given. Then*

(1) *There are only two possibilities for N-$\bar{F}$-type Q-processes: either none exist or there exist infinitely many.*

(2) *A necessary and sufficient condition for the existence of N-F-type Q-processes is that Q is conservative and O-type Q-process is unique, or that Q is nonconservative and no N-$\overline{B\bigcup F}$-type Q-process exists.*

Theorem 14.2.18 *Suppose a Q-matrix is given. Then*

(1) *There are only three possibilities for N-$\bar{B}\bigcap F$-type Q-processes: none exist, only one exists, or there exist infinitely many.*

(2) *If Q is conservative, or Q is nonconservative and the $\bar{F}$-type Q-process is unique, then no N-$\bar{B}\bigcap F$-type Q-process exists. If Q is nonconservative and the N-F-type Q-process is unique, then there exists a unique N-$\bar{B}\bigcap F$-type Q-process. If Q is nonconservative and the N-F-type Q-process is not unique, then there exist infinitely many N-$\bar{B}\bigcap F$-type Q-processes.*

Theorem 14.2.19 *Suppose a Q-matrix is given. Then*

(1) *There are only two possibilities for N-$B\bigcap\bar{F}$-type Q-processes: either none exist or there exist infinitely many.*

(2) *If Q is conservative and the Q-process is unique, or if Q is nonconservative, then there does not exist any N-$B\bigcap\bar{F}$-type Q-process. If Q is conservative and Q-process is not unique, then there exist infinitely many N-$B\bigcap\bar{F}$-type Q-processes.*

Theorem 14.2.20 *Suppose a Q-matrix is given. Then*

(1) *There are three possibilities for N-$B\bigcup F$-type Q-processes: none exist, only one exists, or there exist infinitely many.*

(2) *If Q is nonconservative and Equation* (14.2.2) *only has the zero solution, then no N-$B\bigcup F$-type Q-process exists. If Q is conservative and Q-process is unique, or if Q is nonconservative and Equation* (14.2.2) *has just one linearly independent solution, then N-$B\bigcup F$-type Q-process is unique. If Q is conservative and the Q-process is not unique, or if Q is nonconservative and Equation* (14.2.2) *has more than one linearly independent solution, then there exist infinitely many N-$B\bigcup F$-type Q-processes.*

In what follows we shall only give the proofs of Theorems 14.2.1-14.2.10. The proofs of Theorems 14.2.11-14.2.20 are omitted here since they can be easily accomplished by applying the previous ten theorems and the methods adopted in their proofs.

§ 14.3 Reduction of the construction problem of B-type Q-processes, Doob processes

The aim of this section is to reduce the problem of constructing all the B-type Q-processes in the nonconservative case to that of constructing (O-type) Q-processes in the conservative case. Then, with the results obtained we construct a class of Q-processes--Doob Q-processes. The results presented here are not only used to prove Proposition 14.4.1 and Theorem 14.2.2, but also have independent significance. So we devote a special section to them.

Suppose $P(t) = (p_{ij}(t),\ i,\ j \in E)$ is a B-type Q-process. $X(\omega) = \{x(t,\ \omega),\ t < \sigma(\omega)\}$ is a Q-process with transition probability matrix $P(t)$. Let

$$\hat{x}(t,\ \omega) = \begin{cases} x(t,\ \omega), & t < \sigma(\omega), \\ 0, & t \geqslant \sigma(\omega). \end{cases} \tag{14.3.1}$$

Lemma 14.3.1 $\hat{X}(\omega) = \{\hat{x}(t,\omega),\ t \geqslant 0\}$ *is an honest homogeneous denumerable Markov process. Its transition probability matrix* $\hat{P}(t) = (\hat{p}_{ij}(t),\ i, j \in E \bigcup \{0\})$ *and density matrix* $\hat{Q} = (\hat{q}_{ij},\ i,\ j \in E \bigcup \{0\})$ *are determined uniquely as follows:*

$$\hat{p}_{ij}(t) = \begin{cases} p_{ij}(t), & (i,\ j \in E), \\ 1 - \sum\limits_{k \in E} p_{ik}(t), & (i \in E,\ j = 0), \\ 1, & (i = j = 0), \\ 0, & (i = 0,\ j \in E), \end{cases} \tag{14.3.2}$$

$$\hat{q}_{ij} = \begin{cases} q_{ij}, & (i,\ j \in E), \\ q_i - \sum\limits_{k \neq i} q_{ik}, & (i \in E,\ j = 0), \\ 0, & (i = 0,\ j \in E \bigcup \{0\}). \end{cases} \tag{14.3.3}$$

Proof. It suffices to prove that

$$\hat{q}_{i0} = q_i - \sum_{k \neq i} q_{ik} \qquad (i \in E), \tag{14.3.4}$$

because the remaining conclusions obviously hold from (14.3.4).

Let

$$E_1 = \{i:\ i \in E,\ q_i \neq 0\}, \tag{14.3.5}$$

$$E_2 = \{i: \ i \in E, \ q_i = 0\}. \tag{14.3.6}$$

Equation (14.3.4) obviously holds when $i \in E_2$, so in what follows we shall always assume $i \in E_1$ without further indication.

$\tau^{(1)}(\omega)$ stands for the first point of discontinuity of $X(\omega)$. Hence by [1, II, Theorem 15.6], we have

$$P(x(\tau^{(1)}) = j \mid x_0 = i) = \frac{q_{ij}}{q_i} \qquad (i \neq j). \tag{14.3.7}$$

By virtue of [1, II, Theorem 17.4] and the fact that $P(t)$ is a B-type Q-process, we see

$$P(x(\tau^{(1)}) = +\infty \mid x_0 = i) = 0. \tag{14.3.8}$$

Applying [1, II, Theorem 5.5] enables us to deduce that

$$P(x(\tau^{(1)}) < +\infty \mid x_0 = i) = 1. \tag{14.3.9}$$

By appealing to (14.3.7), (14.3.8) and (14.3.9), we confirm that

$$\begin{aligned} P(\tau^{(1)} &= \sigma, \ \sigma < +\infty \mid x_0 = i) \\ &= P(\tau^{(1)} < +\infty \mid x_0 = i) - \sum_{j \neq i} P(x(\tau^{(1)}) = j \mid x_0 = i) \\ &= 1 - \sum_{j \neq i} \frac{q_{ij}}{q_i} = \frac{q_i - \sum_{j \neq i} q_{ij}}{q_i}. \end{aligned} \tag{14.3.10}$$

From the definition of $\hat{X}(\omega)$, we infer that $\tau^{(1)}(\omega)$ is also the first point of discontinuity of $\hat{X}(\omega)$ (under $\hat{x}_0 = i$), $0 < \hat{q}_i = q_i < +\infty$ and

$$\lim_{t \to 0} \hat{p}_{00}(t) = 1. \tag{14.3.11}$$

It can be seen by (14.3.11) and [1, II, Theorem 15.6] that

$$P(\hat{x}(\tau^{(1)}) = 0 \mid \hat{x}_0 = i) = \frac{\hat{q}_{i0}}{\hat{q}_i}, \tag{14.3.12}$$

but

$$\begin{aligned} P(\hat{x}(\tau^{(1)}) = 0 \mid \hat{x}_0 = i) &= P(\tau^{(1)} = \sigma, \ \sigma < +\infty \mid x_0 = i) \\ &= \frac{q_i - \sum_{k \neq i} q_{ik}}{q_i}. \end{aligned} \tag{14.3.13}$$

From $\hat{q}_i = q_i$, (14.3.12) and (14.3.13), we obtain (14.3.4), and the proof is complete. □

We denote the transformation from Q to $\hat{Q}$ by L, i.e.,

$$\hat{Q} - LQ. \tag{14.3.14}$$

Obviously we have the following: □

Lemma 14.3.2 *$\hat{Q}$ is a conservative Q-matrix.*

Suppose $\hat{P}(t) = (\hat{p}_{ij}(t),\ i,\ j \in E \bigcup \{0\})$ is an arbitrary $\hat{Q}$-process. Let

$$p_{ij}(t) = \hat{p}_{ij}(t) \qquad (i,\ j \in E), \tag{14.3.15}$$

$$P\ (t) = (p_{ij}(t),\ i,\ j \in E), \tag{14.3.16}$$

and denote the transformation from $\hat{P}(t)$ to $P(t)$ by G, i.e.,

$$p(t) = G\hat{P}(t). \tag{14.3.17}$$

Proposition 14.3.1 *Suppose $\hat{P}(t)$ is a $\hat{Q}$-process, then*

$$P(t) = G\hat{P}(t) \tag{14.3.18}$$

is a B-type Q-process. Conversely, suppose $P(t)$ is a B-type Q-process, then there exists at least one $\hat{Q}$-process $\hat{P}(t)$ such that (14.3.18) *holds.*

Proof. By Lemma 14.3.1 it suffices to prove the first half of the proposition.

Suppose $\hat{P}(t) = (\hat{p}_{ij}(t);\ i, j \in E \bigcup \{0\})$ is an arbitrary $\hat{Q}$-process. $P(t) = (p_{ij}(t);\ i,\ j \in E)$ is defined by (14.3.18). We wish to prove

$$P(t) \geqslant 0, \tag{14.3.19}$$

$$P(t)\mathbf{1} \leqslant \mathbf{1}, \tag{14.3.20}$$

$$P(t) \cdot P(s) = P(t+s), \tag{14.3.21}$$

$$\lim_{t \to 0} P(t) = I, \tag{14.3.22}$$

$$P'(0) = Q, \tag{14.3.23}$$

$$P'(t) = QP(t). \tag{14.3.24}$$

Evidently, we need only prove (14.3.21) and (14.3.24). Taking account of (14.3.3), (14.3.15), the fact that $\hat{Q}$ is conservative and

$$\hat{P}_{0j}(t) = 0 \qquad (j \in E) \tag{14.3.25}$$

we immediately obtain (14.3.24). From (14.3.25), we deduce (14.3.21) readily. So $P(t)$ is a B-type Q-process, and the proposition is valid. □

Proposition 14.3.2 *Suppose $\hat{P}(t)$ is an arbitrary N-O-type $\hat{Q}$-process. Then* (14.3.18) *defines a B-type Q-process. Conversely, let $P(t)$ be an arbitrary B-type Q-process, then there exists only one N-O-type $\hat{Q}$-process $\hat{P}(t)$ such that* (14.3.18) *holds.*

Proof. We deduce the proposition immediately from Lemma 14.3.1, Proposition 14.3.1 and the following simple fact, that if $\hat{P}(t) = (\hat{p}_{ij}(t); i, j \in E \bigcup \{0\})$ is a N-O-type $\hat{Q}$-process. then $\hat{p}_{i0}(t) \quad (i \in E)$. and $\hat{p}_{oj}(t) \quad (j \in E \cup \{0\})$ are determined uniquely by $\hat{p}_{ij}(t)$ $(i, j \in E)$. □

We denote the set of all solutions of the equation

$$\left.\begin{aligned} \lambda \mathbf{U} - Q\mathbf{U} = \mathbf{0}_1, \qquad \lambda > 0, \\ \mathbf{0}_1 \leqslant \mathbf{U} \leqslant \mathbf{1} \end{aligned}\right\} \tag{14.3.26}$$

and

$$\left.\begin{aligned} \lambda \hat{\mathbf{U}} - \hat{Q}\hat{\mathbf{U}} = \mathbf{0}_1, \quad \lambda > 0, \\ \mathbf{0}_1 \leqslant \hat{\mathbf{U}} \leqslant \mathbf{1} \end{aligned}\right\} \tag{14.3.27}$$

by D_Q and $D_{\hat{Q}}$, respectively.

Let

$$\mathbf{U} = \begin{pmatrix} u(1) \\ u(2) \\ \vdots \end{pmatrix} \tag{14.3.28}$$

be an element of D_Q. Let

$$\hat{\mathbf{U}} = \begin{pmatrix} \hat{u}(0) \\ \hat{u}(1) \\ \hat{u}(2) \\ \vdots \end{pmatrix} = \begin{pmatrix} 0 \\ u(1) \\ u(2) \\ \vdots \end{pmatrix}. \tag{14.3.29}$$

Then we denote by W the transformation from $\mathbf{U}$ to $\hat{\mathbf{U}}$, i.e.,

$$\hat{\mathbf{U}} = W\mathbf{U}. \tag{14.3.30}$$

Proposition 14.3.3 *The transformation W defined by* (14.3.30) *is one–to–one from D_Q onto $D_{\hat{Q}}$, and $\mathbf{0}_1 = W\mathbf{0}_1$.*

Proof. The following fact can be easily verified, that if

$$\hat{\mathbf{U}} = \begin{pmatrix} \hat{u}(0) \\ \hat{u}(1) \\ \hat{u}(2) \\ \vdots \end{pmatrix} \tag{14.3.31}$$

belongs to $D_{\hat{Q}}$, then $\hat{u}(0) = 0$. Thus, we obtain the proposition immediately. □

Let

$$u_\lambda(i) = 1 - \left(\lambda \sum_{j\in E} p_{ij}^{\min}(\lambda) + \sum_{j\in E} p_{ij}^{\min}\left(q_j - \sum_{k\neq j} q_{jk}\right)\right) \qquad (i \in E), \tag{14.3.32}$$

$$\mathbf{U}_\lambda = \begin{pmatrix} u_\lambda(1) \\ u_\lambda(2) \\ \vdots \end{pmatrix}. \tag{14.3.33}$$

Lemma 14.3.3 *$\mathbf{U}_\lambda$ is the maximal solution of the equation*

$$\left.\begin{aligned} \lambda \mathbf{U} - Q\mathbf{U} = \mathbf{0}_1, \qquad \lambda > 0, \\ \mathbf{0}_1 \leqslant \mathbf{U} \leqslant \mathbf{1}. \end{aligned}\right\} \tag{14.3.34}$$

Hence if (14.3.34) *has a nonzero solution, then*

$$\mathbf{U}_\lambda \not\equiv \mathbf{0}_1. \tag{14.3.35}$$

Proof. It is well known that $\{p_{ij}^{\min}(\lambda), i \in E\}$ is the minimal nonnegative solution of the pseudo-normal system of equations

$$x_i = \sum_{k\neq i} \frac{q_{ik}}{\lambda + q_i} x_k + \frac{\delta_{ij}}{\lambda + q_i} \qquad (i \in E). \tag{14.3.36}$$

Hence we see from Theorem 3.3.2 that $\mathbf{1} - \mathbf{U}_\lambda$ is the minimal nonnegative solution of the system of equations

$$x_i = \sum_{k\neq i} \frac{q_{ik}}{\lambda + q_i} x_k + \frac{\lambda}{\lambda + q_i} + \frac{q_i - \sum_{k\neq i} q_{ik}}{\lambda + q_i} \qquad (i \in E). \tag{14.3.37}$$

Therefore, $\mathbf{U}_\lambda$ is the maximal solution of (14.3.34), and the lemma is established. □

Proposition 14.3.4 *If* (14.3.34) *has a nonzero solution,* $\alpha = (\alpha_1, \alpha_2, \cdots) \geqslant \mathbf{0}_-$, $\alpha \neq \mathbf{0}_-$ *and* $\alpha\mathbf{1} \leqslant \mathbf{1}$, *then*

$$P_\lambda = P_\lambda^{\min} + \mathbf{U}_\lambda \frac{\alpha P_\lambda^{\min}}{1 - \alpha \mathbf{U}_\lambda} \tag{14.3.38}$$

is a B-type Q-process.

Proof. If Q is conservative, it is well known that then (14.3.38) determines a Doob Q-process. If Q is nonconservative, then we denote the minimal $\hat{Q}$-process by $\hat{P}_\lambda^{\min} = (\hat{p}_{ij}^{\min}(\lambda),\ i,\ j \in E \bigcup \{0\})$. Let

$$\hat{\alpha} = (0,\ \alpha_1,\ \alpha_2,\ \cdots), \tag{14.3.39}$$

$$\hat{P}_\lambda = \hat{P}_\lambda^{\min} + (\mathbf{1} - \lambda \hat{P}^{\min}\mathbf{1}) \frac{\hat{\alpha} P_\lambda^{\min}}{1 - \alpha(\mathbf{1} - \lambda \hat{P}^{\min}\mathbf{1})}. \tag{14.3.40}$$

It is well known that $\hat{P}_\lambda$ is a Doob $\hat{Q}$-process. We can easily prove that

$$p_{ij}(\lambda) = \hat{p}_{ij}(\lambda) \qquad (\lambda > 0,\ i,\ j \in E), \tag{14.3.41}$$

hence

$$p_{ij}(t) = \hat{p}_{ij}(t) \qquad (t \geqslant 0,\ i,\ j \in E), \tag{14.3.42}$$

where $p_{ij}(t)$ and $\hat{p}_{ij}(t)$ are the inverse Laplace transforms of $p_{ij}(\lambda)$ and $\hat{p}_{ij}(\lambda)$, respectively.

From the fact that $\hat{P}(t) = (\hat{p}_{ij}(t),\ i,\ j \in E \bigcup \{0\})$ is a $\hat{Q}$-process, (14.3.42) and Proposition 14.3.1, we deduce that $P(t) = (p_{ij}(t),\ i,\ j \in E)$, i.e., P_λ is a B-type Q-process. □

Definition 14.3.1 *If Equation* (14.3.34) *has a nonzero solution, then the B-type Q-process P_λ defined by* (14.3.38) *is called a Doob process.*

Lemma 14.3.4 *Let*

$$\rho = \begin{pmatrix} \rho_1 \\ \rho_2 \\ \vdots \end{pmatrix} \neq \mathbf{0}_1, \qquad \alpha = (\alpha_1,\ \alpha_2, \cdots), \qquad \bar{\alpha} = (\bar{\alpha}_1,\ \bar{\alpha}_2, \cdots).$$

If

$$\rho\alpha = \rho\bar{\alpha}, \tag{14.3.43}$$

then

$$\alpha = \bar{\alpha}. \tag{14.3.44}$$

Proof. Since $\rho \neq \mathbf{0}_1$, there exists a subscript $i_0 \in E$ such that

$$\rho_{i_0} \neq 0. \tag{14.3.45}$$

It follows by (14.3.43) and (14.3.45) that

$$\rho_{i_0}\alpha_j = \rho_{i_0}\bar{\alpha}_j \qquad (j \in E). \tag{14.3.46}$$

Equations (14.3.45) and (14.3.46) imply that

$$\alpha_j = \bar{\alpha}_j \qquad (j \in E), \tag{14.3.47}$$

i.e., (14.3.44) holds, as claimed.

Proposition 14.3.5 *If Equation* (14.3.34) *has a nonzero solution, then for Doob Q-processes*

$$P_\lambda^{(i)} = P_\lambda^{\min} + \mathbf{U}_\lambda \frac{\alpha^{(i)} P_\lambda^{\min}}{1 - \alpha^{(i)}\mathbf{U}_\lambda} \qquad (i = 1,\ 2), \tag{14.3.48}$$

we have

$$P_\lambda^{(1)} \equiv P_\lambda^{(2)} \tag{14.3.49}$$

if and only if there exists a constant r such that

$$\alpha^{(1)} = r\alpha^{(2)}. \tag{14.3.50}$$

Proof. From Lemma 14.3.4, a necessary and sufficient condition for (14.3.49) to hold is that

$$\frac{\alpha^{(1)} P_\lambda^{\min}}{1 - \alpha^{(1)}\mathbf{U}_\lambda} = \frac{\alpha^{(2)} P_\lambda^{\min}}{1 + \alpha^{(2)}\mathbf{U}_\lambda}. \tag{14.3.51}$$

Applying the operator $\lambda I - Q$ to both sides of the above equality from the right and noting the fact that $\alpha^{(i)}\mathbf{1} \leqslant 1$ $(i = 1,\ 2)$, we infer

$$\frac{\alpha^{(1)}}{1 - \alpha^{(1)}\mathbf{U}_\lambda} = \frac{\alpha^{(2)}}{1 - \alpha^{(2)}\mathbf{U}_\lambda}. \tag{14.3.52}$$

Conversely, applying the operator $P_\lambda^{\min}$ to both sides of the above equality, we obtain (14.3.51). So (14.3.51) is equivalent to (14.3.52).

Taking note of (14.3.48), we know that

$$1 - \alpha^{(i)}\mathbf{U}_\lambda \neq 0 \qquad (i = 1,\ 2). \tag{14.3.53}$$

Hence if we let

$$\frac{1-\alpha^{(1)}\mathbf{U}_\lambda}{1-\alpha^{(2)}\mathbf{U}_\lambda}=r, \tag{14.3.54}$$

then (14.3.52) can be transformed into the following equivalent form:

$$\alpha^{(1)}=r\alpha^{(2)}. \tag{14.3.55}$$

The proposition is proved. □

Proposition 14.3.6 *Suppose* (14.3.34) *has a nonzero solution. If let*

$$\alpha^{(n)}=(\underbrace{0,\ 0,\cdots,\ 0}_{(n-1)},\ 1,\ 0,\ \cdots), \tag{14.3.56}$$

then

$$P_\lambda^{(n)}=P_\lambda^{\min}+\mathbf{U}_\lambda\frac{\alpha^{(n)}P_\lambda^{\min}}{1-\alpha^{(n)}\mathbf{U}_\lambda}\qquad(n=1,\ 2,\cdots) \tag{14.3.57}$$

are infinitely many B-type Q-processes, different from each other.

Proof. The proposition follows easily by Propositions 14.3.4 and 14.3.5. □

§ 14.4 Reduction of the construction problem of $B\cap F$-type Q-processes

The aim of this section is to reduce the problem of constructing all the $B\cap F$-type *Q-processes in the nonconservative case to that of constructing all the* F-type Q-processes in the conservative case. The results presented here can not only be used to prove Theorem 12.2.3 but also (like the results in §14.3) have independent significance.

Proposition 14.4.1 *Let* $\hat{P}(t)$ *be an arbitrary F-type Q-process, then*

$$P(t)=G\hat{P}(t) \tag{14.4.1}$$

is a $B\cap F$*-type Q-process. Conversely, let* $P(t)$ *be an arbitrary* $B\cap F$*-type Q-process, then there exists only one F-type* $\hat{Q}$*-process* $\hat{P}(t)$ *such that* (14.4.1) *holds. For the definitions of* $\hat{Q}$ *and G, see* §14.3.

In order to prove Proposition 14.4.1, let us first prove several lemmas.

Lemma 14.4.1 *Suppose* $\hat{P}(t)$ *is an arbitrary F-type* $\hat{Q}$*-process, then* $P(t)$*, defined by* (14.4.1), *is a* $B\cap F$*-type Q-process.*

Proof. From the fact that $\hat{P}(t)=(\hat{p}_{ij}(t),\ i,j\in E)$ is an F-type $\hat{Q}$-process and $\hat{q}_{0j}=0$ $(j\in E)$, we confirm that

$$p'_{ij}(t)=\hat{p}'_{ij}(t)=\sum_{k\in E\cup\{0\}}\hat{p}_{ik}(t)\hat{q}_{kj}=\sum_{k\in E}\hat{p}_{ik}(t)\hat{q}_{kj}+\hat{p}_{i0}(t)\hat{q}_{0j}$$

$$= \sum_{k\in E} p_{ik}(t)q_{kj} \qquad (i,\ j\in E). \tag{14.4.2}$$

Hence from the fact that Q is conservative and Proposition 14.3.1, we get the lemma immediately. □

Lemma 14.4.2 *Suppose $\hat{P}(t) = (\hat{p}_{ij}(t),\ i,\ j\in \bigcup\{0\})$ is an F-type $\hat{Q}$-process, then $\hat{P}(t)$ is determined uniquely by $(\hat{p}_{ij}(t),\ i,\ j\in E)$.*

Proof. We see from $\hat{Q}$ being conservative that

$$\hat{P}'(t) = \hat{P}(t)\hat{Q}, \tag{14.4.3}$$

$$\hat{P}'(t) = \hat{Q}\hat{P}(t). \tag{14.4.4}$$

Then taking account of the definition of $\hat{Q}$, we see

$$\hat{p}'_{i0}(t) = \sum_{k\in E} \hat{p}_{ik}(t)q_{k0} \qquad (i\in E), \tag{14.4.5}$$

$$\hat{p}'_{0j}(t) = 0 \qquad (j\in E\bigcup\{0\}). \tag{14.4.6}$$

If $(\hat{p}_{ij}(t),\ i,\ j\in E)$ is known, adding the initial conditions $\hat{p}_{i0}(0) = 0\ (i\in E)$, $\hat{p}_{00}(0) = 1$ and $\hat{p}_{0j}(0) = 0\ (j\in E)$, then from (14.4.5) and (14.4.6), we may uniquely determine $\hat{p}_{i0}(t)\ (i\in E)$ and $\hat{p}_{0j}(t)\ (j\in E\bigcup\{0\})$. □

Lemma 14.4.3 *Suppose $P(t)$ is an arbitrary $B\bigcap F$-type Q-process, then there exists at least one F-type $\hat{Q}$-process $\hat{P}(t)$ such that* (14.4.1) *holds.*

Proof. Let

$$\hat{p}_{ij}(t) = \begin{cases} p_{ij}(t), & (i,\ j\in E), \\ 1, & (i=j=0), \\ 0, & (i=0,\ j\in E), \\ \displaystyle\int_0^t \Big(\sum_{k\in E} p_{ik}(s)q_{k0}\Big)\, ds & (i\in E,\ j=0), \end{cases} \tag{14.4.7}$$

$$\hat{P}(t) = (\hat{p}_{ij}(t);\ i,\ j\in E\bigcup\{0\}). \tag{14.4.8}$$

We are going to prove that $\hat{P}(t)$, defined by (14.4.8), is an F-type $\hat{Q}$-process such that (14.4.1) holds.

From the proof of the second part of Proposition 14.3.1, we see that $\bar{P}(t) = (\bar{p}_{ij}(t),\ i,\ j\in E)$ is a $\hat{Q}$-process, where

$$\bar{p}_{ij}(t) = \begin{cases} p_{ij}(t), & (i,\ j\in E), \\ 1-\sum_{k\in E} p_{ik}(t), & (i\in E,\ j=0), \\ 1, & (i=j=0), \\ 0, & (i=0,\ j\in E), \end{cases} \tag{14.4.9}$$

hence, from [1, II, §17] and $\bar{q}_{00}=0<+\infty$, we see that

$$\sum_{k\in E\cup\{0\}} \bar{p}_{ik}(t)\bar{q}_{k0} < +\infty \qquad (i\in E,\ 0\leqslant t<+\infty). \tag{14.4.10}$$

It follows by (14.4.9), (14.4.10) and the definition of $\hat{Q}$ that

$$\sum_{k\in E} p_{ik}(t)q_{k0} = \sum_{k\in E\cup\{0\}} \bar{p}_{ik}(t)\hat{q}_{k0} < +\infty \qquad (i\in E,\ 0\leqslant t<+\infty). \tag{14.4.11}$$

Hence $\hat{P}(t)$ is determined uniquely by (14.4.7). From $q_{k0}\geqslant 0$ $(k\in E)$, we have immediately

$$\hat{p}_{ij}(t)\geqslant 0 \qquad (i,\ j\in E\cup\{0\}). \tag{14.4.12}$$

It follows by (14.4.7) that

$$\sum_{k\in E\cup\{0\}} \hat{p}_{0k}(t)=\hat{p}_{00}(t)=1. \tag{14.4.13}$$

From the fact that $\bar{P}(t)$ is a $\hat{Q}$-process, and [1, II, §17], we have

$$\bar{p}'_{i0}(t) \geqslant \sum_{k\in E\cup\{0\}} \bar{p}_{ik}(t)q_{k0} = \sum_{k\in E} p_{ik}(t)q_{k0} \geqslant 0 \qquad (i\in E,\ 0\leqslant t<+\infty). \tag{14.4.14}$$

By (14.4.14) and $\bar{p}_{i0}(0)=0$ $(i\in E)$, we have

$$\bar{p}_{i0}(t) \geqslant \int_0^t \left(\sum_{k\in E} p_{ik}(s)\, q_{k0}\right) ds \qquad (i\in E,\ 0\leqslant t<+\infty). \tag{14.4.15}$$

By (14.4.9) and (14.4.15), we have

$$1-\sum_{k\in E} p_{ik}(t) = p_{i0}(t) \geqslant \int_0^t \left(\sum_{k\in E} p_{ik}(s)q_{k0}\right) ds \qquad (i\in E,\ 0\leqslant t<+\infty). \tag{14.4.16}$$

It follows by (14.4.7) and (14.4.16) that

$$\sum_{k\in E\cup\{0\}} \hat{p}_{ik}(t)\leqslant 1 \qquad (i\in E). \tag{14.4.17}$$

Combining (14.4.12), (14.4.13) and (14.4.17), we obtain

$$\hat{P}(t) \geqslant 0, \tag{14.4.18}$$

$$\hat{P}(t)\mathbf{1} \leqslant \mathbf{1}. \tag{14.4.19}$$

From the fact that $P(t)$ is a $B \cap F$-type Q-process and the definition of $\hat{Q}$, it is easy to verify directly the validity of the following four relations:

$$\hat{P}(t+s) = \hat{P}(t)\hat{P}(s), \tag{14.4.20}$$

$$\lim_{t\to 0} \hat{P}(t) = I, \tag{14.4.21}$$

$$\hat{P}'(0) = \hat{Q}, \tag{14.4.22}$$

$$\hat{P}'(t) = \hat{p}(t)\hat{Q}. \tag{14.4.23}$$

Therefore $\hat{P}(t)$ is an F-type $\hat{Q}$-process. Obviously, from (14.4.7) and the definition of G, we see that (14.4.1) holds.

The lemma is now proved. □

Proposition 14.4.1 follows immediately from Lemmas 14.4.1—14.4.3. □

§ 14.5 Proofs of Theorems 14.2.1—14.2.3

Proof of Theorem 14.2.1. The second part of Theorem 14.2.1 consists of the results due to Reuter [27]. It is well known that the minimal Q-process is a B-type Q-process. So B-type Q-processes always exists. From the second part of our theorem and Proposition 14.3.6, there exist infinitely many B-type Q-processes if B-type Q-process is not unique. Thus the first part of the theorem is established. □

Proof of Theorem 14.2.2 It is well-known that the minimal Q-process is an F-type Q-process. So an F-type Q-process always exists. The latter half of the first part and the former half of the second part of the theorem are also due to Reuter [27]. The latter half of the second part of the theorem is an immediate consequence of the former half of second part and Lemma 12.5.1. Therefore the theorem is valid. □

Proof of Theorem 14.2.3 From Proposition 14.4.1 and Theorem 14.2.2, we obtain the theorem immediately. □

§14.6 Proof and examples of applications of Theorem 14.2.4

Let us first prove the following lemmas.

Lemma 14.6.1 *For all $i \in E$, we have*

$$u_\lambda(i) = 1 - \lambda \sum_{j \in E} p_{ij}^{\min}(\lambda) \downarrow 0 \qquad (\lambda \uparrow + \infty). \tag{14.6.1}$$

Proof. By Lemma 12.2.2, $u_\lambda(i)$ does not increase as λ increases. From the proof of Lemma 12.2.1, we know that $\lambda \sum_{j \in E} p_{ij}^{\min}(\lambda)$ $(i \in E)$ is the minimal nonnegative solution of the pseudo–normal system of equations

$$x_i = \sum_{k \neq i} \frac{q_{ik}}{\lambda + q_i} x_k + \frac{\lambda}{\lambda + q_i} \qquad (i \in E). \tag{14.6.2}$$

Thus

$$\frac{\lambda}{\lambda + q_i} \leqslant \lambda \sum_{j \in E} p_{ij}^{\min}(\lambda) \leqslant 1 \qquad (i \in E), \tag{14.6.3}$$

hence

$$\lambda \sum_{j \in E} p_{ij}^{\min}(\lambda) \uparrow 1 \qquad (i \in E). \tag{14.6.4}$$

Consequently (14.6.1) holds, as desired. □

Lemma 14.6.2

$$\lim_{\lambda \to \infty} \lambda u_\lambda(i) = q_i - \sum_{j \neq i} q_{ij} \qquad (i \in E). \tag{14.6.5}$$

Proof. In view of the proof of Lemma 12.2.2, we see that

$$\lambda u_\lambda(i) = -q_i u_\lambda(i) + \sum_{j \neq i} q_{ij} u_\lambda(j) + q_i - \sum_{j \neq i} q_{ij} \qquad (i \in E). \tag{14.6.6}$$

Hence by Lemma 12.6.1, we easily verify what should be proved. □

Lemma 14.6.3 *Assume* $\mathbf{n}_\lambda = (n_\lambda(1), n_\lambda(2), \cdots)$ *is an arbitrary solution of* (14.2.7), *then for all* $\mu > 0$, *we have*

$$\lim_{\lambda \to \infty} \lambda \sum_{j \in E} n_\lambda(j) \geqslant \mu \sum_{j \in E} n_\mu(j) + \sum_{j \in E} \left(q_i - \sum_{k \neq j} q_{jk} \right) n_\mu(j). \tag{14.6.7}$$

Hence, if

$$\sum_{j \in E} \left(q_j - \sum_{k \neq j} q_{ik} \right) n_\mu(j) = +\infty \tag{14.6.8}$$

for some μ, *then*

$$\lim_{\lambda \to \infty} \lambda \sum_{j \in E} n_\lambda(j) = +\infty. \tag{14.6.9}$$

Proof. From

$$\mathbf{n}_\lambda - \mathbf{n}_\mu + (\lambda - \mu)\mathbf{n}_\mu P_\lambda^{\min} = \mathbf{0}_- \tag{14.6.10}$$

we have

$$\begin{aligned}\lambda \sum_{j\in E} n_\lambda(j) &= \lambda \sum_{j\in E} n_\mu(j) + (\mu - \lambda)\sum_{i\in E} n_\mu(i)\lambda \sum_{j\in E} p_{ij}^{\min}(\lambda)\\ &= \lambda \sum_{j\in E} n_\mu(j) + (\mu - \lambda)\sum_{i\in E} n_\mu(i)(1 - \mu_\lambda(i))\\ &= \mu \sum_{j\in E} n_\mu(j) + (\lambda - \mu)\sum_{i\in E} n_\mu(i) u_\lambda(i)\\ &= \mu \sum_{j\in E} n_\mu(j) + \sum_{i\in E} n_\mu(i)\lambda u_\lambda(i) - \mu \sum_{i\in E} n_\mu(i) u_\lambda(i). \ \square \end{aligned} \tag{14.6.11}$$

By (14.6.11), Lemmas 14.6.1 and 14.6.2, we readily get what should be proved. □

Lemma 14.6.4 *If Q is nonconservative, and Condition* (i) *in Theorem* 14.2.4 *does not hold, then there exist infinitely many N-$B\bigcup F$-type Q-processes.*

Proof. If Condition (i) in Theorem 14.2.4 does not hold, then by Lemma 12.2.4 there exists a row vector $\alpha = (\alpha(1), \alpha(2), \cdots) > \mathbf{0}_-$, independent of λ, such that $\alpha \mathbf{P}_\lambda^{\min} > \mathbf{0}_-$ $(0 < \lambda < +\infty)$ is summable, but α is not summable, so we have

$$\frac{\left(-\sum_{j\in E} q_{ij}\right)\alpha_j}{\sum_{k\in E} \alpha_k} = 0 \qquad (i,\ j \in E). \tag{14.6.12}$$

Hence by [28],

$$P_\lambda = P_\lambda^{\min} + \frac{(\mathbf{1} - \lambda P_\lambda^{\min}\mathbf{1})}{\lambda \alpha P_\lambda^{\min}\mathbf{1}} \cdot \alpha P_\lambda^{\min} \tag{14.6.13}$$

is a Q-process, and it is honest, i.e., it is an N-O-type Q-process.

Now we prove that the Q-process defined by (14.6.13) is not a B-type Q-process. It suffices to show that

$$\lambda P_\lambda - Q P_\lambda - I \neq 0, \tag{14.6.14}$$

where I is the unit matrix. In fact,

$$\begin{aligned}\lambda P_\lambda - Q P_\lambda - I = \lambda\left(P_\lambda^{\min} + \frac{\mathbf{1} - \lambda P_\lambda^{\min}\mathbf{1}}{\lambda \alpha P_\lambda^{\min}\mathbf{1}} \cdot \alpha P_\lambda^{\min}\right)\\ - Q\left(P_\lambda^{\min} + \frac{\mathbf{1} - \lambda P_\lambda^{\min}\mathbf{1}}{\lambda \alpha P_\lambda^{\min}\mathbf{1}} \alpha P_\lambda^{\min}\right) - I = (\lambda P_\lambda^{\min}\end{aligned}$$

$$- QP_\lambda^{\min} - I)$$

$$+ (\lambda(\mathbf{1} - P_\lambda^{\min}\mathbf{1}) - Q(\mathbf{1} - P_\lambda^{\min}\mathbf{1}))\frac{\alpha P_\lambda^{\min}}{\lambda\alpha P_\lambda^{\min}\mathbf{1}}. \quad (14.6.15)$$

Since $P_\lambda^{\min}$ is a B–type Q–process,

$$\lambda P_\lambda^{\min} - QP_\lambda^{\min} - I = 0. \quad (14.6.16)$$

But

$$\alpha P_\lambda^{\min} > \mathbf{0}_-, \quad (14.6.17)$$

$$\lambda\alpha P_\lambda^{\min}\mathbf{1} > 0. \quad (14.6.18)$$

From Condition (i) in Theorem 14.2.4, we have

$$\mathbf{1} - \lambda P_\lambda^{\min}\mathbf{1} \neq \mathbf{0}_1. \quad (14.6.19)$$

From (14.6.6) and the fact that Q is nonconservative, we ascertain

$$\lambda(\mathbf{1} - \lambda P_\lambda^{\min}\mathbf{1}) - Q(\mathbf{1} - \lambda P_\lambda^{\min}\mathbf{1}) = \begin{pmatrix} -\sum_{j\in E} q_{1j} \\ -\sum_{j\in E} q_{2j} \\ \vdots \end{pmatrix} \neq \mathbf{0}_1. \quad (14.6.20)$$

Using (14.6.15)—(14.6.20), we know that (14.6.14) holds. So P_λ is not a B-type Q-process.

Let us now prove that P_λ is not an F-type Q-process either. We need only to prove

$$\lambda P_\lambda - P_\lambda Q - I \neq 0. \quad (14.6.21)$$

In fact,

$$\lambda P_\lambda - P_\lambda Q - I = \lambda\left(P_\lambda^{\min} + \frac{\mathbf{1} - \lambda P_\lambda^{\min}\mathbf{1}}{\lambda\alpha P_\lambda^{\min}\mathbf{1}}\cdot\alpha P_\lambda^{\min}\right)$$

$$-\left(P_\lambda^{\min} + \frac{\mathbf{1} - \lambda P_\lambda^{\min}\mathbf{1}}{\lambda\alpha P_\lambda^{\min}\mathbf{1}}\alpha P_\lambda^{\min}\right)Q - I$$

$$= (\lambda P_\lambda^{\min} - P_\lambda^{\min}Q - I) + \frac{\mathbf{1} - \lambda P_\lambda\mathbf{1}}{\lambda\alpha P_\lambda^{\min}\mathbf{1}}$$

$$.\alpha(\lambda P_\lambda^{\min} - P_\lambda^{\min}Q). \quad (14.6.22)$$

Since $P_\lambda^{\min}$ is an F-type Q-process,

$$\lambda P_\lambda^{\min} - P_\lambda^{\min} Q - I = 0. \tag{14.6.23}$$

By (14.6.22) and (14.6.23), we have

$$\lambda P_\lambda - P_\lambda Q - I = \frac{(\mathbf{1} - \lambda P_\lambda^{\min}\mathbf{1})\alpha}{\lambda \alpha P_\lambda^{\min}\mathbf{1}}. \tag{14.6.24}$$

Equation (14.6.21) follows readily by (14.6.18), (14.6.19), (14.6.24) and $\alpha > \mathbf{0}_-$. So P_λ is not an F-type Q-process either.

Now we know the Q-process P_λ defined by (14.6.13) is an N-$\overline{B \bigcup F}$-type Q-process.

Let

$$\alpha^{(n)}(i) = \begin{cases} \alpha(1) + 2^{-n}, & i = 1, \\ \alpha(i), & i \neq 1, \end{cases} \tag{14.6.25}$$

$$\alpha^{(n)} = (\alpha^{(n)}(1), \alpha^{(n)}(2), \cdots) \qquad (n = 1, 2, \cdots). \tag{14.6.26}$$

Therefore we see from the above assertions that for any natural number n

$$P_\lambda^{(n)} + P_\lambda^{\min} + \frac{\mathbf{1} - \lambda P_\lambda^{\min}\mathbf{1}}{\lambda \alpha^{(n)} P_\lambda \mathbf{1}} \cdot \alpha^{(n)} P_\lambda^{\min} \tag{14.6.27}$$

is an N-O-type Q-process. It is easy to prove by referring to the proofs of Propositions 14.3.6 and 14.3.5 that $P_\lambda^{(n)}$ $(n = 1, 2, \cdots)$ are infinitely many N-$\overline{B \bigcup F}$-type Q-processes different to one another. □

Lemma 14.6.5 *If Q is nonconservative, and Condition* (ii)′ *in Theorem* 14.2.4 *does not hold, then there exist infinitely many N-$\overline{B \bigcup F}$-typc Q-processes.*

Proof. Since Condition (ii)′ does not hold, there exists at least one solution $\mathbf{n}_\lambda = (n_\lambda(1), n_\lambda(2), \cdots)$, among all the solutions of (14.2.7) such that

$$\lim_{\lambda \to \infty} \lambda \sum_{j \in E} n_\lambda(j) = +\infty. \tag{14.6.28}$$

We choose arbitrarily a row vector $\alpha = (\alpha(1), \alpha(2), \cdots) > \mathbf{0}_-$, such that $\alpha P_\lambda^{\min}$ is summable (This is always possible, for example, choosing $\alpha(i) = 2^{-i}$ $(i \in E)$.) From (14.6.28), we have

$$\frac{\left(-\sum_{k\in E} q_{ik}\right)\alpha(j)}{\sum_{k\in E}\alpha(k) + \lim_{\lambda\to\infty}\lambda\sum_{k\in E} n_\lambda(k)} = 0 \qquad (i,\ j\in E). \tag{14.6.29}$$

Hence

$$P_\lambda = P_\lambda^{\min} + \frac{(\mathbf{1} - \lambda P_\lambda^{\min}\mathbf{1})}{\lambda(\alpha P_\lambda^{\min} + \mathbf{n}_\lambda)\mathbf{1}}(\alpha P_\lambda^{\min} + \mathbf{n}_\lambda) \tag{14.6.30}$$

is an N-O-type Q-process by [28]. It is then easy to complete the proof of this lemma by referring to the proof of Lemma 14.6.4. □

Lemma 14.6.6 *If Q is nonconservative and Condition* (ii) *does not hold, then there exist infinitely many N-$\overline{B\bigcup F}$-type Q-processes.*

Proof. This lemma follows readily from Lemmas 14.6.3 and 14.6.5. □

Lemma 14.6.7 *Suppose $X(\omega) = \{x(t,\ \omega),\ t < \sigma(\omega)\}$ is a homogeneous denumerable Markov process which is well–separable, Borel measurable, and its sample function is right-continuous with probability one. Suppose further this process is differentiable, with $Q = (q_{ij})$ as its density matrix, E as its minimal state space, $\tau(\omega)$ as the time of its first jump, then for any $i\in(j\colon q_j \neq 0)$, we have*

$$P(x(\tau) = j \mid x_0 = i) = \frac{q_{ij}}{q_i} \qquad (j\in E), \tag{14.6.31}$$

$$P(x(\tau) = \infty \mid x_0 = i) \leqslant \frac{q_i - \sum_{j\in E} q_{ij}}{q_i}, \tag{14.6.32}$$

$$P(\tau < t \mid x_0 = i,\ x(\tau) = j) = P(\tau < t \mid x_0 = i)$$

$$= \begin{cases} 0, & t < 0 \\ 1 - e^{-q_i t}, & t \geqslant 0 \end{cases} \qquad (j\in E\bigcup\{\infty\}). \tag{14.6.33}$$

Proof. Equation (14.6.31) follows immediately from [1, II, Theorem 15.6] and (14.6.32) follows readily from (14.6.31). Using [1, II, Theorem 15.2], we get (14.6.33) immediately.

Lemma 14.6.8 *Suppose $P_\lambda = (p_{ij}(\lambda),\ i, j\in E)$ is an arbitrary Q-process. Then*

$$0 \leqslant \lambda\sum_{j\in E} p_{ij}(\lambda) - \sum_{k\neq i}\frac{q_{ik}}{\lambda + q_i}\left(\lambda\sum_{j\in E} p_{kj}(\lambda)\right) - \frac{\lambda}{\lambda + q_i}$$

$$\leqslant \frac{q_i - \sum_{j\in E} q_{ij}}{\lambda + q_i} \qquad (\lambda > 0,\ i \in E). \tag{14.6.34}$$

Proof. Suppose $X(\omega) = \{x(t,\ \omega),\ t < \sigma(\omega)\}$ is a Q-process. Assume the Laplace transform of its transition function matrix is $P_\lambda = (p_{ij}(\lambda),\ i, j \in E)$, that this process is well-separable, Borel measurable and that its sample functions are right continuous with probability one. It is well known that such a process $X(\omega)$ with the above properties always exists, hence

$$\int_0^\infty e^{-\lambda t} P(x(t) \in E \mid x_0 = i) dt = \sum_{j\in E} p_{ij}(\lambda) \qquad (i \in E). \tag{14.6.35}$$

We now denote the first jump point of $x(\cdot,\ \omega)$ by $\tau(\omega)$, then

$$\begin{aligned} P(x(t) \in E \mid x_0 = i) &= \sum_{k \neq i} P(x(t) \in E,\ \tau < t,\ x(\tau) = k \mid x_0 = i) \\ &\quad + P(x(t) \in E,\ \tau > t \mid x_0 = i) \\ &\quad + P(x(t) \in E,\ \tau < t,\ x(\tau) = \infty \mid x_0 = i) \quad (i \in E). \end{aligned} \tag{14.6.36}$$

By [1, II, Theorem 16.4], we have

$$\begin{aligned} &P(x(t) \in E,\ \tau < t,\ x(\tau) = k \mid x_0 = i) \\ &= \int_0^t e^{-q_i(t-s)} q_{ik} P(x(s) \in E \mid x_0 = k)\, ds \qquad (i \in E). \end{aligned} \tag{14.6.37}$$

Substituting (14.6.37) into (14.6.36) and using the Laplace transform, we get

$$\begin{aligned} &\sum_{j\in E} p_{ij}(\lambda) - \sum_{k \neq i} \frac{q_{ik}}{\lambda + q_i} \left(\sum_{j \in E} p_{kj}(\lambda) \right) - \frac{1}{\lambda + q_i} \\ &= \int_0^\infty e^{-\lambda t} P(x(t) \in E,\ \tau < t,\ x(\tau) = \infty \mid x_0 = i)\, dt, \qquad (i \in E). \end{aligned} \tag{14.6.38}$$

Obviously we have

$$P(x(t) \in E,\ \tau < t,\ x(\tau) = \infty \mid x_0 = i) = 0, \tag{14.6.39}$$

when $q_i = 0$. Hence

$$\int_0^{\infty} e^{-\lambda t} P(x(t) \in E,\ \tau < t,\ x(\tau) = \infty \mid x_0 = i)\, dt = 0 = \frac{q_i - \sum_{j \neq i} q_{ij}}{\lambda(\lambda + q_i)}$$

$$(i \in \{j:\ q_j = 0\}). \tag{14.6.40}$$

When $q_i \neq 0$, from the nonnegativity of probability and Lemma 14.6.7, we get

$$\begin{aligned} 0 &\leqslant P(x(t) \in E,\ \tau < t,\ x(\tau) = \infty \mid x_0 = i) \\ &\leqslant P(x(\tau) = \infty \mid x_0 = i)\ P(\tau < t \mid x_0 = i,\ x(\tau) = \infty) \\ &\leqslant \frac{q_i - \sum_{j \neq i} q_{ij}}{q_i} P(\tau < t \mid x_0 = i). \end{aligned} \tag{14.6.41}$$

Hence we have

$$\begin{aligned} 0 &\leqslant \int_0^{\infty} e^{-\lambda t} P(x(t) \in E,\ \tau < t,\ x(\tau) = \infty \mid x_0 = i)\, dt \\ &\leqslant \frac{q_i - \sum_{j \neq i} q_{ij}}{\lambda(\lambda + q_i)} \qquad (i \in (j:\ q_j \neq 0)). \end{aligned} \tag{14.6.42}$$

Equations (14.6.38), (14.6.40) and (14.6.42) imply the lemma immediately. □

Lemma 14.6.9 *Suppose Q is nonconservative. If Conditions* (i) *and* (ii) *in Theorem* 14.2.4 *hold simultaneously, then all the Q-processes are of F-type, butnot of $\overline{B \bigcup F}$-type.*

Proof. Suppose $P_\lambda = (p_{ij}(\lambda);\ i, j \in E)$ is an arbitrary Q-process. By referring to §12.3, we have, if Condition (i) holds, that

$$p_{ij}(\lambda) = p_{ij}^{\min}(\lambda) + \sum_{a \in E} p_{ia}^{\min}(\lambda) f_\lambda^{(a)}(j), \qquad f_\lambda^{(a)}(j) \geqslant 0,$$

$$(\lambda > 0,\ i,\ j,\ a \in E). \tag{14.6.43}$$

Let

$$\mathbf{F}_\lambda^{(a)} = (f_\lambda^{(a)}(1),\ f_\lambda^{(a)}(2), \cdots), \tag{14.6.44}$$

thus, by (14.6.43),

$$\lambda \sum_{j \in E} p_{ij}(\lambda) = \lambda \sum_{j \in E} p_{ij}^{\min}(\lambda) + \sum_{a \in E} \lambda p_{ia}^{\min}(\lambda) [\mathbf{F}_\lambda^{(a)},\ \mathbf{1}],$$

$$[\mathbf{F}_\lambda^{(a)},\ \mathbf{1}] \geqslant 0. \tag{14.6.45}$$

From (14.6.45), Theorem 3.3.2 and the fact that $\{p_{ia}^{\min}(\lambda);\ i \in E\}$ is the minimal nonnegative solution of the pseudo-normal system of equations

$$x_i = \sum_{k \neq i} \frac{q_{ik}}{\lambda + q_i} x_k + \frac{\delta_{ia}}{\lambda + q_i} \qquad (i \in E), \tag{14.6.46}$$

we see that $\{\lambda \sum_{j \in E} p_{ij}(\lambda);\ i \in E\}$ is the minimal nonnegative solution of the first-type system of 1-bounded equations

$$x_i = \sum_{k \neq i} \frac{q_{ik}}{\lambda + q_i} x_k + \frac{\lambda}{\lambda + q_i} + \frac{\lambda[\mathbf{F}_\lambda^{(i)}, \mathbf{1}]}{\lambda + q_i} \qquad (i \in E). \tag{14.6.47}$$

Hence

$$\lambda \sum_{j \in E} p_{ij}(\lambda) - \sum_{k \neq i} \frac{q_{ik}}{\lambda + q_i} \left(\lambda \sum_{j \in E} p_{kj}(\lambda) \right) - \frac{\lambda}{\lambda + q_i}$$

$$= \frac{\lambda[\mathbf{F}_\lambda^{(i)}, \mathbf{1}]}{\lambda + q_i} \qquad (i \in E). \tag{14.6.48}$$

By (14.6.48) and Lemma 14.6.8, we get

$$[\mathbf{F}_\lambda^{(i)} \mathbf{1}] \leqslant \frac{q_i - \sum_{j \neq i} q_{ij}}{\lambda} < +\infty \qquad (\lambda > 0,\ i \in E). \tag{14.6.49}$$

Based on [29, Lemma 2.2] and by referring to §12.3, there exist a row vector $\bar{\rho}_\mu^{(a,\lambda)} \geqslant \mathbf{0}_-$ and a row vector $\beta_\lambda^{(a)} \geqslant \mathbf{0}_-$ independent of μ (but dependent on a and λ) such that

$$[\beta_\lambda^{(a)} P_\lambda^{\min}, \mathbf{1}] < +\infty \qquad (\lambda > 0,\ a \in E), \tag{14.6.50}$$

$$\lambda \mathbf{F}_\lambda^{(a)} - \mathbf{F}_\lambda^{(a)} Q = \beta_\lambda^{(a)} \qquad (\lambda > 0,\ a \in E), \tag{14.6.51}$$

$$\mathbf{F}_\lambda^{(a)} = \beta_\lambda^{(a)} P_\lambda^{\min} + \bar{\rho}_\lambda^{(a,\lambda)} \qquad (\lambda > 0,\ a \in E), \tag{14.6.52}$$

$$\beta_\lambda^{(a)} = \beta_\mu^{(a)} + (\mu - \lambda) \sum_{t \in E} [\mathbf{F}_\lambda^{(a)}, \mathbf{P}_\mu^{\min(t)}] \beta_\mu^{(t)}, \tag{14.6.53}$$

and for a and λ fixed $\bar{\rho}_\mu^{(a,\lambda)} = (\bar{\rho}_\mu^{(a,\lambda)}(1), \bar{\rho}_\mu^{(a,\lambda)}(2), \cdots)$ is a solution of the system of equations

$$\left.\begin{array}{l} \mu \rho_\mu - \rho_\mu Q = \mathbf{0}_-, \qquad \mu > 0, \\ 0 \leqslant \rho_\mu, \qquad \rho_\mu \mathbf{1} < +\infty, \\ \rho_\mu - \rho_\nu + (\mu - \nu) \rho_\nu \rho_\mu^{\min} = \mathbf{0}_-, \end{array}\right\} \tag{14.6.54}$$

where

$$\mathbf{P}_{\mu}^{\min(t)} = \begin{pmatrix} p_{1t}^{\min}(\mu) \\ p_{2t}^{\min}(\mu) \\ \vdots \end{pmatrix}. \tag{14.6.55}$$

From (14.6.49) and

$$q_a - \sum_{j \neq a} q_{aj} \geqslant \lambda[\mathbf{F}_{\lambda}^{(a)}, \mathbf{1}] = \lambda[\beta_{\lambda}^{(a)} P_{\lambda}^{\min} + \bar{\rho}_{\lambda}^{(a,\lambda)}, \mathbf{1}]$$

$$\geqslant [\lambda \beta^{(a)} P_{\lambda}^{\min}, \mathbf{1}] = [\beta_{\lambda}^{(a)}, \lambda P_{\lambda}^{\min} \mathbf{1}] \geqslant n_{\lambda}[\beta_{\lambda}^{(a)}, \mathbf{1}], \tag{14.6.56}$$

we have

$$[\beta_{\lambda}^{(a)}, \mathbf{1}] \leqslant \frac{q_a - \sum_{j \neq a} q_{aj}}{n_{\lambda}} < +\infty \qquad (\lambda > 0,\ a \in E). \tag{14.6.57}$$

By referring to §12.3, we can prove

$$[\beta_{\mu}^{(a)}, \mathbf{1}] \downarrow 0 \qquad (\mu \uparrow +\infty), \tag{14.6.58}$$

$$(\mu - \lambda) \sum_{t \in E} [\beta_{\lambda}^{(a)} \mathbf{P}_{\lambda}^{\min}, \mathbf{P}_{\mu}^{\min(t)}] [\beta_{\mu}^{(t)}, \mathbf{1}]$$

$$= \sum_{t \in E} [\beta_{\lambda}^{(a)}, \mathbf{P}_{\lambda}^{\min(t)}] [\beta_{\mu}^{(t)}, \mathbf{1}] - \sum_{t \in E} [\beta_{\lambda}^{(a)}, \mathbf{P}_{\mu}^{\min(t)}] . [\beta_{\mu}^{(t)}, \mathbf{1}] < +\infty. \tag{14.6.59}$$

From Condition (ii) in Theorem 14.2.4 and (14.6.57), we have

$$\sum_{t \in E} \bar{\rho}_{\mu}^{(a,\lambda)}(t) [\beta_{\mu}^{(t)}, \mathbf{1}] \leqslant \frac{1}{n_{\mu}} \sum_{t \in E} \bar{\rho}_{\mu}^{(a,\lambda)}(t) \left(q_t - \sum_{j \neq t} q_{tj} \right) < +\infty. \tag{14.6.60}$$

From (14.6.60) and the fact that $\bar{\rho}_{\mu}^{(a,\lambda)}$ satisfies (14.6.54), we have

$$(\mu - \lambda) \sum_{t \in E} [\bar{\rho}_{\lambda}^{(a,\lambda)}, \mathbf{P}_{\mu}^{\min(t)}] [\beta_{\mu}^{(t)}, \mathbf{1}]$$

$$= \sum_{t \in E} (\mu - \lambda) \bar{\rho}_{\lambda}^{(a,\lambda)} \mathbf{P}_{\mu}^{\min(t)} [\beta_{\mu}^{(t)}, \mathbf{1}]$$

$$= \sum_{t \in E} (\bar{\rho}_{\lambda}^{(a,\lambda)}(t) - \bar{\rho}_{\mu}^{(a,\lambda)}(t)) [\beta_{\mu}^{(t)}, \mathbf{1}]$$

$$= \sum_{t \in E} \bar{\rho}_{\lambda}^{(a,\lambda)}(t) [\beta_{\mu}^{(t)}, \mathbf{1}] - \sum_{t \in E} \bar{\rho}_{\mu}^{(a,\lambda)}(t) [\beta_{\mu}^{(t)}, \mathbf{1}] < +\infty. \tag{14.6.61}$$

It follows by (14.6.53) and (14.6.57) that

$$[\beta_\lambda^{(a)}, \mathbf{1}] = [\beta_\mu^{(a)}, \mathbf{1}] + (\mu - \lambda)\sum_{t\in E}[\mathbf{F}_\lambda^{(a)}, \mathbf{P}_\mu^{\min(t)}]\,[\beta_\mu^{(t)}, \mathbf{1}] < +\infty. \tag{14.6.62}$$

By (14.6.52), (14.6.59), (14.6.61) and (14.6.62), we have

$$\begin{aligned}
[\beta_\lambda^{(a)}, \mathbf{1}] &= [\beta_\mu^{(a)}, \mathbf{1}] + (\mu - \lambda)\sum_{t\in E}[\beta_\lambda^{(a)} P_\lambda^{\min}, \mathbf{P}_\mu^{\min(t)}]\,[\beta_\mu^{(t)}, \mathbf{1}] \\
&\quad + (\mu - \lambda)\sum_{t\in E}[\bar{\rho}_\lambda^{(a,\lambda)}, \mathbf{P}_\mu^{\min(t)}]\,[\beta_\mu^{(t)}, \mathbf{1}] \\
&= [\beta_\mu^{(a)}, \mathbf{1}] + \sum_{t\in E}[\beta_\lambda^{(a)}, \mathbf{P}_\lambda^{\min(t)}]\,[\beta_\mu^{(t)}, \mathbf{1}] \\
&\quad - \sum_{t\in E}[\beta_\lambda^{(a)}, \mathbf{P}_\mu^{\min}(t)]\,[\beta_\mu^{(t)}, \mathbf{1}] + \sum_{t\in E}\bar{\rho}_\lambda^{(a,\lambda)}(t)\,[\beta_\mu^{(t)}, \mathbf{1}] \\
&\quad - \sum_{t\in E}\bar{\rho}_\mu^{(a,\lambda)}(t)\,[\beta_\mu^{(t)}, \mathbf{1}] < +\infty.
\end{aligned} \tag{14.6.63}$$

From the fact that $\rho_\mu^{(a,\lambda)}$ satisfies (14.6.54), we have

$$\bar{\rho}_\mu^{(a,\lambda)} \downarrow \qquad (\mu \uparrow +\infty). \tag{14.6.64}$$

By the definition of $P_\mu^{\min(t)}$, we know

$$P_\mu^{\min(t)} \downarrow \qquad (\mu \uparrow +\infty). \tag{14.6.65}$$

From (14.6.58), (14.6.64), (14.6.65) and (14.6.66), we have

$$[\beta_\lambda^{(a)}, \mathbf{1}] \equiv 0 \qquad (\lambda > 0,\ a \in E). \tag{14.6.66}$$

By $\beta_\lambda^{(a)} \geqslant \mathbf{0}_-$ and (14.6.66) we have

$$\beta_\lambda^{(a)} = \mathbf{0}_1 \qquad (\lambda > 0,\ a \in E). \tag{14.6.67}$$

By (14.6.52) and (14.6.67), we have

$$\mathbf{F}_\lambda^{(a)} = \bar{\rho}_\lambda^{(a,\lambda)} \qquad (\lambda > 0,\ a \in E). \tag{14.6.68}$$

By (14.6.43), (14.6.68) and the fact that $\bar{\rho}_\lambda^{(a,\lambda)}$ satisfies Equation (14.6.54), we see that P_λ is an F-type Q-process. This completes the proof of the lemma.

Lemma 14.6.10 *A necessary and sufficient condition for Condition* (i) *in Theorem* 14.2.4 *to hold is that Conditions* (i)′ *and* (i)″ *in Theorem* 14.2.4 *hold simultaneously.*

Proof. Referring to the proof of Theorem 12.8.1, we get the lemma immediately. □

Lemma 14.6.11 *If Q is nonconservative, and Condition* (i) *in Theorem* 14.2.4 *holds, then a necessary and sufficient condition for*

$$\lim_{\lambda\to\infty} \lambda \sum_{j\in E} n_\lambda(j) < +\infty \tag{14.6.69}$$

is that for all $\mu > 0$, we have

$$\sum_{j\in E}\left(q_j - \sum_{k\neq j} q_{jk}\right) n_\mu(j) < +\infty, \tag{14.6.70}$$

where $\mathbf{n}_\lambda = (n_\lambda(1), n_\lambda(2), \cdots)$ is a solution of the equation system (14.2.7).

Proof. By Lemma 14.6.3, if (14.6.69) holds, then (14.6.70) holds. Conversely, if Condition (i) in Theorem 14.2.4 and (14.6.70) hold, then by Lemma 14.6.9 no N $\cdot\overline{B\bigcup F}$-type Q-process exists. But by Lemma 14.6.5, when (14.6.69) does not hold, there must be an $N\cdot\overline{B\bigcup F}$-type Q-process. Therefore if Condition (i) in Theorem 14.2.4 and (14.6.70) hold, then (14.6.69) must also hold. □

Lemma 14.6.12 *If Q is conservative, no $\overline{B\bigcup F}$-type Q-process exists.*

Proof. It is well known that all the Q-processes in this case are of B-type. So there does not exists any $\overline{B\bigcup F}$-type Q-process. □

Proof of Theorem 14.2.4. From Lemmas 14.6.4, 14.6.6, 14.6.9 ~ 14.6.12, we readily get the theorem.

Corollay 14.6.1 *If Q is nonconservative, and*

$$\sup_{i\in E}\left(q_i - \sum_{j\neq i} q_{ij}\right) \leqslant c < +\infty, \tag{14.6.71}$$

then a necessary and sufficient condition for the nonexistence of $\overline{B\bigcup F}$-type Q-processes is that Condition (i) *in Theorem* 14.2.4 *holds.*

Proof. For any solution of (14.2.7), we now have

$$\sum_{j\in E}\left(q_i - \sum_{k\neq j} q_{ik}\right) n_\lambda(j) \leqslant c\sum_{j\in E} n_\lambda(j) < c\sum_{j\in E} n_\lambda(j) < +\infty, \tag{14.6.72}$$

so condition (ii) in Theorem 14.2.4 holds. Therefore, from Theorem 14.2.4, we readily get what should be proved.

Corollary 14.6.2 *If Q is nonconservative, and*

$$\sup_{i\in \dot{E}} q_i \leqslant c < +\infty, \tag{14.6.73}$$

then a necessary and sufficient condition for the nonexistence of $\overline{B\bigcup F}$-type Q-processes is that the B-type Q-process is unique.

Proof. It is easy to prove that if (14.6.73) holds, then (14.2.8) holds. Therefore by Theorem 14.2.4 and

$$\begin{aligned}\sum_{j\in E} n_\lambda(j)\left(q_j - \sum_{k\neq j} q_{jk}\right) &\leqslant \sum_{j\in \dot{E}} n_\lambda(j) q_i \\ &\leqslant c\sum_{j\in \dot{E}} n_\lambda(j) \leqslant c\sum_{j\in E} n_\lambda(j) < +\infty,\end{aligned} \tag{14.6.74}$$

where $\mathbf{n}_\lambda = (n_\lambda(1), n_\lambda(2), \cdots)$ is any solution of (14.2.7), we get immediately what should be proved. □

Corollary 14.6.3 *If Q is nonconservative and $\dot{E}$ is a finite set, then a necessary and sufficient condition for the nonexistence of $\overline{B\bigcup F}$-type Q-processes is that the B-type Q-process is unique.* □

Corollary 14.6.4 *If Q is nonconservative and no $\overline{B\bigcup F}$-type Q-process exists, then the B-type Q-process is unique.* □

Corollary 14.6.5 *If Q is nonconservative and no $\overline{B\bigcup F}$-type Q-process exists, then all Q-processes are of F-type.* □

§ 14.7 Proofs of Theorems 14.2.5—14.2.10

Proof of Theorem 14.2.5. By Lemma 14.6.10, [24] and Chapter XII, we get the theorem immediately. □

Proof of Theorem 14.2.6. Suppose Q is nonconservative, $P_\lambda = (p_{ij}(\lambda),\ i, j\in E)$ is an arbitrary B-type Q-process, then $\{\lambda\sum_{j\in E} p_{ij}(\lambda),\ i\in E\}$ satisfies the system of equations

$$x_i = \sum_{k\neq i} \frac{q_{ik}}{\lambda + q_i} x_k + \frac{\lambda}{\lambda + q_i} \qquad (i\in E). \tag{14.7.1}$$

Hence

$$\begin{aligned}\lambda\sum_{j\in E} p_{ij}(\lambda) &= \sum_{k\neq i} \frac{q_{ik}}{\lambda + q_i}\left(\lambda\sum_{j\in E} p_{kj}(\lambda)\right) + \frac{\lambda}{\lambda + q_i} \\ &\leqslant \sum_{k\neq i} \frac{q_{ik}}{\lambda + q_i} + \frac{\lambda}{\lambda + q_i}\end{aligned}$$

$$= \frac{\lambda + \sum_{k \neq i} q_{ik}}{\lambda + q_i} \qquad (i \in E). \tag{14.7.2}$$

Since Q is nonconservative, there exists a state $s \in E$ such that

$$\sum_{k \neq s} q_{sk} < q_s, \tag{14.7.3}$$

hence by (14.7.2), we have

$$\lambda \sum_{j \in E} p_{sj}(\lambda) \leqslant \frac{\lambda + \sum_{k \neq s} q_{sk}}{\lambda + q_s} < 1. \tag{14.7.4}$$

So P_λ is not an N-B-type Q-process, and the assertion (2) of Theorem 14.2.6 is valid. □

If Q is conservative and the processes are not unique. It is easy to see that

$$P_\lambda^{(n)} = P_\lambda^{\min} + \frac{1 - \lambda P_\lambda^{\min} \mathbf{1}}{\lambda \alpha^{(n)} P_\lambda^{\min} \mathbf{1}} \cdot \alpha^{(n)} P_\lambda^{\min} \qquad (n = 1, 2, \cdots) \tag{14.7.5}$$

are infinitely many N-B-type Q-processes different to one another, here

$$\alpha^{(n)} = (\underbrace{0, 0, \cdots, 0}_{(n-1)}, 1, 0, 0, \cdots). \tag{14.7.6}$$

Since the other conclusions of the theorem obviously hold, their proofs are omitted.

Proof of Theorem 14.2.7. It is easy to carry out the proof by using some well-known facts and a result from [27], namely, that if the minimal Q-process is honest or (14.2.2) has just one linearly independent solution, then the N-F-type Q-process is unique, and if the minimal Q-process is not honest and (14.2.2) has more than one linearly independent solution, then there exist infinitely many N-F-type Q-processes.

Proof of Theorem 14.2.8. By Theorem 14.2.6 and Theorem 14.2.7, we immediately get the theorem. □

Proof of Theorem 14.2.9. The theorem follows from Theorem 14.2.4, Lemmas 14.6.4 and 14.6.6. □

Proof of Theorem 14.2.10. Note that if Q is conservative, then all the Q-processes are of B-type. The conclusions of Theorem 14.2.10 in the case when Q is conservative can be derived from Theorem 14.2.6.

If Q is nonconservative and the Q-process is unique, then the unique Q-process is the minimal Q-process, hence it is of B-type. Consequently, it can be seen by (2) of Theorem 14.2.6 that it is not honest, i.e., it is not an N-O-type Q-process. So no N-O-type Q-process exists.

If Q is nonconservative and both Conditions (a) and (b) hold, then by Lemma 14.6.9 (since it is seen by [29] that the solution n_λ of Equation (14.2.2) satisfies the equation $\mathbf{n}_\lambda - \mathbf{n}_\mu + (\lambda - \mu)\mathbf{n}_\mu P_\lambda^{\min} = \mathbf{0}_-$ in this case), all the Q-processes are of F-type. Therefore from Condition (b) and Theorem 14.2.7, we deduce that there exists only one N-F-type Q-process, and hence that there exists only one N-O-type Q-process.

If Q is nonconservative and (a) does not hold, then, by Lemma 14.6.4, we see that there exist infinitely many N-O-type Q-processes.

If Q is nonconservative, (14.2.5) holds and (14.2.2) has more than one linearly independent solution, then Theorem 14.2.7 implies that there exist infinitely many N-F-type Q-processes, therefore there exist infinitely many N-O-type Q-processes.

If Q is nonconservative, (14.2.5) holds, and (14.2.2) has just one linearly independent solution $\mathbf{n}_\lambda$ which satisfies (14.2.12), then it can be seen by Lemma 14.6.5 that there exist infinitely many N-O-type Q-processes.

Summing up the above results, we get the theorem immediately. □

Bibliography

[1] K.L.Chung, Markov Chains with Stationary Transition Probabilities, 2nd. Edition, Springer-Verlag, 1967.

[2] Wang Zikun (Wang Tzi-kun), Structure of Birth and Death Processes, *Progress in Math.*, **5:** 2 (1962), 137—170. (*Chinese*)

[3] Wang Zikun (Wang Tzi-kun), Ergodic Property and Zero-one Law for Birth and Death Processes, *Acta Scientiarum Naturalium, Universitatis Nan Kaiensis,* **5:** 5 (1964), 89—94.

[4] Yang Chaoqun, Notes on the Construction Theory for Birth and Death Processes, *Acta Mathematica Sinica,* **15:** 2 (1965), 173—187.

[5] Yang Chaoqun, Properties of Birth and Death Processes, *Progress in Math.*, **9:** 4 (1966), 365—380.

[6] E.B.Dynkin, Markov Processes, Springer-Verlag, 1965.

[7] Shi Renjie, The Replacement of Random Time for Denumerable Markov Processes, *Acta Scientiarum Naturalium, Universitatis Nan Kaiensis,* **5:** 5 (1964), 51—88.

[8] L.V.Kantorovich and B. I. Krylov, Approximate Methods of Advanced Analysis, Fifth Edition, Gostekhizdat, 1962. (in Russian)

[9] F. R. Gantmacher, The Theory of Matrices, New York, Chelsea, 1959.

[10] R. S. Varga, Matrix Iterative Analysis, Prentice-Hall, Inc., 1962.

[11] Yang Chaoqun, Integral Functional of Denumerable Markov Processes and Boundary Property of Bilateral Birth and Death Processes, *Progress in Math.*, **7:** 4 (1964), 397—424 (Chinese)

[12] J. L. Doob, Discrete Potential Theory and Boundaries, *Journ. of Math. and Mech.*, **8:** 3 (1959), 433—458.

[13] G. A. Hunt, Markoff Chains and Martin Boundaries, *Illinois Journ. of Math.*, **4** (1960), 313—340.

[14] H. Kunita, Applications of Martin Boundaries to Instantaneous Return Markov Processes over a Denumerable Space, *J. Math. Soc. Japan*, **14** (1962), 66—100.

[15] E. B. Dynkin, Boundary Theory for Markov Processes (Discrete case), Uspekhi Matematicheskikh Nauk, **24,** Vol. 2 (146), 1969, 3—42. (in Russian)

[16] Guan Zhaozhi, Lecture on Functional Analysis, Advanced Education Press, Beijing, 1958.

[17] Wu Lide, Classification of States of Denumerable Markov Processes, *Acta Mathematica Sinica,* **15:** 1 (1965), 32—41.

[18] Wang Zikun (Wang Tzi-kun), On Distributions of Functions of Birth and Death Processes and Their Applications in the Theory of Queues, *Scientia Sinica,* **X:** 2 (1961), 160—170.

[19] Wu Lide, Distribution of Integral Functional of Homogeneous Denumerable Markov Processes, *Progress in Math.*, **13:** 1 (1963), 86—93.

[20] Wang Zikun (Wang Tzi-kun), The Martin Boundary and Limit Theorems for Excessive Function, *Scientia Sinica,* **XIV:** 8 (1965), 1118—1129.

[21] E. B. Dynkin, Foundations of the Theory of Markov Processes, Pergamon Press, 1960.

[22] Wang Zikun (Wang Tzi-kun), Theory of Stochastic Processes, Science Press, Beijing, 1965.

[23] K. L. Chung, Lectures on Boundary Theory for Markov Chains, Princeton, New Jersey, 1970.

[24] J. L. Doob, Markov Chains-Denumerable Case, *Tran. Am. Math.*, **58** (1945), 455—473.

[25] W. Feller, On the Integro-differential Equations of Purely Discontinuous Markoff Process, *Tran. Am. Math. Soc.*, **48** (1940), 488—515; *ibid.*, **58** (1945), 474·

[26] Hou Zhenting, The Criterion for Uniqueness of a Q Process, *Scientia Sinica*, **2** (1974), 141—159.

[27] G. E. H. Reuter, Denumerable Markov Processes, *Acta Math.*, **97** (1957), 1—46.

[28] G. E. H. Reuter, Denumerable Markov Processes III, *J. London Math. Soc.*, **37** (1962), 64—73.

[29] G. E. H. Reuter, Denumerable Markov Processes II, *J. London Math. Soc.*, **34** (1959), 81—91.

[30] T. E. Harris, The Theory of Branching Processes, Springer-Verlag, 1963.

[31] Hu Dihe, The Construction Theory of Denumerable Markov Processes, *Acta Scientiarum Naturalium, Universitatis Pekinensis,* **2** (1965), 111—143.

[32] Yang Chaoqun, A Class of Birth and Death Processes, *Acta Mathematica Sinica*, **15** (1965), 9—31. (Chinese)

[33] Yang Chaoqun, Boundary Conditions for the System of Kolmogorov Backward Differential Equations, *Acta Mathematica Sinica*, **16:** 4 (1966), 429—452. (Chinese)

[34] G. E. H. Reuter, Denumerable Markov Processes (IV): On C. T. Hou's Uniqueness Theorem for Q-Semigroups, *Z. Wahrscheinlichkeitstheor. verw. Geb.*, **33** (1976), 309—315.

[35] Hou Zhenting, Construction of Sample Function of Homogeneous Denumerable Markov Processes, *Scientia Sinica*, **3** (1975), 259—266. (Chinese)

[36] Hou Zhenting, Guo Qingfeng, Qualitative Theory in Construction Theory of Homogeneous Denumerable Markov Processes, *Scientia Sinica*, **4** (1976), 239—262. (Chinese)

[37] Hou Zhenting, Probabilistic-Analytical Methods in Denumerable Markov Process with Stationary Transition Probabilities, *Kexue Tongbao*, **3** (1973), 115—118. (Chinese)

[38] Hou Zhenting, Construction Theory for Homogeneous Denumerable Markov Processes, *Kexue Tongbao*, **3** (1973). (Chinese)

Index